ISBN 0-8373-0229-3

C-229 CAREER EXAMINATION SERIES

This is your PASSBOOK® for...

Electronic Technician (USPS)

Test Preparation Study Guide

Questions & Answers

NLC

NATIONAL LEARNING CORPORATION

National Learning Corporation

212 Michael Drive, Syosset, New York 11791

(516) 921-8888
Outside N.Y.: 1(800) 645-6337
ORDER FAX: 1(516) 921-8743
www.passbooks.com
email: passbooks @ aol.com
sales @ passbooks.com
info @ passbooks.com

PRINTED IN THE UNITED STATES OF AMERICA

PASSBOOK®

NOTICE

This book is *SOLELY* intended for, is sold *ONLY* to, and its use is *RESTRICTED* to *individual*, bona fide applicants or candidates who qualify by virtue of having seriously filed applications for appropriate license, certificate, professional and/or promotional advancement, higher school matriculation, scholarship, or other legitimate requirements of educational and/or governmental authorities.

This book is *NOT* intended for use, class instruction, tutoring, training, duplication, copying, reprinting, excerption, or adaptation, etc., by:

(1) Other Publishers
(2) Proprietors and/or Instructors of "Coaching" and/or Preparatory Courses
(3) Personnel and/or Training Divisions of commercial, industrial, and governmental organizations
(4) Schools, colleges, or universities and/or their departments and staffs, including teachers and other personnel
(5) Testing Agencies or Bureaus
(6) Study groups which seek by the purchase of a single volume to copy and/or duplicate and/or adapt this material for use by the group as a whole without having purchased individual volumes for each of the members of the group
(7) Et al.

Such persons would be in violation of appropriate Federal and State statutes.

PROVISION OF LICENSING AGREEMENTS. — Recognized educational commercial, industrial, and governmental institutions and organizations, and others legitimately engaged in educational pursuits, including training, testing, and measurement activities, may address a request for a licensing agreement to the copyright owners, who will determine whether, and under what conditions, including fees and charges, the materials in this book may be used by them. In other words, a licensing facility *exists* for the legitimate use of the material in this book on other than an individual basis. However, it is asseverated and affirmed here that the materials in this book *CANNOT* be used without the receipt of the express permission of such a licensing agreement from the Publishers.

NATIONAL LEARNING CORPORATION
212 Michael Drive
Syosset, New York 11791

Inquiries re licensing agreements should be addressed to:

The President
National Learning Corporation
212 Michael Drive
Syosset, New York 11791

PASSBOOK SERIES®

THE *PASSBOOK SERIES®* has been created to prepare applicants and candidates for the ultimate academic battlefield—the examination room.

At some time in our lives, each and every one of us may be required to take an examination—for validation, matriculation, admission, qualification, registration, certification, or licensure.

Based on the assumption that every applicant or candidate has met the basic formal educational standards, has taken the required number of courses, and read the necessary texts, the *PASSBOOK SERIES®* furnishes the one special preparation which may assure passing with confidence, instead of failing with insecurity. Examination questions—together with answers—are furnished as the basic vehicle for study so that the mysteries of the examination and its compounding difficulties may be eliminated or diminished by a sure method.

This book is meant to help you pass your examination provided that you qualify and are serious in your objective.

The entire field is reviewed through the huge store of content information which is succinctly presented through a provocative and challenging approach—the question-and-answer method.

A climate of success is established by furnishing the correct answers at the end of each test.

You soon learn to recognize types of questions, forms of questions, and patterns of questioning. You may even begin to anticipate expected outcomes.

You perceive that many questions are repeated or adapted so that you gain acute insights, which may enable you to score many sure points.

You learn how to confront new questions, or types of questions, and to attack them confidently and work out the correct answers.

You note objectives and emphases, and recognize pitfalls and dangers, so that you may make positive educational adjustments.

Moreover, you are kept fully informed in relation to new concepts, methods, practices, and directions in the field.

You discover that you are actually taking the examination all the time: you are preparing for the examination by "taking" an examination, not by reading extraneous and/or supererogatory textbooks.

In short, this PASSBOOK®, used directedly, should be an important factor in helping you to pass your test.

ELECTRONIC TECHNICIAN (U.S.P.S.)

Test 932 Covers the Following Knowledge, Skills and Abilities

(1) Knowledge of basic mechanics refers to the theory of operation, terminology, usage, and characteristics of basic mechanical principles as they apply to such things as gears, pulleys, cams, pawls, power transmissions, linkages, fasteners, chains, sprockets, and belts; and including hoisting, rigging, roping, pneumatics, and hydraulic devices.

(2) Knowledge of basic electricity refer to the theory, terminology, usage, and characteristics of basic electrical principles such as Ohm's Law, Kirchoff's Law, and magnetism, as they apply to such things as AC-DC circuitry and hardware, relays, switches, and circuit breakers.

(3) Knowledge of basic electronics refers to the theory, terminology, usage, and characteristics of basic electronic principles concerning such things as solid state devices, vacuum tubes, coils, capacitors, resistors, and basic logic circuitry.

(4) Knowledge of digital electronics refers to the terminology, characteristics, symbology, and operation of digital components as used in such things as logic gates, registers, adders, counters, memories, encoders and decoders.

(5) Knowledge of safety procedures and equipment refers to the knowledge of industrial hazards (e.g., mechanical, chemical, electrical, electronic) and procedures and techniques established to avoid injuries to self and others such as lock-out devices, protective clothing, and waste disposal techniques.

(6) Knowledge of basic computer concepts refers to the terminology, usage, and characteristics of digital memory storage/processing devices such as internal memory, input-output peripherals, and familiarity with programming concepts.

(19) Ability to perform basic mathematical computations refers to the ability to perform basic calculations such as addition, subtraction, multiplication and division with whole numbers, fractions and decimals.

(20) Ability to perform more complex mathematics refers to the ability to perform calculations such as basic algebra, geometry, scientific notation, and number conversions, as applied to mechanical, electrical and electronic applications.

(21) Ability to apply theoretical knowledge to practical applications refers to mechanical, electrical and electronic maintenance applications such as inspection, troubleshooting equipment repair and modification, preventive maintenance, and installation of electrical equipment.

(22) Ability to detect patterns refers to the ability to observe and analyze qualitative factors such as number progressions, spatial relationships, and auditory and visual patterns. This includes combining information and determining how a given set of numbers, objects, or sounds are related to each other.

(23) Ability to use written reference materials refers to the ability to locate, read, and comprehend text material such as handbooks, manuals, bulletins, directives, checklists and route sheets.

(26) Ability to follow instructions refers to the ability comprehend and execute written and oral instructions such as work orders, checklists, route sheets, and verbal directions and instructions.

(31) Ability to use hand tools refers to knowledge of, and proficiency with, various hand tools. This ability involves the safe and efficient use and maintenance of such tools as screwdrivers, wrenches, hammers, pliers, chisels, punches, taps, dies, rules, gauges, and alignment tools.

(35) Ability to use technical drawings refers to the ability to read and comprehend technical materials such as diagrams, schematics, flow charts, and blueprints.

(36) Ability to use test equipment refers to the knowledge of, and proficiency with, various types of mechanical, electrical and electronic test equipment such as VOMS, oscilloscopes, circuit tracers, amprobes, and tachometers.

(37) Ability to solder refers to the knowledge of, and ability to safely and effectively apply, the appropriate soldering techniques.

Job Description

Maintenance positions require highly skilled and experienced individuals. All applicants must meet the knowledge, skills, and abilities requirements listed.
Maintenance duties require prolonged standing, walking, climbing, bending, reaching, and stooping. Employees must lift and carry heavy objects on level surfaces, on ladders, and/or on stairways. For positions requiring driving, applicants must have a valid state driver's license, a safe driving record, and at least two years of documented driving experience.

Electronic Technician -- Performs the full range of diagnostic, preventive maintenance, alignment and calibration, and overhaul tasks, on both hardware and software on a variety of mail processing, customer service, and building equipment and systems, applying advanced technical knowledge to solve complex problems.

Sample Questions

Examination 932 – Electronic Technician

The purpose of this booklet is to illustrate the types of questions that will be used in Test 932. The samples will also show how the questions in the test are to be answered.

Test 932 measures 16 knowledge, skills, and abilities (KSAs) used by a variety of maintenance positions. Exhibit A lists the actual KSAs that are measured. However, not all KSAs that are measured in this test are scored for every position. The qualification standard for each position lists the KSAs required for the position. Only those questions that measure KSAs required for the position(s) for which you are applying will be scored for the position(s).

The suggested answers to each question are lettered A, B, C, etc. Select the BEST answer and make a heavy pencil mark in the corresponding space on the Sample Answer Sheet. Each mark must be dense black. Each mark must cover more than half the space and must not extend into neighboring spaces. If the answer to Sample 1 is B, you would mark the Sample Answer Sheet like this:

1. Ⓐ ● Ⓒ Ⓓ Ⓔ

During the test, directions for answering questions in Part I will be given orally, either by a cassette tape, CD, or by the examiner. You are to listen closely to the directions and follow them. To practice for this part of the test, you might have a friend read the directions to you while you mark your answers on the Sample Answer Sheet – Part I. Directions for answering questions in Part II will be completely described in the test booklet you receive at the examination session.

Now practice Part I. When completed, practice Part II and mark your answers on the Sample Answer Sheet – Part II. After recording your answers, compare them with those in the Correct Answers to Sample Questions. If they do not agree, carefully re-read the questions that were missed to get a clear understanding of what each question is asking.

STUDY CAREFULLY BEFORE YOU GO THE EXAMINATION ROOM

Part I

In Part I of the test, you will be told to follow directions by writing in a test booklet and then on an answer sheet. The test booklet will have lines of material like the following five samples. To practice this test, have someone read the instructions to you on page 8 (tear out this page) and follow the instructions accordingly. When they tell you to darken the space on the Sample Answer Sheet, use the one on this page.

Sample Questions

SAMPLE 1. 5 ____________

SAMPLE 2. 1 6 4 3 7

SAMPLE 3. D B A E C

SAMPLE 4. (8 __) (5 __) (2 __) (9 __) (10 __)

SAMPLE 5. (7 __) [6 __] (1 __) [12 __]

Sample Answer Sheet-Part 1

1 Ⓐ Ⓑ Ⓒ Ⓓ Ⓔ	5 Ⓐ Ⓑ Ⓒ Ⓓ Ⓔ	9 Ⓐ Ⓑ Ⓒ Ⓓ Ⓔ
2 Ⓐ Ⓑ Ⓒ Ⓓ Ⓔ	6 Ⓐ Ⓑ Ⓒ Ⓓ Ⓔ	10 Ⓐ Ⓑ Ⓒ Ⓓ Ⓔ
3 Ⓐ Ⓑ Ⓒ Ⓓ Ⓔ	7 Ⓐ Ⓑ Ⓒ Ⓓ Ⓔ	11 Ⓐ Ⓑ Ⓒ Ⓓ Ⓔ
4 Ⓐ Ⓑ Ⓒ Ⓓ Ⓔ	8 Ⓐ Ⓑ Ⓒ Ⓓ Ⓔ	12 Ⓐ Ⓑ Ⓒ Ⓓ Ⓔ

Part I

Instructions to be Read (the words in parentheses should not be read aloud).

You are to follow the instructions that I shall read to you. I cannot repeat them.

Look at the samples. Sample 1 has a number and a line beside it. On the line write an A. (Pause 2 seconds.) Now on the Sample Answer Sheet, find number 5 (pause 2 seconds) and darken the space for the letter you just wrote on the line. (Pause 2 seconds.)

Look at Sample 2. (Pause slightly.) Draw a line under the third number. (Pause 2 seconds.) Now look on the Sample Answer Sheet, find the number under which you just drew a line and darken space B as in baker for that number. (Pause 5 seconds.)

Look at Sample 3. (Pause slightly.) Draw a line under the third letter in the line. (Pause 2 seconds.) Now on your Sample Answer Sheet, find number 9 (pause 2 seconds) and darken the space for the letter under which you drew a line. (Pause 5 seconds.)

Look at the five circles in Sample 4. (Pause slightly.) Each circle has a number and a line in it. Write D as in dog on the blank in the last circle. (Pause 2 seconds.) Now on the Sample Answer Sheet, darken the space for the number-letter combination that is in the circle you just wrote in. (Pause 5 seconds.)

Look at Sample 5. (Pause slightly.) There are two circles and two boxes of different sizes with numbers in them. (Pause slightly.) If 4 is more than 2 and if 5 is less than 3, write A in the smaller circle. (Pause slightly.) Otherwise write C in the larger box. (Pause 2 seconds.) Now on the Sample Answer Sheet, darken the space for the number-letter combination in the circle or box in which you just wrote. (Pause 5 seconds.)

Correct Answers to Sample Questions – Part 1

Now look at the Sample Answer Sheet. You should have darkened spaces 4B, 5A, 9A, 10D, and 12C on the Sample Answer Sheet. (If the person preparing to take the examination made any mistakes, try to help him or her understand why the mistakes are wrong.)

Sample Answer Sheet – Part II

1	Ⓐ Ⓑ Ⓒ Ⓓ Ⓔ	18	Ⓐ Ⓑ Ⓒ Ⓓ Ⓔ
2	Ⓐ Ⓑ Ⓒ Ⓓ Ⓔ	19	Ⓐ Ⓑ Ⓒ Ⓓ Ⓔ
3	Ⓐ Ⓑ Ⓒ Ⓓ Ⓔ	20	Ⓐ Ⓑ Ⓒ Ⓓ Ⓔ
4	Ⓐ Ⓑ Ⓒ Ⓓ Ⓔ	21	Ⓐ Ⓑ Ⓒ Ⓓ Ⓔ
5	Ⓐ Ⓑ Ⓒ Ⓓ Ⓔ	22	Ⓐ Ⓑ Ⓒ Ⓓ Ⓔ
6	Ⓐ Ⓑ Ⓒ Ⓓ Ⓔ	23	Ⓐ Ⓑ Ⓒ Ⓓ Ⓔ
7	Ⓐ Ⓑ Ⓒ Ⓓ Ⓔ	24	Ⓐ Ⓑ Ⓒ Ⓓ Ⓔ
8	Ⓐ Ⓑ Ⓒ Ⓓ Ⓔ	25	Ⓐ Ⓑ Ⓒ Ⓓ Ⓔ
9	Ⓐ Ⓑ Ⓒ Ⓓ Ⓔ	26	Ⓐ Ⓑ Ⓒ Ⓓ Ⓔ
10	Ⓐ Ⓑ Ⓒ Ⓓ Ⓔ	27	Ⓐ Ⓑ Ⓒ Ⓓ Ⓔ
11	Ⓐ Ⓑ Ⓒ Ⓓ Ⓔ	28	Ⓐ Ⓑ Ⓒ Ⓓ Ⓔ
12	Ⓐ Ⓑ Ⓒ Ⓓ Ⓔ	29	Ⓐ Ⓑ Ⓒ Ⓓ Ⓔ
13	Ⓐ Ⓑ Ⓒ Ⓓ Ⓔ	30	Ⓐ Ⓑ Ⓒ Ⓓ Ⓔ
14	Ⓐ Ⓑ Ⓒ Ⓓ Ⓔ	31	Ⓐ Ⓑ Ⓒ Ⓓ Ⓔ
15	Ⓐ Ⓑ Ⓒ Ⓓ Ⓔ	32	Ⓐ Ⓑ Ⓒ Ⓓ Ⓔ
16	Ⓐ Ⓑ Ⓒ Ⓓ Ⓔ	33	Ⓐ Ⓑ Ⓒ Ⓓ Ⓔ
17	Ⓐ Ⓑ Ⓒ Ⓓ Ⓔ	34	Ⓐ Ⓑ Ⓒ Ⓓ Ⓔ

USE THIS PAGE TO RECORD YOUR ANSWERS TO THE QUESTIONS ON PAGES 10 THROUGH 16.

Part II

1. The primary function of a take-up pulley in a belt conveyor is to

   A) carry the belt on the return trip.
   B) track the belt.
   C) maintain proper belt tension.
   D) change the direction of the belt.

2. Which device is used to transfer power and rotary mechanical motion from one shaft to another?

   A) Bearing
   B) Lever
   C) Idler roller
   D) Gear
   E) Bushing

3. Which of the following circuits is shown in Figure III-A-22?

   A) Series circuit
   B) Parallel circuit
   C) Series, parallel circuit
   D) Solid state circuit
   E) None of the above

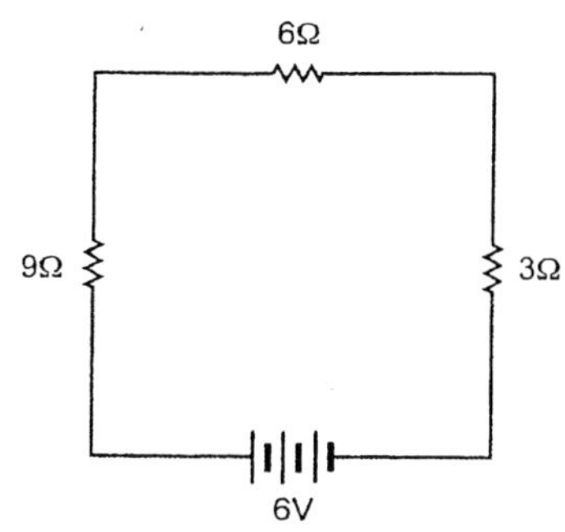

Figure III-A-22

4. A circuit has two resistors of equal value in series. The voltage and current in the circuit are 20 volts and 2 amps respectively. What is the value of EACH resistor?

   A) 5 ohms
   B) 10 ohms
   C) 20 ohms
   D) Not enough information given

5. What is the total net capacitance of two 60-farad capacitors connected in series?

   A) 30 farads
   B) 60 farads
   C) 90 farads
   D) 120 farads
   E) 360 farads

6. Select the Boolean equation that matches the circuit diagram in Figure 79-4-17B.

   A) Z = AB+CD+EF
   B) Z = (A+B) (C+D) (E+F)
   C) Z = A+B+C+D+EF
   D) Z = ABCD (E+F)

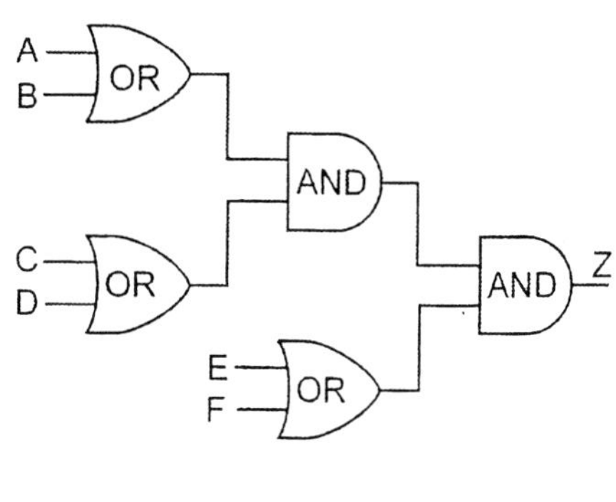

FIGURE 79-4-17B

7. If two 30-mH inductors are connected in series, what is the total net inductance of the combination?

   A) 15 mH
   B) 20 mH
   C) 30 mH
   D) 45 mH
   E) 60 mH

8. In pure binary the decimal number 6 would be expressed as

   A) 001
   B) 011
   C) 110
   D) 111

9. In Figure 75-8-11, which of the following scores will be printed?

   A) All scores > 90 and < 60

   B) All scores < 90

   C) All scores $\leq$ 90 and $\geq$ 60

   D) All scores < 60

10. Crowbars, light bulbs and vacuum bags are to be stored in the cabinet shown in Figure 75-25-1. Considering the balance of weight, what would be the safest arrangement?

    A) Top Drawer – Crowbars
    Middle Drawer - Light Bulbs
    Bottom Drawer - Vacuum bags
    B) Top Drawer – Crowbars
    Middle Drawer - Vacuum bags
    Bottom Drawer - Light Bulbs
    C) Top Drawer - Vacuum Bags
    Middle Drawer – Crowbars
    Bottom Drawer - Light Bulbs
    D) Top Drawer - Vacuum Bags
    Middle Drawer - Light Bulbs
    Bottom Drawer - Crowbars
    E) Top Drawer - Light Bulbs
    Middle Drawer - Vacuum Bags
    Bottom Drawer - Crowbars

11. Which is most appropriate for pulling a heavy load?

    A) Electric lift
    B) Fork lift
    C) Tow Conveyor
    D) Dolly
    E) Pallet truck

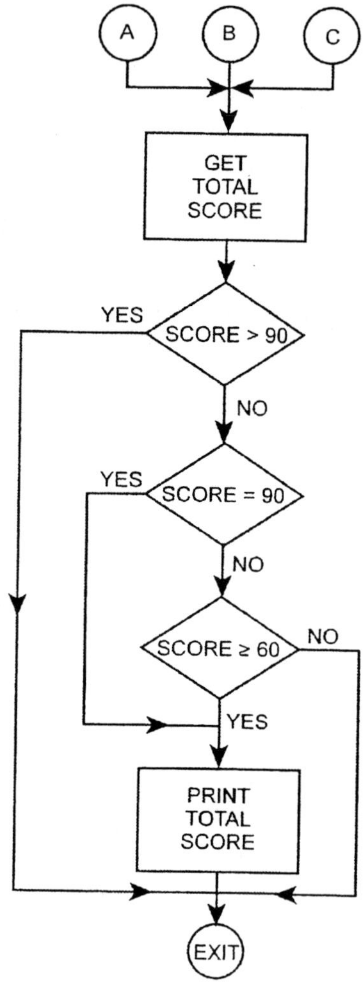

Figure 75-8-11

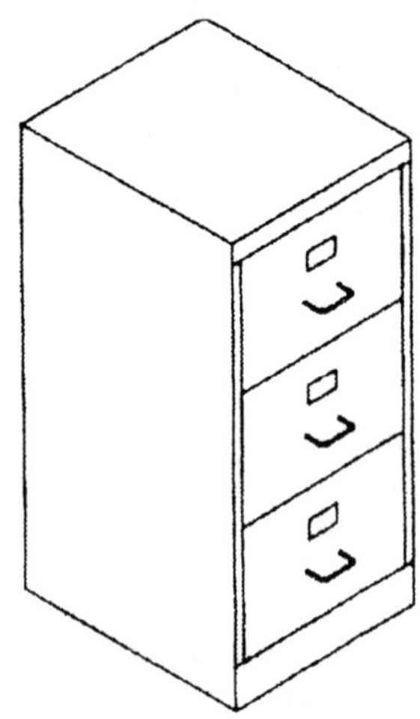
Figure 75-25-1

12. The electrical circuit term "open circuit" refers to a closed loop being opened. When an ohmmeter is connected into this type of circuit, one can expect the meter to

A) read infinity.
B) read infinity and slowly return to ZERO.
C) read ZERO.
D) read ZERO and slowly return to infinity.
E) None of the above

13. Contaminants have caused bearings to fail prematurely. Which pair of the items listed below should be kept away from bearings?

A) Dirt and oil
B) Grease and water
C) Oil and grease
D) Dirt and moisture
E) Water and oil

14. In order to operate a breast drill, which direction should you turn it?

A) Clockwise
B) Counterclockwise
C) Up and down
D) Back and forth
E) Right, then left

15. Which is the correct tool for tightening or loosening a water pipe?

A) Slip joint pliers
B) Household pliers
C) Monkey wrench
D) Water pump pliers
E) Pipe wrench

16. What is one purpose of a chuck key?

A) Open doors
B) Remove drill bits
C) Remove screws
D) Remove set screws
E) Unlock chucks

17. When smoke is generated as a result of using a portable electric drill for cutting holes into a piece of angle iron, one should

A) use a fire watch.
B) cease the drilling operation.
C) use an exhaust fan to remove smoke.
D) use a prescribed coolant solution to reduce friction.
E) call the Fire Department.

18. The primary purpose of soldering is to

A) melt solder to a molten state.
B) heat metal parts to the right temperature to be joined.
C) join metal parts by melting the parts.
D) harden metal.
E) join metal parts.

19. Which of the following statements is correct of a soldering gun?

A) Tip is not replaceable
B) Cannot be used in cramped places
C) Heats only when trigger is pressed
D) Not rated by the number of watts they use
E) Has no light

20. What unit of measurement is read on a dial torque wrench?

A) Pounds
B) Inches
C) Centimeters
D) Foot-pounds
E) Degrees

21. Which instrument is used to test insulation breakdown of a conductor?

A) Ohmmeter
B) Ammeter
C) Megger
D) Wheatstone bridge
E) Voltmeter

22. 1/2 of 1/4 =

A) 1/12
B) 1/8
C) 1/4
D) 1/2
E) 8

23. 2.6 - .5 =

A) 2.0
B) 2.1
C) 3.1
D) 3.3
E) None of the above

24. Simplify the following expression in terms of amps:

563×10^{-6}

A) 563,000,000 amps
B) 563,000 amps
C) .563 amps
D) .000563 amps
E) .000000563 amps

25. Solve the power equation

$P = I^2 R$ for R

A) R = EI

B) $R = I^2P$

C) R = PI

D) $R = P/I^2$

E) R = E/I

26. The product of 3 kilo ohms times 3 micro ohms is

A) 6×10^{-9} ohms

B) 6×10^{-3} ohms

C) 9×10^{3} ohms

D) 9×10^{-6} ohms

E) 9×10^{-3} ohms

In sample question 27 below, select the statement which is most nearly correct according to the paragraph.

"Prior to 1870, a conveyor that made use of rollers was developed for transporting clay. This construction substituted rolling friction at the idler bearing points for the sliding friction of the slider bed. A primitive type of troughing belt conveyor was developed about the same time for the handling of grain. This design was improved during the latter part of the century when the troughing idler was developed."

27. According to the above paragraph, which of the following statements is most nearly correct?

A) The troughing belt conveyor was developed about 1870 to handle clay and grain.

B) Rolling friction construction was replaced by sliding friction construction prior to 1870.

C) In the late nineteenth century, conveyors were improved with the development of the troughing idler.

D) The troughing idler, a significant design improvement for conveyors, was developed in the early nineteenth century.

E) Conveyor belts were invented and developed in the 1800's.

For sample question 28 below, select from the drawings of objects on the right labeled A, B, C, and D, the one that would have the TOP, FRONT, and RIGHT views shown in the drawing at the left

28.

TOP

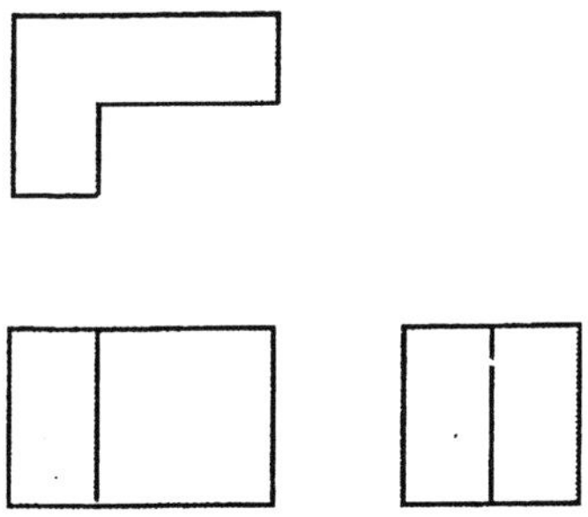

FRONT RIGHT

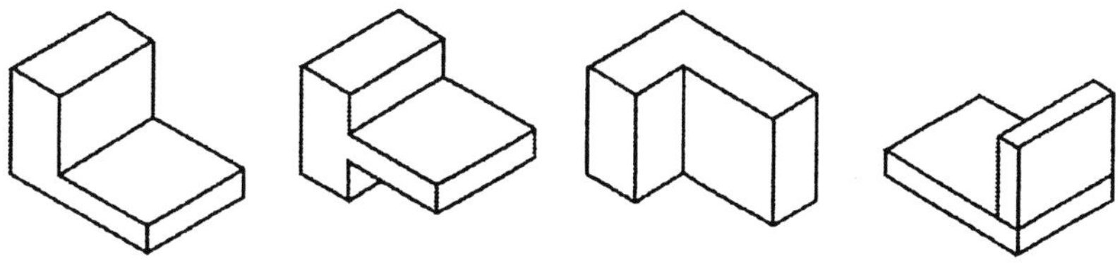

A B C D

In sample question 29 below, there is, on the left, a drawing of a flat piece of paper and, on the right, four figures labeled A, B, C, and D. When the paper is bent on the dotted lines it will form one of the figures on the right. Decide which alternative can be formed from the flat piece.

29.

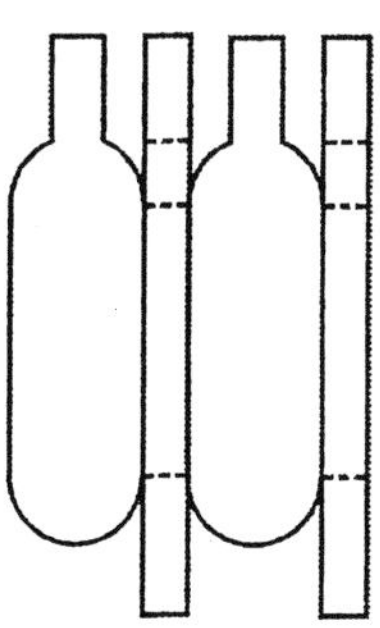

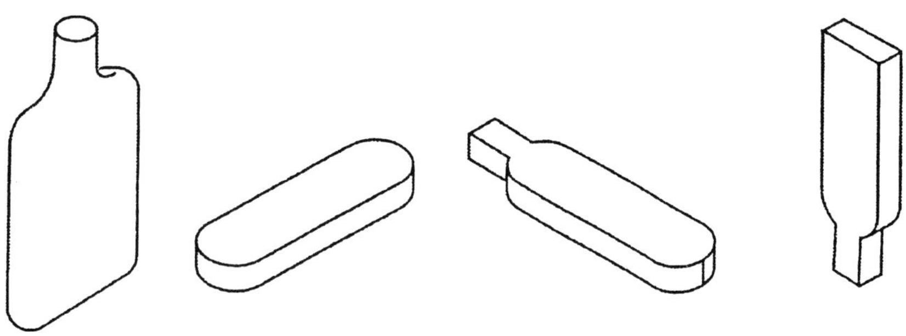

A B C D

In each of the sample questions below, look at the symbols in the first two boxes. Something about the three symbols in the first box makes them alike; something about the two symbols in the other box with the question mark makes them alike. Look for some characteristic that is common to all symbols in the same box, yet makes them different from the symbols in the other box. Among the five answer choices, find the symbol that can best be substituted for the question mark, because it is like the symbols in the second box, and, for the same reason different from those in the first box.

30.

A B C D E

In sample question 30 above, all the symbols in the first box are vertical lines. The second box has two lines, one broken and one solid. Their likeness to each other consists in their being horizontal; and their being horizontal makes them different from the vertical lines in the other box. The answer must be the only one of the five lettered choices that is a horizontal line, either broken or solid. NOTE: There is not supposed to be a series or progression in these symbol questions. If you look for a progression in the first box and the second box, you will be wasting time. Remember, look for a likeness within each box and a difference between the two boxes.

Now do sample questions 31 and 32.

31.

A B C D E

32.

A B C D E

33. In Figure 3-8-6, what is the measurement of Dimension F? Drawing is not actual size.

A) 1 3/4 inches
B) 2 1/4 inches
C) 2 1/2 inches
D) 3 3/4 inches
E) None of the above

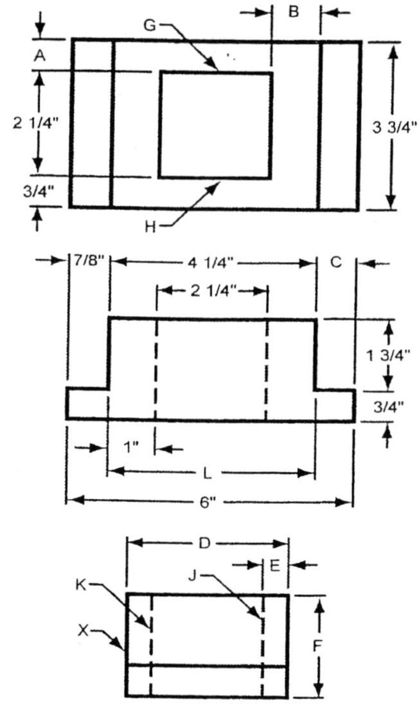

Figure 3-8-6

34. In Figure 160-57, what is the current flow through R_3 when:

V = 50 volts

R_1 = 25 ohms

R_2 = 25 ohms

R3 = 50 ohms

R_4 = 50 ohms

R_5 = 50 ohms

and the current through the entire circuit totals one amp?

A) 0.5 amp
B) 5.0 amps
C) 5.0 milliamps
D) 50.0 milliamps
E) None of the above

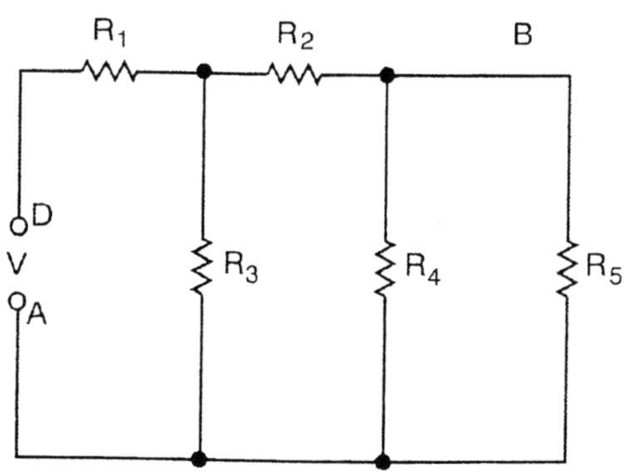

Figure 160-57

Correct Answers to Sample Questions – Part II

1 Ⓐ Ⓑ ● Ⓓ Ⓔ
2 Ⓐ Ⓑ Ⓒ ● Ⓔ
3 ● Ⓑ Ⓒ Ⓓ Ⓔ
4 ● Ⓑ Ⓒ Ⓓ Ⓔ
5 ● Ⓑ Ⓒ Ⓓ Ⓔ
6 Ⓐ ● Ⓒ Ⓓ Ⓔ
7 Ⓐ Ⓑ Ⓒ Ⓓ ●
8 Ⓐ Ⓑ ● Ⓓ Ⓔ
9 Ⓐ Ⓑ ● Ⓓ Ⓔ
10 Ⓐ Ⓑ Ⓒ Ⓓ ●
11 Ⓐ Ⓑ Ⓒ Ⓓ ●
12 ● Ⓑ Ⓒ Ⓓ Ⓔ
13 Ⓐ Ⓑ Ⓒ ● Ⓔ
14 ● Ⓑ Ⓒ Ⓓ Ⓔ
15 Ⓐ Ⓑ Ⓒ Ⓓ ●
16 Ⓐ ● Ⓒ Ⓓ Ⓔ
17 Ⓐ Ⓑ Ⓒ ● Ⓔ
18 Ⓐ Ⓑ Ⓒ Ⓓ ●
19 Ⓐ Ⓑ ● Ⓓ Ⓔ
20 Ⓐ Ⓑ Ⓒ ● Ⓔ
21 Ⓐ Ⓑ ● Ⓓ Ⓔ
22 Ⓐ ● Ⓒ Ⓓ Ⓔ
23 Ⓐ ● Ⓒ Ⓓ Ⓔ
24 Ⓐ Ⓑ Ⓒ ● Ⓔ
25 Ⓐ Ⓑ Ⓒ ● Ⓔ
26 Ⓐ Ⓑ Ⓒ Ⓓ ●
27 Ⓐ Ⓑ ● Ⓓ Ⓔ
28 Ⓐ Ⓑ ● Ⓓ Ⓔ
29 Ⓐ Ⓑ ● Ⓓ Ⓔ
30 Ⓐ Ⓑ ● Ⓓ Ⓔ
31 Ⓐ Ⓑ Ⓒ Ⓓ ●
32 Ⓐ Ⓑ Ⓒ ● Ⓔ
33 Ⓐ Ⓑ ● Ⓓ Ⓔ
34 ● Ⓑ Ⓒ Ⓓ Ⓔ

HOW TO TAKE A TEST

. YOU MUST PASS AN EXAMINATION

A. WHAT EVERY CANDIDATE SHOULD KNOW

Examination applicants often ask us for help in preparing for the written test. What can I study in advance? What kinds of questions will be asked? How will the test be given? How will the papers be graded?

As an applicant for a civil service examination, you may be wondering about some of these things. Our purpose here is to suggest effective methods of advance study and to describe civil service examinations.

Your chances for success on this examination can be increased if you know how to prepare. Those "pre-examination jitters" can be reduced if you know what to expect. You can even experience an adventure in good citizenship if you know why civil service examinations are given.

B. WHY ARE CIVIL SERVICE EXAMINATIONS GIVEN?

Civil service examinations are important to you in two ways. As a citizen, you want public jobs filled by employees who know how to do their work. As a job-seeker, you want a fair chance to compete for that job on an equal footing with other candidates. The best known means of accomplishing this two-fold goal is the competitive examination.

Examinations are widely publicized throughout the nation. They may be administered for jobs in federal, state, city, municipal, town, or village governments or agencies.

Any citizen may apply, with some limitations, such as the age or residence of applicants. Your experience and education may be reviewed to see whether you meet the requirements for the particular examination. When these requirements exist, they are reasonable and are applied consistently to all applicants. Thus, a competitive examination may cause you some uneasiness now, but it is your privilege and safeguard.

C. HOW ARE CIVIL SERVICE EXAMINATIONS DEVELOPED?

Examinations are carefully written by trained technicians who are specialists in the field known as "psychological measurement," in consultation with recognized authorities in the field of work that the test will cover. These experts recommend the subject matter areas or skills to be tested; only those knowledges or skills important to your success on the job are included. The most reliable books and source materials available are used as references. Together, the experts and technicians judge the difficulty level of the questions.

Test technicians know how to phrase questions so that the problem is clearly stated. Their ethics do not permit "trick" or "catch" questions. Questions may have been tried out on sample groups, or subjected to statistical analysis, to determine their usefulness.

Written tests are often used in combination with performance tests, ratings of training and experience, and oral interviews. All of these measures combine to form the best known means of finding the right man for the right job.

II. HOW TO PASS THE WRITTEN TEST

A. NATURE OF THE EXAMINATION

To prepare intelligently for civil service examinations, you should know how they differ from school examinations you have taken. In school you were assigned certain definite pages to read or subjects to cover. The examination questions were quite detailed and usually emphasized memory. Civil service examinations, on the other hand, try to discover your present ability to perform the duties of a position, plus your potentiality to learn these duties. In other words, a civil service examination attempts to predict how successful you will be. Questions cover such a broad area that they cannot be as minute and detailed as school examination questions.

In the public service similar kinds of work, or positions, are grouped together in one "class." This process is known as "position-classification." All the positions in a class are paid according to the salary range for that class. One class title covers all these positions, and they are all tested by the same examination.

B. FOUR BASIC STEPS

1. Study the Announcement.--How, then, can you know what subjects to study? Our best answer is: "Learn as much as possible about the class of positions for which you have applied." The examination will test the knowledge, skills, and abilities needed to do the work.

Your most valuable source of information about the position you want is the official announcement of the examination. This announcement lists the training and experience qualifications. Check these standards and apply only if you come reasonably close to meeting them.

The brief description of the position in the examination announcement offers some clues to the subjects which will be tested. Think about the job itself. Review the duties in your mind. Can you perform them, or are there some in which you are rusty? Fill in the blank spots in your preparation.

Many jurisdictions preview the written test in the examination announcement by including a section called "Knowledge and Abilities Required," "Scope of Examination," or some similar heading. Here you will find out specifically what fields will be tested.

2. Review Your Own Background.-- Once you learn in general what the position is all about, and what you need to know to do the work, ask yourself which subjects you already know fairly well and which need improvement. You may wonder whether to concentrate on improving your strong areas or on building some background in your fields of weakness. When the announcement has specified "some knowledge" or "considerable knowledge," or has used adjectives such as "beginning principles of ...." or "advanced .....methods," you can get a clue as to the number and difficulty of questions to be asked in any given field. More questions, and hence broader coverage, would be included for those subjects which are more important in the work. Now weigh your strengths and weaknesses against the job requirements and prepare accordingly.

3. Determine the Level of the Position.-- Another way to tell how intensively you should prepare is to understand the level of the job for which you are applying. Is it the entering level? In other words, is this the position in which beginners in a field of work are hired? Or is it an intermediate or advanced level? Sometimes this is indicated by such words as "Junior" or "Senior" in the class title.Other jurisdictions use Roman numerals to designate the level: Clerk I,

Clerk II, for example. The word "Supervisor" sometimes appears in the title. If the level is not indicated by the title, check the description of duties. Will you be working under very close supervision, or will you have responsibility for independent decisions in this work?

4. Choose Appropriate Study Materials.-- Now that you know the subjects to be examined and the relative amount of each subject to be covered, you can choose suitable study materials. For beginning level jobs, or even advanced ones, if you have a pronounced weakness in some aspect of your training, read a modern, standard textbook in that field. Be sure it is up-to-date and has general coverage. Such books are normally available at your library, and the librarian will be glad to help you locate one. For entry level positions, questions of appropriate difficulty are chosen -- neither highly advanced questions, nor those too simple. Such questions require careful thought but not advanced training.

If the position for which you are applying is technical or advanced, you will read more advanced, specialized material. If you are already familiar with the basic principles of your field, elementary textbooks would waste your time. Concentrate on advanced textbooks and technical periodicals. Think through the concepts and review difficult problems in your field.

These are all general sources. You can get more ideas on your own initiative, following these leads. For example, training manuals and publications of the government agency which employs workers in your field can be useful, particularly for technical and professional positions. A letter or visit to the government department involved may result in more specific study suggestions, and certainly will provide you with a more definite idea of the exact nature of the position you are seeking.

II. KINDS OF TESTS

Tests are used for purposes other than measuring knowledge and ability to perform specified duties. For some positions, it is equally important to test ability to make adjustments to new situations or to profit from training. In others, basic mental abilities not dependent upon information are essential. Questions which test these things may not appear as pertinent to the duties of the position as those which test for knowledge and information. Yet they are often highly important parts of a fair examination. For very general questions, it is almost impossible to help you direct your study efforts. What we can do is to point out some of the more common of these general abilities needed in public service positions and describe some typical questions.

1. General Information

Broad, general information has been found useful for predicting job success in some kinds of work. This is tested in a variety of ways, from vocabulary lists to questions about current events. Basic background in some field of work, such as sociology or economics, may be sampled in a group of questions. Often these are principles which have become familiar to most persons through "exposure" rather than through formal training. It is difficult to advise you how to study for these questions; being alert to the world around you is our best suggestion.

2. Verbal Ability

An example of an ability needed in many positions is verbal or language ability. Verbal ability is, in brief, the ability to use and understand words. Vocabulary and grammar tests are typical measures of this ability. "Reading comprehension" or "paragraph interpretation" questions are common in many kinds of civil service tests. You are given a paragraph of written material and asked to find its central meaning.

3. Numerical Ability

Number skills can be tested by the familiar arithmetic problem, by checking paired lists of numbers to see which are alike and which are different, or by interpreting charts and graphs. In the latter test, a graph may be printed in the test booklet which you are asked to use as the basis for answering questions.

4. Observation

A popular test for law-enforcement positions is the observation test. A picture is shown to you for several minutes, then taken away. Questions about the picture test your ability to observe both details and larger elements.

5. Following Directions

In many positions in the public service, the employee must be able to carry out written instructions dependably and accurately. You may be given a chart with several columns, each column listing a variety of information. The questions require you to carry out directions involving the information given in the chart.

6. Skills and Aptitudes

Performance tests effectively measure some manual skills and aptitudes. When the skill is one in which you are trained, such as typing or shorthand, you can practice. These tests are often very much like those given in business school or high school courses. For many of the other skills and aptitudes, however, no short-time preparation can be made. Skills and abilities natural to you or that you have developed throughout your lifetime are being tested.

Many of the general questions just described provide all the data needed to answer the questions and ask you to use your reasoning ability to find the answers. Your best preparation for these tests, as well as for tests of facts and ideas, is to be at your physical and mental best. You, no doubt, have your own methods of getting into an exam-taking mood and keeping "in shape." The next section lists some ideas on this subject.

IV. KINDS OF QUESTIONS

Only rarely is the "essay" question, which you answer in narrative form, used in civil service tests. Civil service tests are usually of the short-answer type. Full instructions for answering these questions will be given to you at the examination. But in case this is your first experience with short-answer questions and separate answer sheets, here is what you need to know.

1. Multiple-Choice Questions

Most popular of the short-answer questions is the "multiple-choice" or "best-answer" question. It can be used, for example, to test for factual knowledge, ability to solve problems, or judgment in meeting situations found at work.

A multiple-choice question is normally one of three types:

(1) It can begin with an incomplete statement followed by several possible endings. You are to find the one ending which *best* completes the statement, although some of the others may not be entirely wrong.

(2) It can also be a complete statement in the form of a question which is answered by choosing one of the statements listed.

(3) It can be in the form of a problem -- again you select the best answer.

Here is an example of a multiple-choice question with a discussion which should give you some clues as to the method for choosing the right answer.

SAMPLE QUESTION:

When an employee has a complaint about his assignment, the action which will *best* help him overcome his difficulty is

(A) to discuss his difficulty with his co-workers
(B) to take the problem to the head of the organization
(C) to take the problem to the person who gave him the assignment
(D) to say nothing to anyone about his complaint

In answering this question you should study each of the choices to find which is best. Consider choice (A). Certainly an employee may discuss his complaint with fellow employees, but no change or improvement can result, and the complaint remains unsolved. Choice (B) is a poor choice since the head of the organization probably does not know what assignment you have been given, and taking your problem to him is known as "going over the head" of the supervisor. The supervisor, or person who made the assignment, is the person who can clarify it or correct any injustice. Choice (C) is, therefore, correct. To say nothing, as in choice (D), is unwise. Supervisors have an interest in knowing the problems employees are facing, and the employee is seeking a solution to his problem.

2. True-False Questions

The "true-false" or "right-wrong" form of question is sometimes used. Here a complete statement is given. Your problem is to decide whether the statement is right or wrong.

SAMPLE QUESTION:

A person-to-person long distance telephone call costs less than a station-to-station call to the same city.

This question is wrong, or "false," since person-to-person calls are more expensive.

This is not a complete list of all possible question forms, although most of the others are variations of these common types. You will always get complete directions for answering questions. Be sure you understand *how* to mark your answers -- ask questions until you do.

V. RECORDING YOUR ANSWERS

For an examination with very few applicants, you may be told to record your answers in the test booklet itself. Separate answer sheets are much more common. If this separate answer sheet is to be scored by machine -- and this is often the case -- it is highly important that you mark your answers correctly in order to get credit.

An electric test-scoring machine is often used in civil service offices because of the speed with which papers can be scored. Machine-scored answer sheets must be marked with a special pencil, which will be given to you. This pencil has a high graphite content which responds to the electrical scoring machine. As a matter of fact, stray dots may register as answers, so do not let your pencil rest on the answer sheet while you are pondering the correct answer. Also, if your pencil lead breaks or is otherwise defective, ask for another.

Since the answer sheet will be dropped in a slot in the scoring machine, be careful not to bend the corners or get the paper crumpled.

The answer sheet normally has five vertical columns of numbers, with 30 numbers to a column. These numbers correspond to the question numbers in your test booklet. After each number, going across the page, are four or five pairs of dotted lines. These short dotted lines have small letters or numbers above them. The first two pairs may also have a "T" and "F" above the letters. This indicates that the first two pairs only are to be used if the questions are of the true-false type. If the questions are multiple-choice, disregard this "T" and "F" completely, and pay attention only to the small number or letters.

Answer your questions in the manner of the sample that follows. Proceed in the sequential steps outlined below.

Assume that you are answering question 32, which is:

32. The largest city in the United States is:

A. Washington, D.C. B. New York City C. Chicago
D. Detroit E. San Francisco

1. Choose the answer you think is best.
   New York City is the largest, so choice B is correct.
2. Find the row of dotted lines numbered the same as the question you are answering.
   This is question number 32, so find row number 32.
3. Find the pair of dotted lines corresponding to the answer you have chosen.
   You have chosen answer B, so find the pair of dotted lines marked "B".
4. Make a solid black mark between the dotted lines.
   Go up and down two or three times with your pencil so plenty of graphite rubs off, but do not let the mark get outside or above the dots.

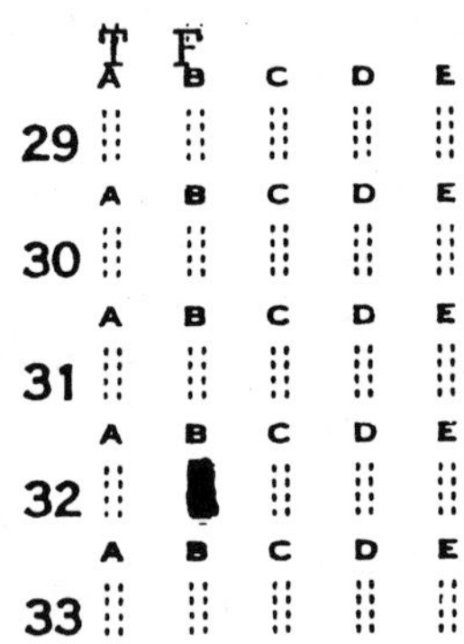

VI. BEFORE THE TEST

Common sense will help you find procedures to follow to get ready for an examination. Too many of us, however, overlook these sensible measures. Indeed, nervousness and fatigue have been found to be the most serious reasons why applicants fail to do their best on civil service tests. Here is a list of reminders.

1. Begin Your Preparation Early

Don't wait until the last minute to go scurrying around for books and materials or to find out what the position is all about.

2. Prepare Continuously

An hour a night for a week is better than an all-night cram session. This has been definitely established. What is more, a night a week for a month will return better dividends than crowding your study into a shorter period of time.

3. Locate the Place of the Examination

You have been sent a notice telling you when and where to report for the examination. If the location is in a different town or otherwise unfamiliar to you, it would be well to inquire the best route and learn something about the building.

4. Relax the Night Before the Test

Allow your mind to rest. Do not study at all that night. Plan some mild recreation or diversion; then go to bed early and get a good night's sleep.

5. Get Up Early Enough to Make a Leisurely Trip to the Place for the Test

Then unforeseen events, traffic snarls, unfamiliar buildings, will not upset you.

6. Dress Comfortably

A written test is not a fashion show. You will be known by number and not by name, so wear something comfortable.

7. Leave Excess Paraphernalia at Home

Shopping bags and odd bundles will get in your way. You need bring only the items mentioned in the official notice sent to you; usually everything you need is provided. Do not bring reference books to the examination. They will only confuse those last minutes and be taken away from you when in the test room.

8. Arrive Somewhat Ahead of Time

If because of transportation schedules you must get there very early, bring a newspaper or magazine to take your mind off yourself while waiting.

9. Locate the Examination Room

When you have found the proper room, you will be directed to the seat or part of the room where you will sit. Sometimes you are given a sheet of instructions to read while you are waiting. Do not fill out any forms until you are told to do so; just read them and be ready.

10. Relax and Prepare to Listen to the Instructions

11. If you have any physical problem that may keep you from doing your best, be sure to tell the test administrator. If you are sick, or in poor health, you really cannot do your best on the test. You can come back and take the test some other time.

AT THE TEST

The day of the test is here and you have the test booklet in your hand. The temptation to get going is very strong. Caution! There is more to success than knowing the right answers. You must know how to identify your papers and understand variations in the type of short-answer question used in this particular examination. Follow these suggestions for maximum results from your efforts:

1. Cooperate with the Monitor

The test administrator has a duty to create a situation in which you can be as much at ease as possible. He will give instructions, tell you when to begin, check to see that you are marking your answer sheet correctly. He is not there to guard you, although he will see that your competitors do not take unfair advantage. He wants to help you do your best.

2. Listen to All Instructions

Don't jump the gun! Wait until you understand all directions. In most civil service tests you get more time than you need to answer the questions. So don't get in a hurry. Read each word of instructions until you clearly understand the meaning. Study the examples. Listen to all announcements. Follow directions. Ask questions if you do not understand what to do.

3. Identify Your Papers

Civil service examinations are usually identified by number only. You will be assigned a number; you must not put your name on your test papers. Be sure to copy your number correctly. Since more than one examination may be given, copy your exact examination title.

4. Plan Your Time

Unless you are told that a test is a "speed" or "rate-of-work" test, speed itself is not usually important. Time enough to answer all the questions will be provided. But this does not mean that you have all day. An overall time limit has been set. Divide the total time (in minutes) by the number of questions to get the approximate time you have for each question.

5. Do Not Linger Over Difficult Questions

If you come across a difficult question, mark it with a paper clip (useful to have along) and come back to it when you have been through the booklet. One caution if you do this -- be sure to skip a number on your answer sheet too. Check often to be sure that you have not lost your place and that you are marking in the row numbered the same as the question you are answering.

6. Read the Questions

Be sure you know what the question asks! Many capable people are unsuccessful because they failed to *read* the questions correctly.

7. Answer All Questions

Unless you have been instructed that a penalty will be deducted for incorrect answers, it is better to guess than to omit a question.

8. Speed Tests

It is often better *not* to guess on speed tests. It has been found that on timed tests people are tempted to spend the last few seconds before time is called in marking answers at random -- without even reading them -- in the hope of picking up a few extra points. To discourage this practice, the instructions may warn you that your score will be "corrected" for guessing. That is, a penalty will be applied. The incorrect answers will be deducted from the correct ones, or some other penalty formula will be used.

9. Review Your Answers

If you finish before time is called, go back to the questions you guessed or omitted to give further thought to them. Review other answers if you have time.

10. Return Your Test Materials

If you are ready to leave before others have finished or time is called, take *all* your materials to the monitor and leave quietly. Never take any test material with you. The monitor can discover whose papers are not complete, and taking a test booklet may be grounds for disqualification.

I. EXAMINATION TECHNIQUES

1. Read the *general* instructions carefully. These are usually printed on the first page of the examination booklet. As a rule, these instructions refer to the timing of the examination; the fact that you should not start work until the signal and must stop work at a signal, etc. If there are any *special* instructions, such as a choice of questions to be answered, make sure that you note this instruction carefully.

2. When you are ready to start work on the examination, that is as soon as the signal has been given, read the instructions to each question booklet, underline any key words or phrases, such as *least, best, outline, describe,* and the like. In this way you will tend to answer as requested rather than discover on reviewing your paper that you *listed without describing,* that you selected the *worst* choice rather than the *best* choice, etc.

3. If the examination is of the objective or so-called multiple-choice type, that is, each question will also give a series of possible answers: A, B, C, or D, and you are called upon to select the best answer and write the letter next to that answer on your answer paper, it is advisable to start answering each question in turn. There may be anywhere from 50 to 100 such questions in the three or four hours allotted and you can see how much time would be taken if you read through all the questions before beginning to answer any. Furthermore, if you come across a question or a group of questions which you know would be difficult to answer, it would undoubtedly affect your handling of all the other questions.

4. If the examination is of the esssay-type and contains but a few questions, it is a moot point as to whether you should read all the questions before starting to answer any one. Of course if you are given a choice, say five out of seven and the like, then it is essential to read all the questions so you can eliminate the two which are most difficult. If, however, you are asked to answer all the questions, there may be danger in trying to answer the easiest one first because you may find that you will spend too much time on it. The best technique is to answer the first question, then proceed to the second, etc.

5. Time your answers. Before the examination begins, write down the time it started, then add the time allowed for the examination and write down the time it must be completed, then divide the time available somewhat as follows:

(a) If 3½ hours are allowed, that would be 210 minutes. If you have 80 objective-type questions, that would be an average of 2½ minutes per question. Allow yourself no more than 2 minutes per question, or a total of 160 minutes, which will permit about 50 minutes to review.

(b) If for the time allotment of 210 minutes, there are 7 essay questions to answer, that would average about 30 minutes a question. Give yourself only 25 minutes per question so that you have about 35 minutes to review.

6. The most important instruction is *to read each question* and make sure you know what is wanted. The second most important instruction is to *time yourself properly* so that you answer every question. The third most important instruction is to *answer every question*. Guess if you have to but include something for each question. Remember that you will receive no credit for a blank and will probably receive some credit if you write something in answer to an essay question. If you guess a letter, say "B" for a multiple-choice question, you may have guessed right. If you leave a blank as the answer to a multiple-choice question, the examiners may respect your feelings but it will not add a point to your score.

7. Suggestions

a. <u>Objective-Type Questions</u>

(1) Examine the question booklet for proper sequence of pages and questions.
(2) Read all instructions carefully.
(3) Skip any question which seems too difficult; return to it after all other questions have been answered.
(4) Apportion your time properly; do not spend too much time on any single question or group of questions.
(5) Note and underline key words -- *all, most, fewest, least, best, worst, same, opposite*.
(6) Pay particular attention to negatives.
(7) Note unusual option, e.g., unduly long, short, complex, different or similar in content to the body of the question.
(8) Observe the use of "hedging" words -- *probably, may, most likely, etc.*
(9) Make sure that your answer is put next to the same number as the question.
(10) Do not second-guess unless you have good reason to believe the second answer is definitely more correct.
(11) Cross out original answer if you decide another answer is more accurate; do not erase.
(12) Answer all questions; guess unless instructed otherwise.
(13) Leave time for review.

b. <u>Essay-Type Questions</u>

(1) Read each question carefully.
(2) Determine exactly what is wanted. Underline key words or phrases.
(3) Decide on outline or paragraph answer.
(4) Include many different points and elements unless asked to develop any one or two points or elements.
(5) Show impartiality by giving pros and cons unless directed to select one side only.
(6) Make and write down any assumptions you find necessary to answer the question.
(7) Watch your English, grammar, punctuation, choice of words.
(8) Time your answers; don't crowd material.

8. Answering the Essay Question

Most essay questions can be answered by framing the specific response around several key words or ideas. Here are a few such key words or ideas:

M's: manpower, materials, methods, money, management;
P's: purpose, program, policy, plan, procedure, practice, problems, pitfalls, personnel, public relations.

a. Six Basic Steps in Handling Problems:
   (1) Preliminary plan and background development
   (2) Collect information, data and facts
   (3) Analyze and interpret information, data and facts
   (4) Analyze and develop solutions as well as make recommendations
   (5) Prepare report and sell recommendations
   (6) Install recommendations and follow up effectiveness

b. Pitfalls to Avoid
   (1) *Taking things for granted*
   A statement of the situation does not necessarily imply that each of the elements is necessarily true; for example, a complaint may be invalid and biased so that all that can be taken for granted is that a complaint has been registered.
   (2) *Considering only one side of a situation*
   Wherever possible, indicate several alternatives and then point out the reasons you selected the best one.
   (3) *Failing to indicate follow-up*
   Whenever your answer indicates action on your part, make certain that you will take proper follow-up action to see how successful your recommendations, procedures, or actions turn out to be.
   (4) *Taking too long in answering any single question*
   Remember to time your answers properly.

IX. AFTER THE TEST

Scoring procedures differ in detail among civil service jurisdictions although the general principles are the same. Whether the papers are hand-scored or graded by the electric scoring machine we have described, they are nearly always graded by number. That is, the person who marks the paper knows only the number -- never the name -- of the applicant. Not until all the papers have been graded will they be matched with names. If other tests, such as training and experience or oral interview ratings have been given, scores will be combined. Different parts of the examination usually have different weights. For example, the written test might count 60 percent of the final grade, and a rating of training and experience 40 percent. In many jurisdictions, veterans will have a certain number of points added to their grades.

After the final grade has been determined, the names are placed in grade order and an eligible list is established. There are various methods for resolving ties between those who get the same final grade: probably the most common is to place first the name of the person whose application was received first. Job offers are made from the eligible list in the order the names appear on it.

You will be notified of your grade and your rank order as soon as all these computations have been made. This will be done as rapidly as possible.

People who are found to meet the requirements in the announcement are called "eligibles." Their names are put on a list of eligibles. An eligible's chances of getting a job depend on how high he stands on this list and how fast agencies are filling jobs from the list.

When a job is to be filled from a list of eligibles, the agency asks for the names of people on the list of eligibles for that job.

When the civil service commission receives this request, it sends to the agency the names of the three people highest on the list. Or, if the job to be filled has specialized requirements, the office sends the agency, from the general list, the names of the top three persons who meet those requirements.

The appointing officer makes a choice from among the three people whose names were sent to him. If the selected person accepts the appointment, the names of the others are put back on the list to be considered for future openings.

That is the rule in hiring from all kinds of eligible lists, whether they are for typist, carpenter, chemist, or something else. For every vacancy, the appointing officer has his choice of any one of the top three eligibles on the list. This explains why the person whose name is on top of the list sometimes does not get an appointment when some of the persons lower on the list do. If the appointing officer chooses the No.2 or No.3 eligible, the No.1 eligible does not get a job at once, but stays on the list until he is appointed or the list is terminated.

X. HOW TO PASS THE INTERVIEW TEST

The examination for which you applied requires an oral interview test. You have already taken the written test and you are now being called for the interview test -- the final part of the formal examination.

You may think that it is not possible to prepare for an interview test and that there are no procedures to follow during an interview.

Our purpose is to point out some things you can do in advance that will help you and some good rules to follow and pitfalls to avoid while you are being interviewed.

A. WHAT IS AN INTERVIEW SUPPOSED TO TEST?

The written examination is designed to test the technical knowledge and competence of the candidate; the oral is designed to evaluate intangible qualities, not readily measured otherwise, and to establish a list showing the relative fitness of each candidate, *as measured against his competitors,* for the position sought. Scoring is not on the basis of "right" or "wrong," but on a sliding scale of values ranging from "not passable" to "outstanding." As a matter of fact, it is possible to achieve a relatively low score without a single "incorrect" answer because of evident weakness in the qualities being measured,

Occasionally, an examination may consist entirely of an oral test -- either an individual or a group oral. In such cases, information is sought concerning the technical knowledges and abilities of the candidate, since there has been no written examination for this purpose. More commonly, however, an oral test is used to supplement a written examination.

B. WHO CONDUCTS INTERVIEWS?

The composition of oral boards varies among different jurisdictions. In nearly all, a representative of the personnel department serves as chairman. One of the members of the board may be a representative of the department in which the candidate would work. In some cases, "outside experts" are used, and, frequently, a business man or some other representative of the general public is asked to serve. Labor and management or other special groups may be represented. The aim is to secure the services of experts in the appropriate field.

However the board is composed, it is a good idea (and not at all improper or unethical) to ascertain in advance of the interview who the members are and what groups they represent. When you are introduced to them, you will have some idea of their backgrounds and interests, and at least you will not stutter and stammer over their names.

C. WHAT TO DO BEFORE THE INTERVIEW

While knowledge about the board members is useful and takes some of the surprise element out of the interview, there is other preparation which is more substantive. It *is* possible to prepare for an oral -- in several ways:

1. Keep a Copy of Your Application and Review it Carefully Before the Interview

   This may be the only document before the oral board, and the starting point of the interview. Know what experience and education you have listed there, and the sequence and dates of it. Sometimes the board will ask *you* to review the highlights of your experience for them; you should not have to hem and haw doing it.

2. Study the Class Specification and the Examination Announcement

   Usually, the oral board has one or both of these to guide them. The qualities,characteristics,or knowledges required by the position sought are stated in these documents. They offer valuable clues as to the nature of the oral interview. For example, if the job involves supervisory responsibilities, the announcement will usually indicate that knowledge of modern supervisory methods and the qualifications of the candidate as a supervisor will be tested. If so, you can expect such questions, frequently in the form of a hypothetical situation which you are expected to solve. *Never* go into an oral without knowledge of the duties and responsibilities of the job you seek.

3. Think Through Each Qualification Required

   Try to visualize the kind of questions *you* would ask if you were a board member. How well could you answer them? Try especially to appraise your own knowledge and background in each area, *measured against the job sought,* and identify any areas in which you are weak. Be critical and realistic -- do not flatter yourself.

4. Do Some General Reading in Areas in Which You Feel You May be Weak

   For example, if the job involves supervision and your past experience has *not,* some general reading in supervisory methods and practices, particularly in the field of human relations, might be useful. *Do not* study agency procedures or detailed manuals. The oral board will be testing your understanding and capacity, *not* your memory.

5. Get a Good Night's Sleep and Watch Your General Health and Mental Attitude

   You will want a clear head at the interview. Take care of a cold or other minor ailment, and, of course, *no hangovers.*

D. WHAT TO DO THE DAY OF THE INTERVIEW

Now comes the day of the interview itself. Give yourself plenty of time to get there. Plan to arrive somewhat ahead of the scheduled time, particularly if your appointment is in the fore part of the day. If a previous candidate fails to appear, the board might be ready for you a bit early. By early afternoon an oral board is almost invariably behind schedule if there are many candidates, and you may have to wait. Take along a book or magazine to read, or your application to review. But leave any extraneous material in the waiting room when you go in for your interview. In any event, relax and compose yourself.

The matter of dress is important. The board is forming impressions about you -- from your experience, your manners, your attitudes, and from your appearance. Give your personal appearance careful attention. Dress your *best*, but not your flashiest. Choose conservative, appropriate clothing, and be sure it and you are immaculate. This is a business interview, and your appearance should indicate that you regard it as such. Besides, being well-groomed and properly dressed will help boost your confidence.

Sooner or later, someone will call your name and escort you into the interview room. *This is it.* From here on you are on your own. It is too late for any more preparation. But, remember, you asked for this opportunity to prove your fitness, and you are here because your request was granted.

E. WHAT HAPPENS WHEN YOU GO IN?

The usual sequence of events will be as follows: The clerk (who is often the board stenographer) will introduce you to the chairman of the oral board, who will introduce you to each other member of the board. Acknowledge the introductions before you sit down. Do not be surprised if you find a microphone facing you or a stenotypist sitting by. Oral interviews are usually recorded, in the event of an appeal or other review.

Usually the chairman of the board will open the interview by reviewing the highlights of your education and work experience from your application -- primarily for the benefit of the other members of the board, as well as to get the material into the record. Do not interrupt or comment unless there is an error or significant misinterpretation; if so, do not hesitate. But do not quibble about insignificant matters. Usually, also, he will ask you some question about your education, your experience, or your present job -- partly to get you started talking, to establish the interviewing "rapport." He may start the actual questioning, or turn it over to one of the other members. Frequently each member undertakes the questioning on a particular area, one in which he is perhaps most competent. So you can expect each member to participate in the examination. And because the time is limited, you may expect some rather abrupt switches in the direction the questioning takes. Do not be upset by it. Normally, a board member will not pursue a single line of questioning unless he discovers a particular strength or weakness.

After each member has participated, the chairman will usually ask whether any member has any further questions, then will ask you if you have anything you wish to add. Unless you are expecting this question, it may floor you. Or worse, it may start you off on an extended, extemporaneous speech. The board is not usually seeking more information. The question is principally to offer you a last opportunity to present further qualifications or to indicate that you have nothing to add. So, if you feel that a significant qualification or characteristic has been overlooked, it is proper to point it out in a sentence or so. Do not compliment the board on the thoroughness of their examination -- they have been sketchy, and you know it. If you wish, merely say, "No thank you, I have nothing further to add." This is a point where you can "talk yourself out" of a good impression or fail to present an important bit of information. *Remember, you close the interview yourself.*

The chairman will then say, "That is all, Mr. Smith, thank you." Do not be startled; the interview is over, and quicker than you think. Say, "Thank you and good morning," gather up your belongings and take your leave. Save your sigh of relief for the other side of the door.

F. HOW TO PUT YOUR BEST FOOT FORWARD

Throughout all this process, you may feel that the board individually and collectively is trying to pierce your defenses, to seek out your hidden weaknesses, and to embarrass and confuse you. Actually, this is not true. They are obliged to make an appraisal of your qualifications for the job you are seeking, and they *want to see you in your best light.* Remember, they must interview all candidates and a noncooperative candidate may become a failure in spite of their best efforts to bring out his qualifications. Here are fifteen(15) suggestions that will help you:

1. Be Natural. Keep Your Attitude Confident, But Not Cocky

If *you* are not confident that you can do the job, do not exexpect the *board* to be. Do not apologize for your weaknesses, try to bring out your strong points. The board is interested in a positive, not a negative presentation. Cockiness will antagonize any board member, and make him wonder if you are covering up a weakness by a false show of strength.

2. Get Comfortable, But Don't Lounge or Sprawl

Sit erectly but not stiffly. A careless posture may lead the board to conclude you are careless in other things, or at least that you are not impressed by the importance of the occasion to you. Either conclusion is natural, even if incorrect. Do not fuss with your clothing, or with a pencil or an ashtray. Your hands may occasionally be useful to emphasize a point; do not let them become a point of distraction.

3. Do Not Wisecrack or Make Small Talk

This is a serious situation, and your attitude should show that you consider it as such. Further, the time of the board is limited; they do not want to waste it, and neither should you.

4. Do Not Exaggerate Your Experience or Abilities

In the first place, from information in the application, from other interviews and other sources, the board may know more about you than you think; in the second place, you probably will not get away with it in the first place. An experienced board is rather adept at spotting such a situation. Do not take the chance.

5. If You Know a Member of the Board, Do Not Make a Point of It, Yet Do Not Hide It.

Certainly you are not fooling him, and probably not the other members of the board. Do not try to take advantage of your acquaintanceship -- it will probably do you little good.

6. Do Not Dominate the Interview

Let the board do that. They will give you the clues -- do not assume that you have to do all the talking. Realize that the board has a number of questions to ask you, and do not try to take up all the interview time by showing off your extensive knowledge of the answer to the first one.

7. Be Attentive

You only have twenty minutes or so, and you should keep your attention at its sharpest throughout. When a member is addressing a problem or a question to you, give him your undivided attention. Address your reply principally to him, but do not exclude the other members of the board.

8. Do Not Interrupt

A board member may be stating a problem for you to analyze. He will ask you a question when the time comes. Let him state the problem, and wait for the question.

9. Make Sure You Understand the Question

Do not try to answer until you are sure what the question is. If it is not clear, restate it in your own words or ask the board member to clarify it for you. But do not haggle about minor elements.

10. Reply Promptly But Not Hastily

A common entry on oral board rating sheets is "candidate responded readily," or "candidate hesitated in replies." Respond as promptly and quickly as you can, but do not jump to a hasty, ill-considered answer.

11. Do Not Be Peremptory in Your Answers

A brief answer is proper -- but do not fire your answer back. That is a losing game from your point of view. The board member can probably ask questions much faster than you can answer them.

12. Do Not Try To Create the Answer You Think the Board Member Wants

He is interested in what kind of mind you have and how it works -- not in playing games. Furthermore, he can usually spot this practice and will usually grade you down on it.

13. Do Not Switch Sides in Your Reply Merely to Agree With a Board Member

Frequently, a member will take a contrary position merely to draw you out and to see if you are willing and able to defend your point of view. Do not start a debate, yet do not surrender a good position. If a position is worth taking, it is worth defending.

1 Do Not Be Afraid to Admit an Error in Judgment if You Are Shown to Be Wrong

The board knows that you are forced to reply without any opportunity for careful consideration. Your answer may be demonstrably wrong. If so, admit it and get on with the interview.

15. Do Not Dwell at Length on Your Present Job

The opening question may relate to your present assignment. Answer the question but do not go into an extended discussion. You are being examined for a *new* job, not your present one. As a matter of fact, try to phrase *all* your answers in terms of the job for which you are being examined.

G. BASIS OF RATING

Probably you will forget most of these "do's" and "don'ts" when you walk into the oral interview room. Even remembering them all will not insure you a passing grade. Perhaps you did not have the qualifications in the first place. But remembering them *will* help you to put your best foot forward, without treading on the toes of the board members.

Rumor and popular opinion to the contrary notwithstanding, an oral board wants you to make the best appearance possible. They know you are under pressure -- but they also want to see how you respond to it as a guide to what your reaction would be under the pressures of the job you seek. They will be influenced by the degree of poise you display, the personal traits you show, and the manner in which you respond.

EXAMINATION SECTION

FOLLOWING ORAL DIRECTIONS

COMMENTARY

A large part of any job is listening to the supervisor and following his instructions. Since it is important that each employee do exactly as he is instructed, this test is used to make sure that each applicant can and will listen carefully and follow through without extra supervision.

The directions in the test are not hard to follow, but you must listen carefully and do exactly what you are told to do.

In order to do this practice section, you must have a friend who will read the directions to you. *Do not read the material in this section yourself; if you do, you will lose the value of this practice.*

DESCRIPTION OF THE TEST

FOLLOWING ORAL DIRECTIONS - SAMPLE QUESTIONS

The directions are to be read at the rate of 80 words per minute. Since not everybody speaks at this speed, your friend should practice reading the 1-minute practice that follows until he can read it in exactly 1 minute whenever he wants to. He will also need a watch with a second hand. Give the 1-Minute Practice box to your friend to use. (Each friend who is helping you will have to use it to practice, so don't throw it away.)

FOR THE PERSON WHO WILL READ THE FOLLOWING ORAL DIRECTIONS TEST TO YOU

The directions should be read at about 80 words per minute. Practice reading aloud the material in the box below until you can do it in exactly 1 minute. This will give you a feel for the way you should read the test material.

1-MINUTE PRACTICE

(This is for practice in reading aloud. It is not the sample test.)

Look at line 20 in your work booklet. There are two circles and two boxes of different sizes with numbers in them. If 7 is less than 3 and if 2 is smaller than 4, write a G in the larger circle. Otherwise write B as in baker in the smaller box. Now on your Code Sheet darken the space for the number-letter combination in the box or circle.

When your friend reads the directions to you, listen carefully and do what he says. If you fall behind and miss a direction, don't get excited. Let that one go and listen for the next one. Since B and D sound very much alike, he will say "B as in baker" when he means B and "D as in dog" when he means D.

He will tell you some things to do with the 5 sample questions below. Then, when he tells you to darken a box on the Sample Answer Sheet, use the one on this page.

SAMPLE QUESTIONS

QUESTION 1. 5____

QUESTION 2. 1 6 4 3 7

QUESTION 3. D B A E C

QUESTION 4. 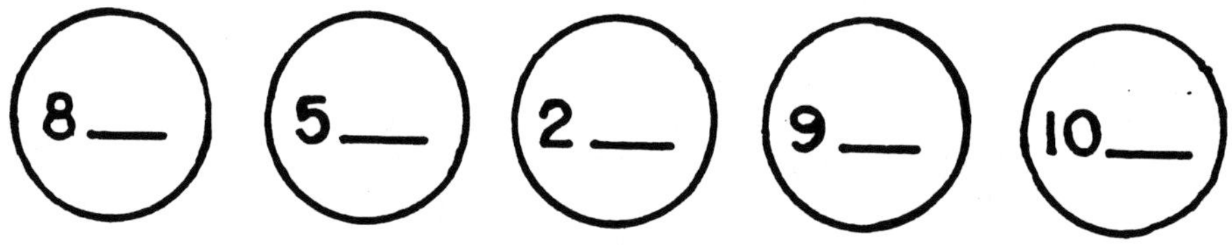

QUESTION 5. 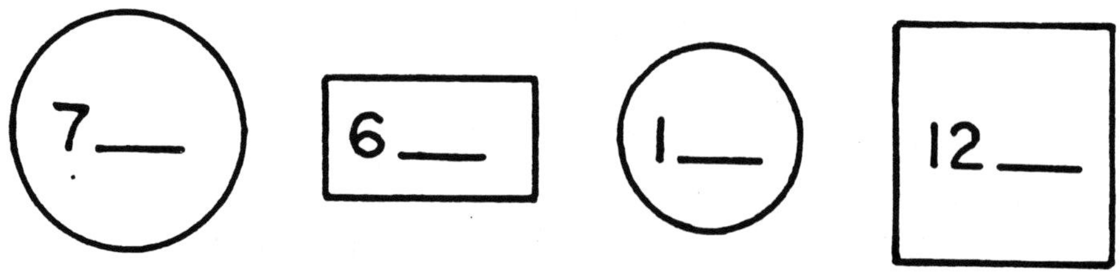

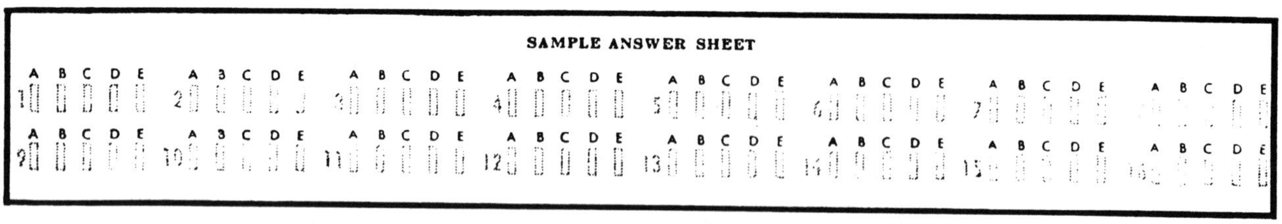

DIRECTIONS to be read. (The words in parentheses should *not* be read aloud. They tell you how long you should pause at the various spots. You should time the pauses with a watch with a second hand. The instruction "Pause slightly" means that you should stop long enough to take a breath.) You should not repeat any directions.

QUESTIONS ON THE SAMPLE

You are to follow the instructions that I read to you. I cannot repeat them.

Look at the Sample Questions. Question 1 has a number and a line beside it. On the line write an A.(Pause 2 seconds.) Now on the Sample Answer Sheet, find number 5 (pause 2 seconds) and darken the box for the letter you just wrote on the line. (Pause 5 seconds.)

Look at Question 2. (Pause slightly.) Draw a line under the third number. (Pause 2 seconds.) Now on the Sample Answer Sheet, find the number under which you just drew a line and darken box B as in baker for that number. (Pause 5 seconds.)

Look at the letters in Question 3. (Pause slightly.) Draw a line under the third letter in the line. (Pause 2 seconds.) Now on your answer sheet, find number 9 (pause 2 seconds) and darken the box for the letter under which you drew a line. (Pause 5 seconds.)

Look at the five circles in Question 4. (Pause slightly.) Each circle has a number and a line in it. Write D as in dog on the blank in the last circle. (Pause 2 seconds.) Now on the Sample Answer Sheet, darken the space for the number-letter combination that is in the circle you just wrote in. (Pause 5 seconds.)

Look at Question 5. (Pause slightly.) There are two circles and two boxes of different sizes with numbers in them. (Pause slightly.) If 4 is more than 2 and if 5 is less than 3, write A in the smaller circle. (Pause slightly.) Otherwise write C in the larger box. (Pause 2 seconds.) Now on the Sample Answer Sheet, darken the space for the number-letter combination in the box or circle in which you just wrote. (Pause 5 seconds.)

Now look at the Sample Answer Sheet. (Pause slightly.) You should have darkened spaces 4B, 5A, 9A, 10D, and 12C on the Sample Answer Sheet.

SUGGESTIONS FOR DOING THE TEST OF FOLLOWING ORAL DIRECTIONS

* Listen carefully to the directions.
* Do exactly what the examiner tells you to do.
* Do not try to get ahead of the examiner.
* If you missed an instruction, wait for the next one.
* Make sure that you darken ONLY one box for each number on the answer sheet.

TEST 1

NOTE: In the examination, the examiner will read aloud directions for you to follow. A sample of directions is given below. The directions are not the same as the directions in the test, but they are somewhat alike. You should have a sheet of lined paper and a pencil as well as the Answer Sheet before you begin.

DIRECTIONS:
1. Fold your lined paper into 4 columns. (Pause for examinee to do this.)
2. In the first column, on the first line, write the number 4. (Pause)
3. On the second line in the same column, write the number 15.
4. Next line, write 12. (Pause)
5. Now go to column 2.
6. Write 35 on the first line (Pause), 26 on the next line, (Pause), and 38 on the third line. (Pause)
7. In column 3, write 11 on the first line (Pause), 18 on the next line (Pause) and 6 last.
8. In column 4, write 16 on the first line next to 4, (Pause), 32 next (Pause) and 19 last.
9. The first number in the first column is 4.
10. Write the letter C next to 4, so it reads 4C. (Pause)
11. The first number in the second column is 35.
12. Write the same letter next to it, so it reads 35C. (Pause)
13. Write C next to the other numbers on the first line, so they read 11C (Pause) and 16C. (Pause)
14. Write the letter A next to each number on the second line, so they read 15A, 26A, etc. (Pause)
15. Write the letter B as in Boy next to each number on the third line. (Pause)
16. Now, take the Answer Sheet you cut out.
17. It has numbers from 1 to 40, and letter spaces.
18. You will mark one space for certain numbers.
19. See how D has been marked for number 1.
20. You will make the same kind of black mark where I tell you. (Pause)
21. Mark 2E. That is, make a black mark at space E for number 2. (Pause)
22. Mark 9C. (Pause)
23. Mark 26C. (Pause)
24. Mark B as in Boy for 15, 16, and 20. (Pause)
25. Mark E for 12, 29, 34, and 39. (Pause)
26. Remember you should *NOT* have more than one mark for any number.

27. If I call a *SECOND* letter for a number where you already have a letter, do *NOT* mark the new letter. Instead, mark the letter A for the number below it.
28. Now I call 2D. You should *NOT* mark 2D, because you have already marked 2E. Instead, mark A for the next number.
29. The next number to 2 is 3. So, you should mark 3A. (Pause)
30. Remember to mark A for the *NEXT* number to the one I call if I call a number where you already have a mark.
31. Now I call 28C. (Pause)
32. Next, 9B. (Pause)
33. 17C. (Pause)
34. 12D. (Pause)
35. 26E and 29D. (Pause)
36. Now, take the sheet of lined paper on which you wrote letters and numbers. (Pause)
37. You will mark the space on your answer sheet for each number and letter you wrote. For example, the first is 4C, so you will mark 4 C on your answer sheet.
38. Do *NOT* start until I tell you.
39. Remember: if you have a mark *ALREADY MADE* for a number, do *NOT* mark another letter. If there is already a mark for a number, make *NO* new mark at all.
40. Start to mark, now!

KEY (CORRECT ANSWERS)

1.		11.	C	21.		31.	
2.	E	12.	E	22.		32.	A
3.	A	13.	A	23.		33.	
4.	C	14.		24.		34.	E
5.		15.	B	25.		35.	C
6.	B	16.	B	26.	C	36.	
7.		17.	C	27.	A	37.	
8.		18.	A	28.	C	38.	B
9.	C	19.	B	29.	E	39.	E
10.	A	20.	B	30.	A	40.	

NOTE: ANY OTHER MARK COUNTS AS WRONG. YOU LOSE CREDIT FOR EACH WRONG MARK.

TEST 2

DIRECTIONS: In the test that follows the examiner will read directions aloud and you will mark your answer sheet as directed.

1. "Mark E for 82, 83, 85, (slight pause) 78, and 102. (Pause)
2. "Mark C for 107, 110, and 103. (Pause)
3. "Mark D as in dog for 101, 110, (slight pause) 76, and 85. (Pause)

"For the next set of questions, mark space E and also mark the letter I call, unless E is already marked. If E is already marked for that number, do not make any mark for that number.

4. "Mark B as in boy for 106, 78, (slight pause) 80, and 84. (Pause)
5. "Mark A for 108, 104, 83, and 109. (Pause)
6. "Mark C for 79, 102, (slight pause) and 77."

KEY (CORRECT ANSWERS)

76.	D	86.		96.		106.	B,E
77.	C,E	87.		97.		107.	C
78.	E	88.		98.		108.	A,E
79.	C,E	89.		99.		109.	A,E
80.	B,E	90.		100.		110.	C,D
81.		91.		101.	D		
82.	E	92.		102.	E		
83.	E	93.		103.	C		
84.	B,E	94.		104.	A,E		
85.	D,E	95.		105.			

NOTE: ANY OTHER MARK COUNTS AS WRONG. YOU LOSE CREDIT FOR EACH WRONG MARK.

TEST 3

DIRECTIONS:

1. "Mark B as in boy for 29, 12, 17, 38, 8. (Pause)
2. "Mark D as in dog for 13, 6, 24, 5. (Pause)
3. "Mark A for 40, 27, 1, 15, 9. (Pause)
4. "Mark E for 13, 39, 31, 4, and 10. (Pause)

"For the next set of questions, mark space E and also mark the letter I call, unless E is already marked. If E is already marked for that number, do *NOT* make any mark for that number.

5. "Mark D as in dog for 12, 9, 19, 23, 2. (Pause)
6. "Mark C for 31, 37, 4, 39. (Pause)
7. "Mark B as in boy for 21, 16, 7, 10, and 26."

KEY (CORRECT ANSWERS)

1.	A	11.		21.	B,E	31.	E
2.	D,E	12.	B,D,E	22.		32.	
3.		13.	D,E	23.	D,E	33.	
4.	E	14.		24.	D	34.	
5.	D	15.	A	25.		35.	
6.	D	16.	B,E	26.	B,E	36.	
7.	B,E	17.	B	27.	A	37.	C,E
8.	B	18.		28.		38.	B
9.	A,D,E	19.	D,E	29.	B	39.	E
10.	E	20.		30.		40.	A

NOTE: ANY OTHER MARK COUNTS AS WRONG. YOU LOSE CREDIT FOR EACH WRONG MARK.

TEST 4

DIRECTIONS:

1. "Mark A for 59, 33, 44, 66, and 75. (Pause)
2. "Mark B as in boy for 69, 42, 31, and 72. (Pause)
3. "Mark E for 35, 64, 58, 47, and 61. (Pause)

"For the next set of questions, mark space B and also mark the letter I call, unless B is already marked. If B is already marked for that number, do *NOT* mark the new letter. Instead, mark the letter B for the number below it.

4. "Mark D as in dog for 32, 41, 70, and 63. (Pause)
5. "Mark C for 44, 48, 37, 74, and 37. (Pause)
6. "Mark E for 72, 67, 60, 42, and 46. (Pause)
7. "Mark A for 34, 56, 67, 38, and 71."

KEY (CORRECT ANSWERS)

No.	Answer	No.	Answer	No.	Answer
31.	B	46.	B,E	61.	E
32.	B,D	47.	E	62.	
33.	A	48.	B,C	63.	B,D
34.	A,B	49.		64.	E
35.	E	50.		65.	
36.		51.		66.	A
37.	B,C	52.		67.	B,E
38.	B	53.		68.	B
39.	B	54.		69.	B
40.		55.		70.	B,D
41.	B,D	56.	A,B	71.	A,B
42.	B	57.		72.	B
43.	B	58.	E	73.	B
44.	A,B,C	59.	A	74.	B,C
45.		60.	B,E	75.	A

NOTE: ANY OTHER MARK COUNTS AS WRONG. YOU LOSE CREDIT FOR EACH WRONG MARK.

TEST 5

DIRECTIONS:

1. "Mark C for 73, 96, 84, and 80. (Pause)
2. "Mark D as in dog for 68, 88, 99, 91, 78, and 67. (Pause)
3. "Mark E for 70, 93, 82, 75, and 92. (Pause)
4. "Mark B as in boy for 87, 69, 77, 98, and 71. (Pause)

"For the next set of questions, mark space C and also mark the letter I call, unless C is already marked. If C is already marked for that number, do *NOT* mark the new letter. Instead mark the letter A for the number below it.

5. "Mark D as in dog for 72, 89, 92, and 84. (Pause)
6. "Mark A for 66, 95, 77, and 73. (Pause)
7. "Mark B as in boy for 75, 83, 88, 90, 96, 100, and 94."

KEY (CORRECT ANSWERS)

66.	A,C	76.		86.		96.	C
67.	D	77.	A,B,C	87.	B	97.	A
68.	D	78.	D	88.	B,C,D	98.	B
69.	B	79.	C,D	89.	C,D	99.	D
70.	E	80.	C	90.	B,C	100.	B,C
71.	B	81.		91.	D		
72.	C,D	82.	C,D,E	92.	E		
73.	C	83.	B,C	93.	E		
74.	A	84.	C	94.	B,C		
75.	B,C,E	85.	A	95.	A,C		

NOTE: ANY OTHER MARK COUNTS AS WRONG. YOU LOSE CREDIT FOR EACH WRONG MARK.

TEST 6

DIRECTIONS:

1. "Mark E for 50, 37, 19, 24, and 11. (Pause)
2. "Mark B as in boy for 16, 22, 40, and 31. (Pause)
3. "Mark D as in dog for 24, 40, 49, 33, and 17. (Pause)

"For the next set of questions, mark space D as in dog and also mark the letter I call, unless D is already marked. If D is already marked for that number, do *NOT* mark the new letter. Instead mark the letter E for the number above it.

4. "Mark C for 12, 21, 42, and 29. (Pause)
5. "Mark A for 19, 49, 24, 15, 47, and 40. (Pause)
6. "Mark E for 41, 34, 29, and 17."

KEY (CORRECT ANSWERS)

10.		20.		30.		40.	B,D
11.	E	21.	C,D	31.	B	41.	D,E
12.	C,D	22.	B	32.		42.	C,D
13.		23.	E	33.	D	43.	
14.		24.	D,E	34.	D,E	44.	
15.	A,D	25.		35.		45.	
16.	B,E	26.		36.		46.	
17.	D	27.		37.	E	47.	A,D
18.		28.	E	38.		48.	E
19.	A,D,E	29.	C,D	39.	E	49.	D
						50.	E

NOTE: ANY OTHER MARK COUNTS AS WRONG. YOU LOSE CREDIT FOR EACH WRONG MARK.

TEST 7

DIRECTIONS:
1. "Mark D as in dog for 79, 51, 69, 42, and 64. (Pause)
2. "Mark A for 44, 62, 51, 59, 50, 42, 76, and 67. (Pause)
3. "Mark C for 64, 73, 80, 49, 55, and 62. (Pause)

"For the next set of questions, mark space A and also the letter I call, unless A is already marked. If A is already marked for that number, do *NOT* mark the new letter. Instead mark the letter E for that number.

4. "Mark E for 74, 68, 41, 77, and 58. (Pause)
5. "Mark B as in boy for 67, 60, 78, 44, and 76. (Pause)
6. "Mark C for 60, 51, 48, 69, 56, 66, and 79."

KEY (CORRECT ANSWERS)

41.	A,E	51.	A,D,E	61.		71.	
42.	A,D	52.		62.	A,C	72.	
43.		53.		63.		73.	C
44.	A,E	54.		64.	C,D	74.	A,E
45.		55.	C	65.		75.	
46.		56.	A,C	66.	A,C	76.	A,E
47.		57.		67.	A,E	77.	A,E
48.	A,C	58.	A,E	68.	A,E	78.	A,B
49.	C	59.	A	69.	A,C,D	79.	A,C,D
50.	A	60.	A,B,E	70.		80.	C

NOTE: ANY OTHER MARK COUNTS AS WRONG. YOU LOSE CREDIT FOR EACH WRONG MARK.

TEST 8

DIRECTIONS:

1. "Mark C for 37, 8, 29, 23, and 46. (Pause)
2. "Mark E for 50, 40, 28, 3, and 29. (Pause)
3. "Mark B as in boy for 38, 26, 23, 45, 47, and 35. (Pause)

"For the next set of questions, mark space C and also the letter I call, unless C is already marked. If C is already marked for that number, do *NOT* mark the new letter. Instead mark the letter B for the number that is two below it.

4. "Mark D as in dog for 48, 14, 8, 23, 33, 18, and 34. (Pause)
5. "Mark A for 42, 2, 16, 43, and 29. (Pause)
6. "Mark E for 4, 41, 48, and 15."

KEY (CORRECT ANSWERS)

1.		16.	A,C	31.	B	46.	C
2.	A,C	17.		32.		47.	B
3.	E	18.	C,D	33.	C,D	48.	C,D
4.	C,E	19.		34.	C,D	49.	
5.		20.		35.	B	50.	B,E
6.		21.		36.			
7.		22.		37.	C		
8.	C	23.	B,C	38.	B		
9.		24.		39.			
10.	B	25.	B	40.	E		
11.		26.	B	41.	C,E		
12.		27.		42.	A,C		
13.		28.	E	43.	A,C		
14.	C,D	29.	C,E	44.			
15.	C,E	30.		45.	B		

NOTE: ANY OTHER MARK COUNTS AS WRONG. YOU LOSE CREDIT FOR EACH WRONG MARK.

TEST 9

DIRECTIONS:

1. "Mark A for 87, 56, 95, 98, 99, 54, 63, and 59. (Pause)
2. "Mark D as in dog for 84, 100, 57, 68, 87, and 60. (Pause)
3. "Mark C for 70, 52, 69, 96, 78, 84, 58, 53, 68, and 76. (Pause)

"For the next set of questions, mark space A and also mark the letter I call, unless A is already marked. If A is already marked for that number, do *NOT* mark the new letter. Instead mark the letter E for the number that is two above it.

4. "Mark B as in boy for 89, 51, 66, 73, 62, and 98. (Pause)
5. "Mark E for 55, 71, 90, 87, 65, 99, and 66. (Pause)
6. "Mark D as in dog for 75, 91, 80, 54, 89, and 95."

KEY (CORRECT ANSWERS)

51.	A,B	66.	A,B	81.		96.	C,E
52.	C,E	67.		82.		97.	E
53.	C	68.	C,D	83.		98.	A
54.	A	69.	C	84.	C,D	99.	A
55.	A,E	70.	C	85.	E	100.	D
56.	A	71.	A,E	86.			
57.	D	72.		87.	A,D,E		
58.	C	73.	A,B	88.			
59.	A	74.		89.	A,B		
60.	D	75.	A,D	90.	A,E		
61.		76.	C	91.	A,D		
62.	A,B	77.		92.			
63.	A	78.	C	93.	E		
64.	E	79.		94.			
65.	A,E	80.	A,D	95.	A		

NOTE: ANY OTHER MARK COUNTS AS WRONG. YOU LOSE CREDIT FOR EACH WRONG MARK.

TEST 10

DIRECTIONS:

1. "Mark E for 87, 12, 93, 29, 9, 94, 16, 33, 21, 59, 67, 43, and 17. (Pause)
2. "Mark C for 82, 7, 63, 37, 97, 55, 39, 5, 47, and 25. (Pause)
3. "Mark B as in boy for 89, 66, 77, 35, 92, 18, 54, 13, 71, and 30. (Pause)

"For the next set of questions, mark space E and also mark the letter I call unless E is already marked. If E is already marked for that number, do *NOT* mark the new letter. Instead mark the letter D for the number that is <u>three above</u> it and the letter A for the number that is <u>three below</u> it.

4. "Mark A for 91, 62, 14, 87, and 33. (Pause)
5. "Mark B as in boy for 51, 11, 98, 51, 68, and 9. (Pause)
6. Mark C for 56, 41, 28, 94, 43, and 29."

KEY (CORRECT ANSWERS)

1.		26.	D	51.	D,E	76.	
2.		27.		52.		77.	B
3.		28.	C,E	53.		78.	
4.		29.	E	54.	A,B	79.	
5.	C	30.	B,D	55.	C	80.	
6.	D	31.		56.	C,E	81.	
7.	C	32.	A	57.		82.	C
8.		33.	E	58.		83.	
9.	E	34.		59.	E	84.	D
10.		35.	B	60.		85.	
11.	D,E	36.	A	61.		86.	
12.	A,E	37.	C	62.	A,E	87.	E
13.	B	38.		63.	C	88.	
14.	A,E	39.	C	64.		89.	B
15.		40.	D	65.		90.	A
16.	E	41.	C,E	66.	B	91.	A,D,E
17.	E	42.		67.	E	92.	B
18.	B	43.	E	68.	D,E	93.	E
19.		44.		69.		94.	E
20.		45.		70.		95.	
21.	E	46.	A	71.	B	96.	
22.		47.	C	72.		97.	A,C
23.		48.	D	73.		98.	D,E
24.		49.		74.		99.	
25.	C	50.		75.		100.	

EXAMINATION SECTION

TEST 1

DIRECTIONS: Each question or incomplete statement is followed by several suggested answers or completions. Select the one that BEST answers the question or completes the statement. *PRINT THE LETTER OF THE CORRECT ANSWER IN THE SPACE AT THE RIGHT.*

1. Reversing the current flow through a coil 1.___
   A. reduces the amount of flux produced
   B. reverses its two-pole field
   C. reduces power consumed
   D. reduces eddy currents

2. The unit of magnetic flux is 2.___
   A. ampere-turn B. gilbert
   C. intensity D. maxwell

3. Flux in a magnetic circuit compares in an electrical circuit with 3.___
   A. voltage B. current
   C. resistance D. opposition

4. If a compass is placed in the vicinity of a conductor carrying d.c., the needle aligns itself 4.___
   A. in the direction of current flow in the conductor
   B. at right angles to the conductor
   C. in the general direction of the north pole
   D. in the general direction of the south pole

5. The left-hand rule for coils states: Grasping the 5.___
   A. coil with the left hand, with fingers pointing in the direction of the magnetic field, the middle finger will point to the north pole
   B. coil with the left hand, thumb pointing in the direction of the conductor movement, the fingers will point in the direction of the magnetic field
   C. conductor with the left hand, fingers pointing in the direction of the north pole, the thumb will indicate current flow
   D. coil in the left hand, fingers wrapped around in the direction of electron flow, the thumb will point towards north pole

6. The word *permeability* indicates the 6.___
   A. amount of reluctance of one centimeter-cube of air
   B. number of turns of an air core
   C. ability of a substance to conduct magnetic lines of force
   D. m.m.f. required to produce one gilbert

7. If an iron core is inserted part way into a coil and current is applied to the coil, the core will be drawn into the coil in an effort to 7.___

A. increase reluctance of the magnetic circuit
B. decrease the length of the magnetic circuit
C. reduce the permeability of the circuit
D. increase the residual effect

8. Current flow in a straight conductor produces 8.___
A. hysteresis
B. magnetic lines of force
C. a north pole
D. permeability

9. The induction of a coil depends upon the 9.___
A. number of turns, current flow, type of core material, and ratio of coil's length to its width
B. current, flux, and core material
C. direction of rotation, flux, and speed
D. permeability, reluctance, gausses, and maxwells

10. The ability of a magnetic substance to hold its magnetism after the magnetizing force has been removed is 10.___
A. residual
B. retentivity
C. hysteresis
D. permeability

11. The solenoid and plunger type of magnet is employed in various forms to control 11.___
A. electrical equipment
B. hysteresis losses
C. permeability
D. all of the above

12. When using the left-hand rule for a conductor, 12.___
A. grasp the conductor with fingers pointing in the direction of current flow and thumb will indicate direction of field
B. the compass will point in the direction of current flow and fingers will indicate direction of magnetic field
C. grasp the conductor with the thumb extended in the direction of electron flow, the fingers will point in the direction of the magnetic lines of force
D. reverse the current flow and thumb will point in the direction of the magnetic field

13. The magnetic circuit is determined by 13.___
A. lines of force leaving the south pole and entering the north pole externally
B. permeability to current and the north pole
C. maxwells to length in centimeters and the south pole
D. the complete path taken by magnetic lines leaving the north pole, and returning to the north pole

14. In what material is retentivity MOST apparent? 14.___
A. Hard steel
B. Soft iron
C. Copper
D. Wood

15. The armature type of electromagnet 15.___
A. has a movable iron bar in its magnetic circuit
B. uses a permanent magnet core
C. has a hard steel core and a pivoted bar of copper
D. has a movable iron core

16. When two parallel conductors carrying current in the same direction are placed side by side, the fields produced by both 16.___
   A. cancel each other's field
   B. push each other apart
   C. form a north pole
   D. encircle each other, drawing the conductors together

17. The equivalent Ohm's law formula for magnetic circuits is 17.___
   A. $I = \frac{e}{r}$ B. $R = \frac{I}{E}$ C. $\Phi = \frac{E}{R}$ D. $P = \frac{I}{E}$

18. Which of the following materials incurs low hysteresis loss when used for transformer cores and similar applications? 18.___
   A. Hard steel B. Annealed steel
   C. Soft iron D. Cast steel

19. The magnetic field around a current-carrying wire 19.___
   A. is parallel to the current flow in the conductor
   B. exists at all points along its length
   C. exists only at the beginning of electron movement
   D. moves in the direction of current flow

20. Direct current flows through a coil of wire which has an iron core. When the iron core is removed and all other factors remain unchanged, the total number of lines of force through the coil will change because 20.___
   A. changing the core material affects m.m.f.
   B. permeance has been increased
   C. reluctance has been increased
   D. the permeability of the magnetic circuit has been increased

KEY (CORRECT ANSWERS)

1. B	11. A
2. D	12. C
3. B	13. D
4. B	14. A
5. D	15. A
6. C	16. D
7. B	17. C
8. B	18. C
9. A	19. B
10. B	20. C

TEST 2

DIRECTIONS: Each question or incomplete statement is followed by several suggested answers or completions. Select the one that BEST answers the question or completes the statement. *PRINT THE LETTER OF THE CORRECT ANSWER IN THE SPACE AT THE RIGHT.*

1. In formula form, angular velocity as applied to voltage or current is 1.___
   A. 2ω B. 2πω C. 2πF D. 2F

2. How much larger is the maximum voltage value than the effective voltage value? 2.___
   A. 1.414 B. 1.111 C. 0.707 D. 0.637

3. How many radians are in a complete circle? 3.___
   A. 0.707 B. 1.11 C. 3.14 D. 6.28

4. Magnitude of a vector is denoted by the 4.___
   A. time axis
   B. length of the line
   C. length of the line, drawn to scale
   D. length of the time axis, drawn to scale

5. Which of the following BEST defines a cycle? 5.___
   A. One alternation B. Two alternations
   C. Three alternations D. Four alternations

6. The MOST important value associated with sine waves of voltage or current are 6.___
   A. instantaneous B. maximum
   C. average D. all of the above

7. The MAXIMUM value of a single sine wave of a-c voltage is 141.4 volts. What is the effective value? ____ volts. 7.___
   A. 63.7 B. 70.7 C. 99.97 D. 141.4

8. Direction of a simple vector 8.___
   A. cannot be determined
   B. will indicate the reference point
   C. is indicated by an arrow at one end of the line
   D. is indicated by the magnitude

9. Frequency as applied to alternating current or voltage is the number of ____ per second of time. 9.___
   A. r.p.m. B. cycles
   C. poles D. volts generated

10. The effective value of an a-c voltage or current of sine waveform is defined in terms of the equivalent d.c. "_____ effect". 10.___
    A. voltage B. current
    C. heating D. cooling

11. The average value of a single sine wave of a-c current is 63.7 amps. What is the r.m.s. value? ____ amps. 11.____
   A. 63.7 B. 70.7 C. 141.4 D. 156.95

12. The instantaneous value of induced voltage of a single-phase generator 12.____
   A. will never be zero
   B. depends on the sine of the angle
   C. will always be the same
   D. depends on the cosine of the angle

13. The period of a-c voltage of sine waveform is the time required for one complete 13.____
   A. r.p.m. B. alternation
   C. cycle D. degree

14. The "root-mean-square" value is also known as the ____ value. 14.____
   A. average B. peak
   C. instantaneous D. effective

15. Two voltages of the same frequency "120 cycles per second" have their positive maximum values displaced by 90^0. What is the time difference between the phases? ____ sec. 15.____
   A. 0.00208 B. 0.00416 C. 0.02080 D. 0.04160

16. Positive rotation of a vector is 16.____
   A. clockwise
   B. counterclockwise
   C. either clockwise or counterclockwise
   D. neither clockwise nor counterclockwise

17. The effective generated voltage of a single-phase a-c generator may be obtained by the formula 17.____
   A. $E = 1.111\ \Phi\ fD10^{-8}$ B. $E = 2.22\ \Phi\ fN^{-6}$
   C. $E = 1.111\ \Phi\ FD^{-6}$ D. $E = 2.22\ \Phi\ fN10^{-8}$

18. Most a-c voltmeters and ammeters are calibrated in ____ values. 18.____
   A. r.m.s. B. average
   C. peaks D. instantaneous

19. A radian is ALWAYS equal to 19.____
   A. the circumference of the circle
   B. the radius of the circle
   C. π
   D. 60 degrees

20. Angular velocity is symbolized by the Greek letter 20.____
   A. gamma γ B. mu μ C. omega ω D. beta β

21. If an active conductor's length is doubled, the 21.____
   A. pole length is cut in half
   B. flux per pole is cut in half
   C. generated voltage is doubled
   D. generated voltage is tripled

22. The "form factor" in most cases refers to the ratio of 22.___
    A. average value to maximum value
    B. instantaneous value to maximum value
    C. r.m.s. value to effective value
    D. r.m.s. value to average value

23. Conductor size in an a-c circuit is based on ____ current . 23.___
    A. instantaneous B. maximum
    C. average D. effective

24. The term "angular velocity," as applied to electricity, refers to the number of 24.___
    A. radians per second a voltage vector rotates
    B. radians per minute a voltage vector rotates
    C. degrees averaged together
    D. degrees subtracted from the radians

25. Vectors indicate 25.___
    A. direction
    B. magnitude
    C. time, direction, and magnitude
    D. magnitude and direction

KEY (CORRECT ANSWERS)

1. C
2. A
3. D
4. C
5. B
6. D
7. C
8. C
9. B
10. C
11. B
12. B
13. C
14. D
15. B
16. B
17. D
18. A
19. B
20. C
21. C
22. D
23. D
24. A
25. D

TEST 3

DIRECTIONS: Each question or incomplete statement is followed by several suggested answers or completions. Select the one that BEST answers the question or completes the statement. *PRINT THE LETTER OF THE CORRECT ANSWER IN THE SPACE AT THE RIGHT.*

1. Inductance is the property of an electric circuit that 1.____
   A. opposes any change in the applied voltage through that circuit
   B. opposes any change in the current through that circuit
   C. aids any change in the applied voltage through that circuit
   D. aids any change in the current through that circuit

2. A coil has a resistance of 22 ohms, and, in 0.1 second after the switch is closed, the current has reached 63.2% of its final value. 2.____
   The value of inductance (L) is _____ henries.
   A. 0.22   B. 0.44   C. 2.20   D. 4.40

3. When a capacitor's counter e.m.f. has risen to equal the battery voltage, the current is 3.____
   A. maximum in the circuit
   B. at 63.2% of its maximum Ohm's law value in the circuit
   C. zero in the circuit
   D. at 36.8% of its maximum Ohm's law value in the circuit

4. The total capacitance of two 100-microfarad capacitors connected in series is _____ microfarads. 4.____
   A. 25   B. 50   C. 100   D. 200

5. When the magnetic field is collapsing, the induced e.m.f. 5.____
   A. aids and tends to prolong the impressed current
   B. opposes the impressed current
   C. has no effect upon the impressed current
   D. causes the impressed current to go immediately to zero

6. To increase the time required for the current to reach its maximum Ohm's law value in a series resistance and inductance circuit, it is necessary to 6.____
   A. increase the resistance
   B. decrease the resistance
   C. decrease the inductance
   D. increase the applied voltage

7. In a simple capacitor, the plate area is 2 square inches, the dielectric material is mica, and the distance between the plates is 0.01 inch. To increase the capacitance, 7.____
   A. increase the distance between the plates
   B. change dielectric material from mica to paraffin paper
   C. decrease the plate area
   D. change dielectric material from mica to flint glass

8. What is the total capacitance of two 200-microfarad capacitors connected in series with each other and both in parallel with a 50-microfarad capacitor? _____ microfarads. 8.___
   A. 50 B. 250 C. 100 D. 150

9. The unit of inductance is the 9.___
   A. ohm B. farad C. henry D. coulomb

10. The coefficient of coupling of two coils is 10.___
   A. increased by turning one coil axis at right angles to the other coil axis
   B. greatly increased with the addition of a soft iron core
   C. generally higher in an air core circuit than in an iron core circuit
   D. almost unity on a diamagnetic core

11. Factors that determine the capacitance of parallel-electrode capacitors are 11.___
   A. area of the plates and the type of dielectric
   B. thickness of the plates, type and thickness of the dielectric
   C. type and thickness of the dielectric and the plate area
   D. thickness of the dielectric, area of the plate, and the direction of the current flow

12. The time constant of a resistance-capacitance circuit is 12.___
   A. the time required to discharge a capacitor to 36.8% of its final voltage
   B. the time required to charge a capacitor to 36.8% of its maximum voltage
   C. equal to the circuit capacitance divided by the circuit resistance
   D. equal to the circuit resistance divided by the circuit capacitance

13. The voltage due to self-induction 13.___
   A. can only occur when the conductor is wound in the form of a coil
   B. is produced when the strength of the magnetic field changes
   C. is produced by moving a conductor through a magnetic field
   D. cannot occur in a direct-current circuit

14. If two coils are positioned with respect to each other so as to have unity coefficient of coupling and the inductance of coil A equals coil B, what is the mutual inductance between the two coils? 14.___
   A. Equal to the inductance of coil A
   B. One-half of the inductance of coil A
   C. Double the inductance of coil A
   D. Equal to the square root of the inductance of coil A

15. The capacitance of a capacitor is inversely proportional to the 15.____
   A. frequency of the applied voltage
   B. dielectric constant
   C. active plate area
   D. distance between the plates

16. A 40-microfarad capacitor in series with 2,000 ohms is connected to a 200-volt source. The time constant is ____ second. 16.____
   A. 0.05 B. 0.08 C. 0.50 D. 0.80

17. To increase the inductance of a coil, 17.____
   A. increase the permeability of the core material
   B. increase the length of the coil
   C. increase the magnitude of the current flow through the coil
   D. decrease the cross-sectional area of the coil

18. What is the total inductance (L_t) of two coils connected in parallel when the coefficient of coupling is zero? 18.____
   A. $L_t = L_1 + L_2$
   B. $L_t = L_1$
   C. $L_t = \frac{1}{L_1} + \frac{1}{L_2}$
   D. $L_t = \frac{L_1 L_2}{L_1 + L_2}$

19. The charge that is stored in a capacitor depends upon 19.____
   A. voltage divided by the capacitance
   B. voltage multiplied by the capacitance
   C. capacitance divided by the voltage
   D. capacitance multiplied by the resistance

20. When selecting a capacitor for use in a circuit, its working voltage should be AT LEAST 20.____
   A. 10% greater than the highest voltage to be applied to it
   B. equal to the highest voltage to be applied to it
   C. 30% greater than the highest voltage to be applied to it
   D. 50% greater than the highest voltage to be applied to it

21. A battery, a switch, a resistor, and a coil are connected in series. What happens at the instant the switch is closed? The 21.____
   A. voltage across the resistor is maximum
   B. voltage across the coil is maximum
   C. current in the circuit is maximum
   D. voltage across the resistor and the coil are minimum

22. Capacitance is the property of an electric circuit that 22.____
   A. opposes any change of current in the circuit
   B. opposes any change of voltage in the circuit
   C. is not affected by a change of voltage
   D. aids any change of current in the circuit

23. When capacitors are connected in parallel, the resulting capacitance is the 23.___
   A. sum of the individual capacitances
   B. reciprocal of the sum of reciprocals of the individual capacitors
   C. sum of the individual reciprocals of the capacitors
   D. reciprocal of the sum of capacitances

24. The time constant of an inductive-resistive circuit is 24.___
   A. that time required for a direct current to rise one-half of its maximum value
   B. a method of determining how much current would flow at the end of one second
   C. the product of the resistance and the inductance of a circuit
   D. the time in seconds required for the current to rise to 63.2% of its final Ohm's law value

25. When capacitors are connected in series, the resulting capacitance is found by 25.___
   A. a rule similar to that for combining parallel resistors
   B. a rule similar to that for combining resistors in series
   C. dividing the applied voltage by the total resistance of series circuit
   D. multiplying the applied voltage by the total resistance of series circuit

---

KEY (CORRECT ANSWERS)

1. B
2. C
3. C
4. B
5. A

6. B
7. D
8. D
9. C
10. B

11. C
12. A
13. B
14. A
15. D

16. B
17. A
18. D
19. B
20. D

21. B
22. B
23. A
24. D
25. A

TEST 4

DIRECTIONS: Each question or incomplete statement is followed by several suggested answers or completions. Select the one that BEST answers the question or completes the statement. *PRINT THE LETTER OF THE CORRECT ANSWER IN THE SPACE AT THE RIGHT.*

1. When frequency is increased in an inductive circuit, the current flow will 1.____
   A. increase because of greater inductive voltage
   B. increase because of less inductive voltage
   C. decrease because of greater inductive voltage
   D. decrease because of less inductive voltage

2. The unit of measurement for reactive power is the 2.____
   A. Volt Amp (VA) B. Volt Amp Wattage (VAW)
   C. Volt Amp Reactive (VAR) D. Volt Amp Kilo Watt (VAKW)

3. The prime factors that determine inductive reactance are 3.____
   A. frequency and current
   B. frequency and inducted voltage
   C. inductance and frequency
   D. inductance and voltage

4. What are the prime factors affecting capacitive reactance? 4.____
   A. Frequency and capacitance
   B. Frequency and inductance
   C. Frequency and resistance
   D. Frequency squared

5. The reaction of an inductor to any change in current is known as inductive 5.____
   A. reactance B. voltage
   C. current D. resistance

6. What two quantities comprise the total power of an a-c circuit? 6.____
   A. Apparent power and VARS
   B. True power and VARS
   C. True power and VAWS
   D. True power and power factor

7. Power factor is the 7.____
   A. percentage of apparent power expended in heat
   B. amount of apparent power in an a-c circuit
   C. amount of apparent power plus true power
   D. efficiency of a circuit expressed as an angle

8. The unit of measurement for inductive reactance is the 8.____
   A. ohm B. volt C. ampere D. henry

9. What is the power factor of a pure capacitive circuit? 9.___
   A. 70.7% B. 86.6% C. 100% D. 0%

10. The self-induced voltage in a coil depends on the 10.___
   A. voltage applied
   B. inductive reactance
   C. c.e.m.f.
   D. current change in the coil

11. What is the path for current flow through a capacitor? 11.___
   A. Does not actually flow through
   B. From negative plate to positive plate
   C. From positive plate to negative plate
   D. From dielectric to plates

12. Power is described as the rate at which 12.___
   A. work is being done
   B. energy is being used
   C. energy is being expended
   D. all of the above are produced

13. What is the power characteristic of a positive inductive circuit? 13.___
   A. Positive power is the largest.
   B. Positive and negative power are equal.
   C. Negative power is the largest.
   D. Negative power is the smallest.

14. The vector sum of X_C and R in a series R-C circuit is the total 14.___
   A. opposition to voltage in the circuit
   B. capacitive reactance in the circuit
   C. resistance in the circuit
   D. opposition to current flow in the circuit

15. Self-induced voltage in an inductor will ALWAYS oppose 15.___
   A. the applied voltage by 180 degrees
   B. circuit current
   C. the applied voltage by 90 degrees
   D. the circuit current by 180 degrees

16. What is the value of apparent current flow in a capacitive circuit when it is FIRST energized? 16.___
   A. No current flow in a capacitive circuit
   B. Minimum
   C. Maximum
   D. Zero

17. What is the phase relationship between current and voltage in a resistive circuit? 17.___
   A. 90° phase difference
   B. 0° phase difference
   C. Out of phase
   D. Cannot be determined without numerical value of R

18. Voltage and current are considered to be out of phase with each other in a pure inductive circuit by what amount? Current 18.___
   A. leads voltage by 90^0 B. lags voltage by 90^0
   C. leads voltage by 180^0 D. lags voltage by 180^0

19. What is the phase relationship between current and voltage in a pure capacitive circuit? 19.___
   A. No phase angle
   B. Current lags voltage by 90^0
   C. Voltage leads current by 90^0
   D. Voltage lags current by 90^0

20. The phase angle between current and voltage in a circuit containing both resistive and inductive elements is 20.___
   A. greater than 0^0 but less than 90^0
   B. a constant 45^0
   C. 90^0 at all times
   D. 0^0 because X_L and R are equal

21. In the formula for capacitive reactance, the symbol "C" represents capacitance in 21.___
   A. farads B. micromicrofarads
   C. microfarads D. ohms

22. What is the effect on capacitor voltage when frequency is increased? Voltage 22.___
   A. increases B. decreases
   C. not affected D. is minimum

23. What is impedance? 23.___
   A. Opposition to current flow in an a-c circuit created by resistance and reactances
   B. Resistance of an a-c circuit
   C. Vector sum of voltage drops in an a-c circuit
   D. Current divided by voltage in an a-c circuit

24. When voltages are out of phase, the total voltage can be found by adding 24.___
   A. arithmetically B. vectorially
   C. peak value D. average values

25. The number of radians per second traversed by an alternating-voltage vector is CLOSELY related to its 25.___
   A. current sine wave B. voltage sine wave
   C. frequency D. none of the above

KEY (CORRECT ANSWERS)

1. C	6. B	11. A	16. C	21. A
2. C	7. A	12. D	17. B	22. B
3. C	8. A	13. B	18. B	23. A
4. A	9. D	14. D	19. D	24. B
5. A	10. D	15. A	20. A	25. C

TEST 5

DIRECTIONS: Each question or incomplete statement is followed by several suggested answers or completions. Select the one that BEST answers the question or completes the statement. *PRINT THE LETTER OF THE CORRECT ANSWER IN THE SPACE AT THE RIGHT.*

1. When resistive, inductive, and capacitive elements are connected in series, 1.___
   A. they lose their individual characteristics
   B. their individual characteristics remain unchanged
   C. the inductance component is always the largest
   D. the capacitive component is always the largest

2. In a circuit containing both X_C and X_L, if the difference between X_C and X_L increases, the total 2.___
   A. impedance decreases
   B. impedance increases
   C. impedance remains the same
   D. resistance increases

3. In a 60-cycle a-c circuit, with an inductor of 0.053 microhenry and a resistance of 50 ohms connected in series, the impedance would be _____ ohms. 3.___
   A. 539 B. 5.39 C. 53.9 D. 19.97

4. The effective resistance of a circuit may be defined as the ratio of the 4.___
   A. true power absorbed by the circuit to the square of the effective current flowing
   B. apparent power absorbed by the circuit to the square of the effective current flowing
   C. true power absorbed by the circuit to the square of the total resistance in the circuit
   D. apparent power absorbed by the circuit to the square of the total resistance in the circuit

5. In a series circuit, when X_C and X_L are equal, 5.___
   A. line voltage leads line current by an unknown $\angle\theta$
   B. line current leads line voltage by an unknown $\angle\theta$
   C. total impedance is minimum
   D. total impedance is maximum

6. To find true power, use the formula 6.___
   A. T.P. = E x l B. T.P. = E x l x cos $\angle\theta$
   C. T.P. = $E^2 \div \cos \angle\theta$ D. T.P. = E x l x sin $\angle\theta$

7. A low-loss inductor has a low 7.___
   A. X_L at high frequencies B. current at low frequencies
   C. inductance D. resistance

8. The MOST inefficient method of voltage reduction, from the standpoint of power loss, is a(n) 8.____
   A. capacitor in series with the load
   B. inductor in series with the load
   C. capacitor and an inductor in series with the load
   D. resistor in series with the load

9. The formula for finding the loss factor of a capacitor is the dielectric constant 9.____
   A. times the power factor
   B. divided by the power factor
   C. times the apparent power
   D. times the true power

10. A phase difference between E and I causes 10.____
   A. true power to increase
   B. apparent power to increase
   C. true power to decrease
   D. apparent power to decrease

11. The energy component of a current flowing in an R-L circuit is the current flowing through the 11.____
   A. inductor and resistor   B. inductor
   C. resistor   D. power source

12. Corona loss is the result of 12.____
   A. emission of electrons from the surface of a conductor
   B. electron collision inside a conductor
   C. overheating a high-frequency conductor
   D. none of the above

13. To find total true power in a parallel circuit, first find the true power in each branch and then 13.____
   A. add the power in all branches arithmetically
   B. add the power in all branches using the parallelogram method
   C. multiply the voltage times the current
   D. multiply the square of the current times the resistance

14. The hypotenuse of the power triangle represents 14.____
   A. true power   B. VARS
   C. apparent power   D. none of the above

15. To find a circuit's power factor in percent, use the formula 15.____
   A. $P.F. = \frac{AP}{TP} \times 100$ B. $P.F. = \cos \underline{/\theta}$
   C. P.F. = angle in degrees x 100
   D. $P.F. = \frac{E \times I \times \cos \underline{/\theta}}{E \times I} \times 100$

16. Corona loss can be held to a satisfactory low value by 16.____
   A. avoiding sharp points, bends, and turns
   B. using low voltages
   C. using large diameter conductors
   D. all of the above

17. In a parallel circuit with a lagging power factor, to improve the power factor, 17.___
   A. increase the inductance
   B. put an inductor in parallel with the rest of the circuit
   C. put a capacitor in parallel with the rest of the circuit
   D. decrease the resistance

18. The non-energy component of an a-c circuit is 18.___
   A. true power
   B. reactive power
   C. apparent power
   D. none of the above

19. The formula for computing R - L - C in series is 19.___
   A. $Z = R_T + X_L + X_C$
   B. $Z = R + X_L - X_C$
   C. $Z^2 = R^2 + (X_L{}^2 - X_C{}^2)$
   D. $Z = R^2 + (X_L{}^2 - X_C{}^2)$

20. One radian is equal to 20.___
   A. $\frac{180}{2\pi}$
   B. 2π
   C. 1.414
   D. $\frac{360}{2\pi}$

21. Skin effect describes the tendency of 21.___
   A. d-c conductors to carry the circuit current on their surfaces
   B. a-c conductors to carry the circuit current on their surfaces
   C. both a-c and d-c conductors to carry the circuit current on their surfaces
   D. none of the above

22. The current in an a-c parallel circuit varies 22.___
   A. inversely with the cos $\angle\theta$
   B. directly with the cos $\angle\theta$
   C. inversely with the sine $\angle\theta$
   D. directly with the sine $\angle\theta$

23. In a circuit containing X_C and X_L, if X_C is larger, 23.___
   A. the voltage would lead the current
   B. the current would lead the voltage
   C. there would be no phase difference
   D. true power is maximum

24. The effective resistance of a capacitor dissipated in heat in the 24.___
   A. positive capacitor plate
   B. negative capacitor plate
   C. dielectric
   D. leads connecting the capacitor in the circuit

25. When loops of wire are formed into a coil and current is passed through this coil, 25.___
   A. the result is a permanent magnet
   B. each turn will cancel the magnetic field of each adjacent turn
   C. the magnetic field of each turn of wire links with the fields of adjacent turns
   D. permeability decreases to minimum

KEY (CORRECT ANSWERS)

1. B
2. B
3. C
4. A
5. C

6. B
7. D
8. D
9. A
10. C

11. C
12. A
13. A
14. C
15. D

16. D
17. C
18. B
19. C
20. D

21. B
22. A
23. B
24. C
25. C

TEST 6

DIRECTIONS: Each question or incomplete statement is followed by several suggested answers or completions. Select the one that BEST answers the question or completes the statement. *PRINT THE LETTER OF THE CORRECT ANSWER IN THE SPACE AT THE RIGHT.*

1. In a D'Arsonval type meter, the PRIMARY purpose of the iron core is to 1.___
   A. dampen coil movement
   B. concentrate the flux between the core and the pole piece
   C. concentrate the flux between the core and the hairsprings
   D. weaken the flux between the core and the pole piece

2. The D'Arsonval movement operates on the principle of 2.___
   A. mutual induction
   B. repulsion
   C. magnetic repulsion and attraction
   D. magnetic attraction

3. In order to measure large currents with the D'Arsonval ammeter, 3.___
   A. the meter must be made more rugged
   B. an increased number of turns must be put on the moving coil
   C. shunts are used
   D. the pointer must be lengthened

4. The PRIMARY purpose of the megger is to measure 4.___
   A. milliohms B. kilohms
   C. megohms D. microhms

5. A moving iron-vane meter works on the principle of 5.___
   A. induction B. repulsion
   C. conduction D. attraction

6. When hairsprings are used in a D'Arsonval type meter, they 6.___
   A. serve as conductors
   B. are wound oppositely and provide restoring forces
   C. are attached to the ends of the bobbin
   D. produce all of the above

7. The metal ribbons attached to the moving coil in the galvanometer have no force on them when the 7.___
   A. north pole of the moving coil is close to the south pole of the horseshoe magnet
   B. north pole of the moving coil is as close as possible to the north pole of the horseshoe magnet
   C. north pole of the moving coil is 90° from the north pole of the horseshoe magnet
   D. south pole of the moving coil is as close as possible to the south pole of the horseshoe magnet

8. When shunts are used with an ammeter, they 8.____
   A. must be located within the case
   B. must be located outside the case
   C. may be located either inside or outside the case
   D. reduce the accuracy of the movement

9. The power supply for a megger comes from 9.____
   A. two flashlight batteries
   B. a hand-driven generator
   C. any 115 v. wall plug
   D. the aircraft battery

10. The iron-vane meter can be used on 10.____
   A. a.c. only
   B. a.c. and d.c.
   C. d.c. only
   D. rectified a.c.

11. The Weston meter uses the principle of operation of the 11.____
   A. D'Arsonval galvanometer
   B. Thompson incline coil
   C. iron vane
   D. V.T.V.M.

12. Current-measuring instruments MUST always be connected in 12.____
   A. parallel with a circuit
   B. series with a circuit
   C. series-parallel with a circuit
   D. delta with the shunt

13. If the pointer fails to come back to zero when the megger is not in use, 13.____
   A. the megger is out of calibration
   B. this is normal operation
   C. the hairsprings are burned out
   D. the pointer is stuck

14. When using ammeters, 14.____
   A. reverse polarity must be used
   B. + or - polarity can be used
   C. polarity should be observed
   D. regardless of polarity, the instrument cannot be damaged because it is grounded

15. For measuring resistances of multimillions of ohms, use a(n) 15.____
   A. TS 297 multimeter
   B. ohmmeter with high scales
   C. megger
   D. combustion volt ammeter

16. Dampening is accomplished in the iron-vane meter by 16.____
   A. hairsprings
   B. the hermetically sealed case
   C. an aluminum bobbin
   D. an aluminum vane

17. Balance springs on each end of the shaft of the Weston ammeter 17.___
  A. are used to carry current to the moving coil
  B. are not balanced due to temperature change
  C. provide a turning force for the pointer
  D. are factory adjusted and must not be re-adjusted

18. The electrodynamometer-type meter employs 18.___
  A. two permanent magnets
  B. one permanent magnet and one electromagnet
  C. two fixed coils and one movable permanent magnet
  D. none of the above

19. With the ohmmeter setting on R x 1, it takes 0.01 ma. to deflect the pointer to half scale. If the meter were set on R x 100, how much would it take to deflect the pointer to half scale? _____ ma. 19.___
  A. 0.1 B. 0.01 C. 0.001 D. 1.0

20. Electrodynamometer-type meters 20.___
  A. never utilize shunts
  B. utilize four coils, all of which are movable
  C. are seldom used in the laboratory because they are not accurate enough
  D. are not as sensitive as the D'Arsonval meter

21. Meggers provided aboard ships are usually rated at _____ volts. 21.___
  A. 250 B. 500 C. 750 D. 1,000

22. The voltage developed in the thermocouple-type meter depends on the 22.___
  A. material of which the wires are made
  B. direction of current flow
  C. frequency of the heater voltage
  D. type of meter movement

23. A multimeter contains a 23.___
  A. voltmeter and wattmeter
  B. voltmeter and frequency meter
  C. voltmeter, ohmmeter, and milliammeter
  D. voltmeter, ammeter, and ohmmeter

24. In the iron-vane power factor meter, the spiral springs 24.___
  A. are eliminated B. return the pointer to zero
  C. oppose torque D. carry current to coils A and B

25. Instruments that are used to measure voltage and current regardless of circuit ratings are 25.___
  A. instrument transformers B. autotransformers
  C. electrodynamometers D. D'Arsonvals

KEY (CORRECT ANSWERS)

1.	B	11.	A
2.	C	12.	B
3.	C	13.	B
4.	C	14.	C
5.	B	15.	C
6.	D	16.	D
7.	B	17.	A
8.	C	18.	D
9.	B	19.	B
10.	B	20.	D

21.	B
22.	A
23.	C
24.	A
25.	A

EXAMINATION SECTION
TEST 1

DIRECTIONS: Each question or incomplete statement is followed by several suggested answers or completions. Select the one that BEST answers the question or completes the statement. *PRINT THE LETTER OF THE CORRECT ANSWER IN THE SPACE AT THE RIGHT.*

1. When control current is sufficient to saturate the core, inductance ____ and reactance ____. 1.____
   A. increases; decreases B. decreases; increases
   C. increases; increases D. decreases; decreases

2. The power handling capacity of magnetic amplifiers has been improved PRIMARILY because of the development of 2.____
   A. dry-disk rectifiers B. high-quality steels
   C. electron amplifier tubes D. high-wattage resistors

3. When the inductive reactance of the load winding is decreased, load ____ and load voltage ____. 3.____
   A. current rises; drops
   B. current drops; rises
   C. circuit impedance drops;rises
   D. circuit impedance rises;rises

4. Doubling the number of turns on a coil will 4.____
   A. double the inductance
   B. decrease the inductance by one-half
   C. increase the inductance as the square of the turns
   D. have no effect on inductance

5. Which of the following is a disadvantage of a magnetic amplifier? 5.____
   A. High efficiency
   B. Ruggedness
   C. It has no moving parts
   D. It is not useful at high frequencies

6. When a magnetic amplifier is operating at the knee of the magnetization curve, a small increase in control current will 6.____
   A. cause a large increase in flux density
   B. desaturate the core
   C. increase permeability of the core
   D. decrease inductance of the load winding

7. Transformer action between the a-c load windings and d-c control winding is 7.____
   A. necessary to increase amplification
   B. desired in order to reduce the required control current
   C. desired to increase response time
   D. not desired

8. Which of the following is considered an advantage of a magnetic-amplifier circuit over a vacuum-tube circuit? 8.___
   A. Little distortion
   B. Frequency response time
   C. Shock resistance
   D. Impedance matching

9. Current flow in the load circuit can be increased by 9.___
   A. *increasing* the impedance of the circuit
   B. *decreasing* the permeability of the core
   C. *increasing* the permeability of the core
   D. *decreasing* the control current

10. The BASIC control action of a magnetic amplifier depends upon 10.___
   A. variations in the load impedance
   B. changes in inductance
   C. type of core material
   D. construction of the core

11. Permeability of a substance is defined as the 11.___
   A. ease with which it conducts magnetic lines of flux
   B. opposition it offers to a-c current flow
   C. opposition it offers to d-c current flow
   D. ease with which it retains magnetic properties

12. Which of the following is an advantage of a magnetic amplifier? 12.___
   A. It has a time delay associated with magnetic effects
   B. The output waveform is not an exact reproduction of the input waveform
   C. No warm-up time
   D. It cannot handle low-level signals

13. Rectifiers are placed in the control circuits to prohibit current flow in the control winding during the 13.___
   A. warm-up time
   B. gating half cycle
   C. reset half cycle
   D. complete cycle

14. Silicon steel cores were not satisfactory for saturable reactors because of ____ saturation flux density and ____ hysteresis losses. 14.___
   A. low; high
   B. high; high
   C. low; low
   D. high; low

15. An important advantage of controlling circuit current by an adjustable inductor is 15.___
   A. low circuit power factor
   B. high circuit power factor
   C. absence of heat loss in the control element
   D. high heat loss in the control element

16. Operating the reactor core in the region of saturation for a portion of each cycle will cause the circuit gain to 16.___
   A. hold steady
   B. fluctuate rapidly
   C. decrease
   D. increase

KEY (CORRECT ANSWERS)

1. D
2. B
3. C
4. C

5. D
6. D
7. D
8. C

9. B
10. B
11. A
12. C

13. B
14. A
15. C
16. D

TEST 2

DIRECTIONS: Each question or incomplete statement is followed by several suggested answers or completions. Select the one that BEST answers the question or completes the statement. *PRINT THE LETTER OF THE CORRECT ANSWER IN THE SPACE AT THE RIGHT.*

1. The direction of rotation of a capacitor motor can be reversed by 1.____
   A. reversing connection to the noncapacitor phase
   B. reversing connection to the capacitor phase
   C. reversing source power connections
   D. shifting the capacitor from one phase to the other

2. A synchro transmitter is connected to a synchro motor 2.____
   A. mechanically B. magnetically
   C. directly D. electrically

3. If S2 and either S1 or S3 are reversed at the receiver, the rotor will 3.____
   A. reverse direction with no error
   B. be 180° in error
   C. be 120° in error
   D. not be affected

4. A BASIC synchro system is used to 4.____
   A. transmit position information
   B. control drive motors
   C. control large amounts of current
   D. transfer energy

5. To reverse the direction of rotation of the rotor of the receiver (with no error), reverse 5.____
   A. S1 and S2 B. R1 and R2
   C. S2 and S3 D. S1 and S3

6. A control transformer synchro uses a ____ rotor. 6.____
   A. squirrel cage B. drum wound
   C. salient pole D. lap wound

7. The purpose of the compensating windings is to create a magnetomotive force to ____ current m.m.f. 7.____
   A. counterbalance the armature load
   B. aid the armature load
   C. counterbalance the control field
   D. aid the control field

8. The a-c servomotor is a ____ motor. 8.____
   A. 3Φ induction B. 1Φ repulsion induction
   C. 1Φ induction D. 1Φ series

9. A synchro is comparable to a 9.____
   A. single-phase transformer B. generator
   C. synchronous motor D. three-phase transformer

10. Synchro generators and motors are usually NOT interchangeable because 10.____
   A. they are not electrically identical
   B. they are not mechanically identical
   C. the generator is larger than the motor
   D. the generator operates at a higher voltage

11. The essential components of a servomechanism are ____ controller and ____. 11.____
   A. input; transmitter
   B. input; receiver
   C. output; transmitter
   D. output; input controller

12. The differential synchro transmitter uses a rotor having ____ coil(s). 12.____
   A. a single
   B. three separately connected
   C. three wye-connected
   D. three delta-connected

13. With both rotors in the same position and maximum voltage induced in winding S2, there is ____ current flow in ____. 13.____
   A. *minimum*; the system
   B. *maximum*; the system
   C. *minimum*; stator windings
   D. *maximum*; stator windings

14. Driving the follow-up center tap rotor from the electrical zero position causes a voltage to be induced into the rotor by the excited stator. 14.____
This voltage will be ____ with the voltage generated by the ____ controller.
   A. in phase; input B. out of phase; output
   C. in phase; output D. out of phase; input

15. The generator and motor rotors are connected 15.____
   A. in parallel B. in series
   C. to separate sources D. by mutual inductance

16. The stator voltages are 16.____
   A. in phase and subtractive
   B. out of phase and subtractive
   C. in phase and additive
   D. out of phase and additive

17. The motor rotor will follow the generator rotor because of 17.____
   A. magnetic coupling B. induced e.m.f.
   C. c.e.m.f. D. generator movement

18. The windings of the synchro stators are displaced from each other by 18.___
A. 45° B. 90° C. 120° D. 240°

19. An amplidyne is a(n) 19.___
A. a-c amplifier B. d-c amplifier
C. ac-dc amplifier D. d-c servomotor

20. The rotor of a synchro MUST continuously draw current to 20.___
A. produce heat
B. maintain a magnetized rotor
C. set up a reference voltage
D. produce an induced voltage in the stator

21. Synchro systems will be at a null when induced e.m.f. s are 21.___
A. in phase and equal B. out of phase and equal
C. in phase and not equal D. out of phase and not equal

22. The amplidyne drive motor is USUALLY a ____ motor. 22.___
A. d-c compound
B. d-c series
C. one-phase a-c induction
D. three-phase a-c induction

KEY (CORRECT ANSWERS)

1. D
2. D
3. C
4. A
5. D

6. B
7. A
8. D
9. A
10. B

11. D
12. C
13. A
14. B
15. A

16. A
17. A
18. C
19. B
20. B

21. A
22. D

TEST 3

DIRECTIONS: Each question or incomplete statement is followed by several suggested answers or completions. Select the one that BEST answers the question or completes the statement. *PRINT THE LETTER OF THE CORRECT ANSWER IN THE SPACE AT THE RIGHT.*

1. A molded capacitor on which the top row of dots (left to right) is silver, brown, and gray and the bottom row (left to right) is yellow, gold, and orange will have a value of _____ mfd. 1.____
   A. 0.018 B. 18 C. 0.183 D. 183

2. Electrical drawings which show the electric wiring for a building or other structure are called 2.____
   A. electrical layouts
   B. elementary wiring diagrams
   C. schematic wiring diagrams
   D. isometric wiring diagrams

3. The electrical symbol used to indicate a potentiometer is 3.____

   A.  B. 

   C.  D. 

4. The lower left dot in a five-dot capacitor indicates the 4.____
   A. multiplier B. tolerance
   C. working voltage D. characteristic

5. Small power transformer primary leads are color coded 5.____
   A. red B. black C. green D. yellow

6. The temperature coefficient of a ceramic capacitor is indicated by the 6.____
   A. first dot B. last dot C. first band D. last band

7. A carbon resistor color coded red, blue, green, and gold would indicate a tolerance of 7.____
   A. 2% B. 5% C. 10% D. 20%

8. A carbon resistor color coded brown, black, and silver has a resistance value of 8.____
   A. 0.01Ω B. 0.1Ω C. 1Ω D. 10Ω

9. The secondary high voltage windings of a small power transformer are color coded 9.____
   A. black B. green C. white D. red

10. Before making any adjustments on a faulty circuit, the maintenance man should 10.___
   A. check all resistors and capacitors in the circuit
   B. make sure he understands what the circuit is supposed to do in normal operation
   C. check the circuit power supply
   D. observe the circuit's faulty operation

11. The value of a radial carbon resistor with a red end, blue body, and an orange dot would be 11.___
   A. 26,000Ω ± 20% B. 6,300Ω ± 20%
   C. 3,600Ω ± 20% D. 62,000Ω ± 20%

12. The drawings MOST often used in maintenance work are 12.___
   A. master electrical drawings and schematics
   B. schematics and wiring diagrams
   C. wiring diagrams and isometric drawings
   D. isometric drawings and schematics

13. In a standard circuit drawing 1 milli-ampere, the power rating of a 4-megohm resistor should be _____ watts. 13.___
   A. 3 B. 4 C. 5 D. 6

14. A molded capacitor on which the top row of dots (left to right) is red, violet, and yellow and the bottom row of dots (left to right) is green, silver, and orange will have a voltage rating of ____ volts. 14.___
   A. 200 B. 300 C. 400 D. 500

15. The type of diagram that is consulted to find location and interconnection of parts is called a _____ diagram. 15.___
   A. schematic B. master C. wiring D. pictorial

16. In what direction does current flow in a circuit? From 16.___
   A. left to right
   B. right to left
   C. the negative terminal to the positive
   D. the positive terminal to the negative

KEY (CORRECT ANSWERS)

1. A	9. D
2. A	10. B
3. C	11. D
4. A	12. B
5. B	13. D
6. C	14. D
7. B	15. C
8. B	16. C

EXAMINATION SECTION

TEST 1

DIRECTIONS: Each question or incomplete statement is followed by several suggested answers or completions. Select the one that BEST answers the question or completes the statement. *PRINT THE LETTER OF THE CORRECT ANSWER IN THE SPACE AT THE RIGHT.*

1. A piece of equipment listed as drawing 100 watts is plugged into a 24 volt DC circuit. 1.___
   The MINIMUM size fuse which would handle this load is _____ amps.
   A. 2 B. 3 C. 4 D. 5

2. A resistor of 1000 ohms has 3 milliamperes passing through it. 2.___
   The voltage drop across the resistor is _____ volts.
   A. 3 B. 6 C. 15 D. 300

3. A certain resistor has three colored bands around it. The one nearest the end is green, the next one is orange, and the next one is red. 3.___
   The value of this register is _____ ohms.
   A. 74 B. 270 C. 5300 D. 64,000

4. An alternating voltage is applied to a capacitor. 4.___
   As the frequency of this voltage is increased, the impedance of the capacitor
   A. increases
   B. decreases
   C. remains the same
   D. increases or decreases depending on its construction

5. The one of the following that is NOT a part of a transistor is the 5.___
   A. emitter B. collector C. base D. grid

6. A 0.2 ufd capacitor is connected in series with a 0.1 ufd capacitor. 6.___
   The resultant capacity is _____ ufd.
   A. 0.067 B. 0.67 C. 0.15 D. 0.3

7. The term *Hertz* means the same as 7.___
   A. degrees Centigrade B. degrees Fahrenheit
   C. revolutions per minute D. cycles per second

8. In an electrolytic condenser, the dielectric material is 8.___
   A. mylar B. aluminum oxide
   C. paper D. sodium chloride

9. The amount by which a transformer will step up or step down a voltage is determined by its 9.___
   A. inductance B. resistance
   C. magnetic flux D. turns ratio

10. The electrolyte in a lead plate storage battery (such as that used in cars) is 10.___
   A. aluminum hydroxide B. sulfuric acid
   C. hydrochloric acid D. sodium chloride

11. A diode in an electronic circuit is used to 11.___
   A. amplify B. oscillate C. attenuate D. rectify

12. The MAIN function of a filter in a power supply is to 12.___
   A. increase the voltage
   B. decrease the load
   C. smooth out the peaks of the ripple frequency
   D. protect the power transformer

13. The expression *pH* as applied to a liquid refers to its 13.___
   A. salinity B. specific gravity
   C. viscosity D. acidity/alkalinity

14. The speed of a synchronous motor is controlled by 14.___
   A. the voltage applied to it
   B. the frequency of the alternating current applied to it
   C. a mechanical governor
   D. the current it draws

15. The capacitance of a condenser is measured in 15.___
   A. oersteds B. ohms C. henrys D. farads

16. The power lost in a 20-ohm resistor, with 0.25 amperes passing through it, is ______ watts. 16.___
   A. 0.04 B. 0.4 C. 1.25 D. 5

17. When soldering a transistor into a circuit, it is good practice to clamp a pair of long-nosed pliers on the lead between the transistor and the end being soldered. 17.___
This is done to
   A. prevent the lead from moving
   B. prevent burning the fingers
   C. ground the transistor
   D. prevent the soldering iron's heat from reaching the transistor

18. The commutator of a motor should 18.___
   A. not be lubricated
   B. be lubricated with light oil
   C. be lubricated with heavy grease
   D. be lubricated with hypoid oil

19. The band of wavelengths of visible light covers 19.___
   A. 20-50 centimeters B. 10-50 meters
   C. 400-700 millimicrons D. 400-700 millimeters

20. The heat reaching the earth from the sun is transmitted by 20.____
A. ions
B. convection
C. radiation
D. cosmic rays

21. A *thermistor* is a 21.____
A. type of thermometer
B. high power transistor
C. water heating device
D. resistor with a negative temperature coefficient

22. In an AC circuit, the term *power factor* refers to the 22.____
A. horsepower
B. BTU per watt
C. ratio of the resistance to the impedance
D. kilowatts per horsepower

23. 23.____

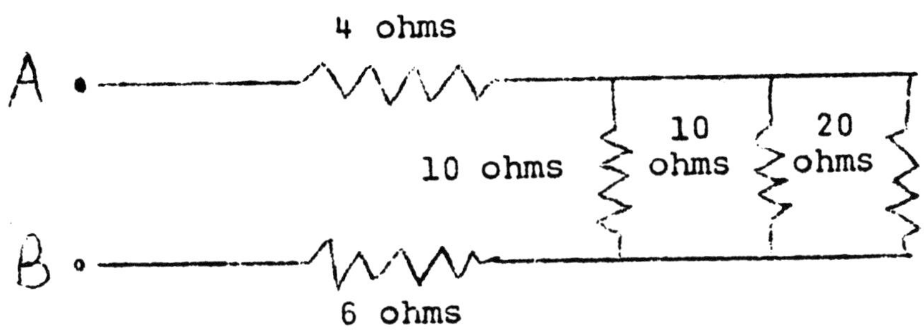

In the above circuit, the TOTAL resistance between points A and B is ______ ohms.
A. 5
B. 14
C. 20
D. 45

24. Of the four gases listed below, the one that is NOT an air pollutant is 24.____
A. carbon dioxide
B. carbon monoxide
C. sulfur dioxide
D. hydrogen sulfide

25. The term *milli-roentgen* refers to a unit of 25.____
A. x-ray radiation
B. ultraviolet radiation
C. reluctance
D. inductance

26. An AC motor drawing 12 amps is plugged into a 15-amp circuit. The starting surge of the motor, however, is 18 amps. 26.____
The PROPER type of fuse to be used in this situation is
A. varistor
B. thermistor
C. fast-blow
D. slow-blow

27. Degrees Kelvin is numerically equal to degrees 27.____
A. Fahrenheit - 15
B. Centigrade + 27
C. Fahrenheit + 135
D. Centigrade + 273

28. In the term *micromicrofarads*, the prefix *micromicro* means multiply by 28.___
A. 10^6 B. 10^3 C. 10^{-12} D. 10^{-6}

29. One horsepower is equivalent to 29.___
A. 276 joules B. 746 kilowatts
C. 746 watts D. 291 calories

30. Laminated iron or steel is generally used instead of solid metal in the construction of the field and armature cores in motors and generators. 30.___
The reason for this is to
A. reduce eddy current losses
B. increase the voltage
C. decrease the flux
D. reduce the cost

31. The instrument used to measure current flow is called a(n) 31.___
A. wattmeter B. voltmeter
C. ammeter D. wavemeter

32. Reversing the polarity of the voltage applied to a mica condenser will 32.___
A. destroy it B. increase its capacity
C. decrease its capacity D. have no effect on it

33. The *decibel* is the unit used for expressing 33.___
A. light levels
B. DC voltage
C. AC current
D. the ratio between two quantities of either electrical or sound energy

34. In a three-phase Y-connected AC power system, the voltage from leg to ground is 120 volts. 34.___
The voltage between each pair of hot legs is _____ volts.
A. 160 B. 180 C. 208 D. 240

35. An hygrometer is an instrument which measures 35.___
A. humidity B. temperature
C. specific gravity D. luminosity

36. The impedance ratio of a transformer varies _____ the turns ratio. 36.___
A. directly with B. as the square of
C. as the square root of D. inversely with

37. Two resistors are connected in series. The current through these resistors is 3 amperes. Resistance #1 has a value of fifty ohms; resistance #2 has a voltage drop of fifty volts across its terminals. 37.___
The TOTAL impressed voltage (across both resistors) is ____ volts.
A. 100 B. 150 C. 200 D. 250

38. The piece of equipment that should be used to obtain more than one voltage from a fixed voltage direct current source is a(n) 38.____
   A. multitap transformer
   B. resistance-type voltage divider
   C. autotransformer
   D. copper oxide rectifier

39. The ratio of peak to effective (rms) voltage value of a sine wave is 39.____
   A. 2 to 1 B. 1 to 2 C. .707 to 1 D. 1.414 to 1

40. Two coils are connected in series. 40.____
If there is no mutual inductance between the coils, the TOTAL inductance of the two coils is the ______ inductances.
   A. sum of the individual
   B. product of the individual
   C. product of the square roots of the two
   D. sum of the squares of the individual

41. The impedance of a coil with zero resistance is called the 41.____
   A. reluctance B. conductance
   C. inductive reactance D. flux

42. The ratio of the energy stored to the energy lost in a coil over a period of one cycle is called its 42.____
   A. efficiency B. *Q*
   C. reactance D. resistance

43. In a vacuum tube, the current is carried by 43.____
   A. ions B. neutrons C. electrons D. molecules

44. The device used to vary the intensity of an incandescent light on a 120V AC circuit is a 44.____
   A. variable capacitor
   B. silicon controlled rectifier
   C. copper oxide rectifier
   D. rf amplifier

45. High power transistors must be mounted on *heat sinks*. 45.____
The purpose of the heat sinks is to
   A. improve voltage regulation
   B. increase the transistors' output
   C. keep the transistors warm
   D. keep the transistors cool

46. The one of the following materials that has the HIGHEST conductivity is 46.____
   A. iron B. zinc C. copper D. silver

47. The unit used to express the alternating current impedance of a circuit is the 47.____
   A. mho B. farad C. ohm D. rel

48. A certain resistor has four colored bands on it. The fourth band is gold. 48.___
This means that the resistor
A. is wirewound B. is non-inductive
C. has a ± 20% tolerance D. has a ± 5% tolerance

49. An amplifier has an output voltage waveform that does not exactly follow that of the input voltage. 49.___
This type of distortion is called _____ distortion.
A. modular B. frequency
C. resonance D. amplitude

50. A parallel circuit, resonant at 1000 khz, has its value of capacity doubled and its value of inductance halved. 50.___
Its resonant frequency now is _____ khz.
A. 500 B. 1000 C. 1500 D. 2000

KEY (CORRECT ANSWERS)

1. D	11. D	21. D	31. C	41. C
2. A	12. C	22. C	32. D	42. B
3. C	13. D	23. B	33. D	43. C
4. B	14. B	24. A	34. C	44. B
5. D	15. D	25. A	35. A	45. D
6. A	16. C	26. D	36. B	46. D
7. D	17. D	27. D	37. C	47. C
8. B	18. A	28. C	38. B	48. D
9. D	19. C	29. C	39. D	49. D
10. B	20. C	30. A	40. A	50. B

TEST 2

DIRECTIONS: Each question or incomplete statement is followed by several suggested answers or completions. Select the one that BEST answers the question or completes the statement. *PRINT THE LETTER OF THE CORRECT ANSWER IN THE SPACE AT THE RIGHT.*

1. A voltmeter which reads 100V full scale has a specified accuracy of 3%. It is hooked across a circuit and reads 97 volts. 1.___
The TRUE voltage can be assumed to be somewhere between
   A. 96.7 and 97.3   B. 94 and 100
   C. 96.07 and 97.03   D. 95.5 and 98.5

2. The product of 127.2 and .0037 is 2.___
   A. 4706.4   B. 470.64   C. .47064   D. .0047064

3. The wind velocity at a certain location was measured four times in a 24-hour period. The readings were 32 mph, 10 mph, 16 mph, and 2 mph. 3.___
The AVERAGE wind velocity for that day was _____ mph.
   A. 24   B. 20   C. 15   D. 13

4. When 280 is divided by .014, the answer is 4.___
   A. .002   B. 20   C. 200   D. 20,000

5. The square root of 289 is 5.___
   A. 1.7   B. 9.7   C. 17   D. 144.5

6. The watts drawn by a resistive load is to be determined. To do this, a voltmeter (10V full scale) is connected across the load, and an ammeter (10 amps full scale) is connected in series with the load. Both instruments are specified as having 1% (full scale) accuracy. The voltmeter reads 9.2V; the ammeter reads 8.3 amps. 6.___
The MOST valid value for the watts drawn is _____ watts.
   A. 76   B. 76.36   C. 76.4   D. 80

7. The formula for converting degrees Centigrade to degrees Fahrenheit is: °F = (9/5)·(°C) + 32. 7.___
A temperature of 25°C is equal to
   A. 102.6°F   B. 85°F   C. 77°F   D. 43°F

8. The prefix *kilo* means 8.___
   A. multiply by one million
   B. divide by one million
   C. multiply by one thousand
   D. divide by one hundred

9. 2^8 is equal to 9.___
   A. 512   B. 256   C. 124   D. 82

10. The prefix *milli* means 10.___
   A. multiply by 100
   B. divide by one thousand
   C. divide by one million
   D. multiply by one million

11. If 1/X = 1/20 + 1/20 + 1/40, the value of X is 11.___
   A. .125 B. 8 C. 16 D. 20

12. 2×10^6 multiplied by 4×10^{-6} equals 12.___
   A. 8 B. 8×10^{-12} C. 8×10^{12} D. 8×10^3

13. 1 inch equals ______ cm. 13.___
   A. 0.62 B. 2.54 C. 3.94 D. 16.2

14. 1 kgm equals 14.___
   A. 2.2 lbs. B. 17.3 oz. C. 0.52 lbs. D. 12 oz.

15. 1 liter equals 15.___
   A. 3.78 quarts
   B. 1.057 quarts
   C. 1.39 pints
   D. .067 gallons

16. A circle has a radius of 10 inches. 16.___
Its circumference is ______ inches.
   A. 72.3 B. 62.8 C. 31.4 D. 25

17. A right angle triangle has sides measuring 3 inches and 4 inches; its hypotenuse is 5 inches. 17.___
The area of this triangle is ______ square inches.
   A. 6 B. 20 C. 15 D. 60

18. A square has an area of 81 square inches. 18.___
The length of each side is ______ inches.
   A. 7.9 B. 9 C. 11 D. 17

19. A bottle contains 11 pints of liquid. To this bottle 1.32 pints is then added. 19.___
This is an increase of
   A. 6% B. 9% C. 12% D. 16%

20. A week ago a storage battery read 12.4V. Today its voltage is 8.1% less. 20.___
Its voltage is now
   A. 11.4 B. 10.8 C. 9.3 D. 10.2

21. The advantage of a vacuum tube voltmeter over a regular voltmeter is that it 21.___
   A. operates on batteries
   B. operates on 120V AC
   C. has a low input impedance
   D. has a high input impedance

22. A g_m tube tester measures a vacuum tube's 22.___
   A. capacitance
   B. resistance
   C. emission
   D. transconductance

23. A cathode ray tube is used in a(n) 23.____
A. audio amplifier
B. radio frequency amplifier
C. oscilloscope
D. volt-ohm-milliammeter

24. A voltmeter is described as having *1000 ohms per volt*. The current required to produce full scale deflection is 24.____
A. 1 milliampere
B. 1 ampere
C. 20 milliamperes
D. 0.05 milliamperes

25. The PRIMARY use of a test oscilloscope is to 25.____
A. analyze complex waveforms
B. measure resistance
C. measure capacitance
D. measure DC voltages

26. A spectrophotometer is an instrument that measures 26.____
A. photographic film density
B. the amount of light of a particular wavelength
C. the amount of airborne dust
D. x-ray radiation

27. The test instrument generally known as a *multitester* will measure, among other things, 27.____
A. temperature
B. beta radiation
C. AC watts
D. DC milliamperes

28. A lightmeter used in measuring incident light gives readings in 28.____
A. footcandles
B. candlepower
C. lumens
D. foot-lamberts

29. A selenium photocell is a type known as photo- 29.____
A. emissive
B. resistive
C. voltaic
D. transistive

30. In wiring electronic circuits, the solder GENERALLY used is ______ solder. 30.____
A. silver
B. acid core
C. aluminum
D. rosin core

31. An unconscious victim of electric shock should be orally administered 31.____
A. nothing
B. coffee
C. alcohol
D. aromatic apirits of ammonia

32. Persons operating x-ray equipment should wear 32.____
A. safety goggles
B. insulating gloves
C. a lead-coated apron and gloves
D. a surgical mask

33. Harmful radiation is emitted by the element 33.___
A. neon B. lithium C. platinum D. radium

34. When a victim of electrical shock or near drowning is given artificial respiration and he does not appear to respond, the treatment should continue for at least 34.___
A. four hours
B. fifteen minutes
C. five minutes
D. fifteen hours

35. A person maintaining high voltage equipment should avoid wearing 35.___
A. long hair
B. sneakers
C. rings and metallic watchbands
D. eyeglasses

36. Portable AC equipment is often equipped with a three-wire cable and a three-prong male plug. 36.___
The reason for this is to prevent
A. radiation
B. electric shock
C. oscillation
D. ground currents

37. Smoke is seen issuing from a piece of electronic equipment. 37.___
The FIRST thing that should be done is to
A. call the fire department
B. pour water on it
C. look for a fire extinguisher
D. shut off the power

38. A match should not be used when inspecting the electrolyte level in a lead-acid battery because the cells emit 38.___
A. nitrogen
B. hydrogen
C. carbon dioxide
D. sulfur dioxide

39. A person feels nauseated, his mental capacity has been lowered, and he has a severe throbbing headache. It is suspected that he has been poisoned by gas, but there is no apparent odor. 39.___
The poisonous gas is MOST likely to be
A. sulfur dioxide
B. hydrogen cyanide
C. carbon monoxide
D. chlorine

40. The purpose of an interlock on a piece of electronic equipment is to 40.___
A. prevent theft of the vacuum tubes
B. prevent electrical shock to maintenance personnel
C. prevent rf radiation
D. keep the equipment cool

41. An alternating voltage is applied to an inductance. 41.___
As the frequency of the voltage is decreased, the impedance of the inductance
A. decreases
B. increases
C. follows the alternating voltage
D. remains the same

42. A 0.25 ufd condenser is connected in parallel with a 0.50 ufd condenser. 42.___
The resultant capacity is _____ ufd.
A. 0.167 B. 0.37 C. 0.75 D. 2.5

43. The electrolyte in a carbon-zinc dry cell is 43.___
A. sulfuric acid B. ammonium chloride
C. lithium chloride D. sodium chloride

44. A 5000-ohm resistor has a voltage of 25 volts applied to it. 44.___
The current drawn by the resistor is
A. 5 milliamperes B. 5 amperes
C. 75 milliamperes D. 1.25 milliamperes

45. A certain resistor has three colored bands around it. The one nearest the end is red, the next one is gray, and the next one is yellow. 45.___
The value of the resistor is
A. 2.7 megaohms B. 280,000 ohms
C. 3270 ohms D. 449 ohms

Questions 46-50.

DIRECTIONS: Questions 46 through 50 are to be answered on the basis of the following paragraph.

The second half of the twin triode acts as a phase modulator. The rf output of the crystal oscillator is impressed on the phase-modulator grid by means of a blocking condenser. The cathode circuit is provided with a large amount of degeneration by an un-bypassed cathode resistor. Because of this degenerative feed-back, the transconductance of the triode is abnormally low, so low that the plate current is affected as much by the direct grid-plate capacitance as by the transconductance. The two effects result in plate current vectors almost 180° apart, and the total plate current is the resultant of the two components. In phase, it will be about 90° removed from the phase of the voltage impressed on the grid.

46. As used in the above paragraph, the word *impressed* means MOST NEARLY 46.___
A. applied B. blocked C. changed D. detached

47. As used in the above paragraph, the word *components* refers to the 47.___
A. blocking condenser and cathode resistor
B. twin triode
C. plate current vectors
D. grid-plate capacitance

48. According to the above paragraph, degenerative feed-back is obtained by means of 48.___
A. a crystal oscillator
B. the plate voltage
C. an un-bypassed cathode resistor
D. a blocking condenser

49. According to the above paragraph, the cathode resistor is 49.___
   A. very large
   B. not bypassed
   C. in series with an inductance
   D. shunted by a blocking condenser

50. According to the above paragraph, the phase angle between the grid voltage and the total plate current is APPROXIMATELY 50.___
   A. 180° B. 90° C. 270° D. zero

KEY (CORRECT ANSWERS)

1. B	11. B	21. D	31. A	41. A
2. C	12. C	22. D	32. C	42. C
3. C	13. B	23. C	33. D	43. B
4. D	14. A	24. A	34. A	44. A
5. C	15. B	25. A	35. C	45. B
6. A	16. B	26. B	36. B	46. A
7. C	17. A	27. D	37. D	47. C
8. C	18. B	28. A	38. B	48. C
9. B	19. C	29. C	39. C	49. B
10. B	20. A	30. D	40. B	50. B

EXAMINATION SECTION
TEST 1

DIRECTIONS: Each question or incomplete statement is followed by several suggested answers or completions. Select the one that BEST answers the question or completes the statement. *PRINT THE LETTER OF THE CORRECT ANSWER IN THE SPACE AT THE RIGHT.*

1. If a nichrome wire 2 meters long has a resistance of 10 ohms, the resistance of another nichrome wire 1 meter long and with a cross-sectional area half that of the longer wire is, in ohms, 1.___
   A. 5 B. 10 C. 20 D. 40

2. Two nichrome wires of exactly the same composition have the same weight, but one is 5 times as long as the other. If the resistance of the shorter wire is R, the resistance of the other is 2.___
   A. R B. 5R C. 25R D. 50R

3. A magnet pole has a strength of 400 units. The magnetic field intensity in air due to this pole and at a distance 4 cm from this pole is _____ oersteds. 3.___
   A. 25 B. 50 C. 100 D. 25π

4. The number of revolutions per minute that a 6-pole alternator must make to produce a frequency of 60 cycles/sec is 4.___
   A. 1080 B. 1200 C. 2160 D. 21,600

5. A voltmeter having a full scale deflection of 10 volts has an internal resistance of 1000 ohms. To convert this instrument to a voltmeter having a full scale deflection of 300 volts requires, in ohms, a multiplier of 5.___
   A. 3000 B. 6000 C. 9000 D. 12,000

6. Which one of the following scientists is NOT directly connected with the invention and development of the transistor? 6.___
   A. John Bardeen
   B. William B. Shockley
   C. Polykarp Kusch
   D. Walter H. Brattain

7. If there is a current of 0.1 ampere through a lamp for 100 seconds, the number of coulombs passing through in that time is 7.___
   A. 0.1 B. 1 C. 10 D. 100

8. If the core of an electromagnet is made of 2 pieces of iron, each running the full length of the electromagnet, when a direct current is sent through the coil, the 2 core pieces will 8.___
   A. attract each other
   B. repel each other
   C. attract and repel alternately
   D. have no effect on each other

9. If a 25-watt, 120-volt lamp and a 100-watt, 12-volt lamp are connected in a series to a 120-volt source, 9.___
   A. both lamps light normally
   B. neither lamp lights
   C. the 100-watt lamp is brighter
   D. the 25-watt lamp is brighter

10. A 30-watt, 120-volt resistor is connected to a 120-volt, 60-cycle source. 10.___
The maximum current flow in the lamp, in amperes, is APPROXIMATELY
   A. 0.18 B. 0.25 C. 0.35 D. 0.4

11. A 0-20 milliampere meter has a resistance of 20 ohms. To convert this meter to a voltmeter with a range of 0-10 volts, one should connect a resistance of APPROXIMATELY ______ ohms in _____. 11.___
   A. 200; series B. 200; parallel
   C. 500; series D. 500; parallel

12. The current flow through a galvanometer is 10^{-5} milliamperes and produces a deflection of 1 scale division. If the resistance of the moving coil is 200 ohms, the voltage across the coil, in volts, is 12.___
   A. 2×10^{-3} B. 5×10^{-10} C. 5×10^{-5} D. 2×10^{-6}

13. A capacitor using a dielectric whose coefficient is 5 has a capacitance of A. 13.___
An identical capacitor using a dielectric whose coefficient is 20 will have a capacitance equal to
   A. 2A B. 4A C. 10A D. 100A

14. When resonance occurs in a circuit supplied with an alternating voltage, the 14.___
   A. impedance equals zero
   B. inductance equals the reciprocal of the capacitance
   C. capacitance equals the inductance
   D. inductive reactance equals the capacitive reactance

15. An electric heating coil of 6 ohms resistance is connected across a 120-volt line for 10 minutes. 15.___
The energy liberated, in joules, in this period of time equals
   A. 7.2×10^3 B. 14.4×10^5
   C. 25.8×10^4 D. 43.2×10^4

16. If a triode has its plate current increased 20 milliamperes when the plate voltage is increased from 50 to 90 volts and the plate current is also increased 20 milliamperes when the grid potential changes 4 volts, the amplification factor of the tube is 16.___
   A. 1.8 B. 5.0 C. 10.0 D. 40.0

17. A charge of 30 coulombs passes through a wire in 3 seconds. The current flow in this wire, in amperes, equals 17.___
A. 3.3 B. 10 C. 30 D. 90

18. The heat developed by 5 amperes flowing through a resistance of 4 ohms is 18.___
A. 20 calories
B. 24 calories per second
C. 100 calories
D. 4.8 calories per degree

19. 1 e.s.u. of potential difference equals 19.___
A. 1 volt
B. 300 joules/coulomb
C. 10^8 e.m.u. of potential
D. 4.187 volts

20. Two electrical condensers having capacitances of 6 and 12 microfarads, respectively, are connected in series. The TOTAL capacitance, in microfarads, of this combination is 20.___
A. 2 B. 4 C. 9 D. 18

21. The voltage induced in a coil with an inductance of 0.25 henries when the current decreased uniformly from 2 amperes to zero amperes in 1/16 second is 21.___
A. 4 B. 8 C. 16 D. 24

22. A transformer placed on DC is LIKELY to burn out because of the absence of 22.___
A. a fuse
B. voltage regulation
C. hysteresis
D. inductive reactance

23. If the effective AC voltage of a given circuit is 100, the maximum voltage is CLOSEST to which one of the following? 23.___
A. 0 B. 71 C. 141 D. 173

24. One milliampere produces full-scale deflection in a galvanometer whose internal resistance is 50 ohms. To convert this instrument into an ammeter whose full-scale deflection is 1 amp, it should be shunted with a resistance, in ohms, CLOSEST to which one of the following? 24.___
A. 0.005 B. 0.05 C. 0.5 D. 5.0

25. The results of the Millikan oil drop experiment lead to the conclusion that 25.___
A. electric charges are negative
B. electric charges are due to a transfer of electrons
C. there is a fundamental unit of charge
D. there are no isolated magnetic poles

26. A 60-watt, 120-volt incandescent lamp has a resistance, in ohms, of 26.___
A. 0.5 B. 2.0 C. 60 D. 240

27. A 0-5 amp ammeter reads full scale when its 2-ohm movable coil has a voltage of 0.01 applied across it. 27.___
The shunt has a resistance, in ohms, CLOSEST to which one of the following?
A. 0.001 B. 0.002 C. 250 D. 500

28. The heat in kilocalories (1 kilocalorie = 4200 joules) developed by a 60 watt lamp in one hour is APPROXIMATELY 28.___
A. 36 B. 50 C. 60 D. 95

29. Assume that a circuit consisting of a coil and a capacitor is adjusted to give resonance. 29.___
If some turns are removed from the coil, then resonance can be restored by
A. increasing the frequency
B. decreasing the capacitance
C. decreasing the frequency
D. decreasing the inductance

30. A series AC circuit contains a resistance of 40 ohms, an inductive reactance of 100 ohms, and a capacitive reactance of 70 ohms. 30.___
The impedance of this circuit is, in ohms,
A. 50 B. 110 C. 150 D. 210

31. A series AC circuit contains a 20-ohm resistance, a 40-ohm resistance, a 60-ohm inductive reactance, and an 80-ohm capacitive reactance. 31.___
The GREATEST amount of heat per second produced by any of these will be produced by the
A. 20-ohm resistance B. 40-ohm resistance
C. 60-ohm reactance D. 80-ohm reactance

32. If a sheet of glass is slipped between the plates of an air capacitor, the capacitance of the combination 32.___
A. drops to zero
B. is reduced to approximately one-half of the original value
C. increases
D. has a value which depends on the charge of the capacitor

33. The resistance of a piece of wire is 16 ohms. 33.___
The resistance, in ohms, of a piece of the same wire twice as long and twice the diameter is
A. 8 B. 16 C. 32 D. 64

34. A circuit containing resistance, capacitance, and inductance is in resonance when supplied with an alternating current of a given frequency. 34.___
An increase in the frequency of this current will
A. decrease the inductive reactance
B. increase the capacitive reactance
C. increase the total current in the circuit
D. decrease the impedance

35. The heat developed by a 5 ampere current flowing through a resistance of 4 ohms is 35.____
A. 20 calories
B. 24 calories/sec
C. 100 calories
D. 4.8 calories per degree

36. A capacitor stores 50 joules of energy when charged by a 5000 volt source. 36.____
The capacitance of the capacitor, in microfarads, is
A. .02
B. .04
C. 2.0
D. 4.0

37. The amplification factor of a triode is represented BEST by which one of the following expressions? 37.____
The
A. change in grid voltage divided by the change in filament voltage
B. plate current
C. change in plate voltage divided by the change in grid voltage needed to produce that change in plate current
D. change in plate voltage divided by the change in plate current needed to produce that change in grid voltage

38. Kirchhoff's Laws with regard to the current in an electrical circuit states that 38.____
A. current is directly proportional to voltage
B. the algebraic sum of the currents at a junction equals zero
C. current is inversely proportional to electrical resistance
D. the total current in a parallel circuit equals the sum of the currents in the individual parallel branches

39. A Fleming valve or diode performs which one of the following functions? 39.____
A. Converts rectified DC to AC
B. Converts AC to half-wave rectified DC
C. Converts AC to smooth DC
D. Steps up low voltage AC to high voltage AC

40. A capacitor discharges at a certain frequency through a circuit containing an inductance. 40.____
If the capacitance is multiplied by four, the oscillation frequency is multiplied by
A. $\frac{1}{4}$
B. $\frac{1}{2}$
C. 2
D. 4

41. A coil of 100 turns is wound on an iron core. The coil is connected to an AC source. Suitable connections are made to display applied voltage and circuit current on an oscilloscope. 41.____
It is found that the current _____ the voltage by _____.
A. leads; 90°
B. lags behind; 90°
C. leads; an acute angle
D. lags behind; an acute angle

42. The mixing of an audio frequency electric current with a radio frequency carrier in a broadcasting station is known as 42.___
A. oscillation
B. modulation
C. amplification
D. rectification

43. An electron is accelerated through a potential difference of 20,000 volts. 43.___
Its gain in kinetic energy is 20,000
A. volts
B. joules
C. electron volts
D. ergs

44. The radius of the circular path of a charged particle moving at right angles to a uniform magnetic field is DIRECTLY proportional to the 44.___
A. momentum of the particle
B. flux density
C. charge on the particle
D. wavelength of its radiation

45. The direction of an induced current is always such that the magnetic field belonging to it tends to oppose the change in the strength of the magnetic field belonging to the primary current. 45.___
This law was FIRST enunciated by
A. Ampere
B. Faraday
C. Henry
D. Lenz

46. A split ring commutator will be found on a 46.___
A. synchronous motor
B. AC generator
C. DC motor
D. induction-repulsion type of motor

47. A ballistic galvanometer is used MAINLY to measure which one of the following? 47.___
A. Electric charge
B. Electric current
C. EMF
D. Resistance

48. If a calibrated oscilloscope shows a sinusoidal current having a peak to peak value of 2.0 amperes, the effective value of the current as measured by an ammeter and expressed in amperes would be 48.___
A. 0.50
B. 0.71
C. 1.0
D. 2.0

49. Of the following, the pair of functions of a vacuum tube that are MOST closely allied are 49.___
A. rectification and amplification
B. oscillation and rectification
C. amplification and detection
D. detection and rectification

50. If a thin sheet of metal is placed halfway between the two plates of a parallel-plate, air-dielectric capacitor, the capacitance is 50.____
 A. quadrupled
 B. doubled
 C. decreased to 1/2 of the original value
 D. decreased to 1/4

KEY (CORRECT ANSWERS)

1. B	11. C	21. B	31. B	41. D
2. C	12. D	22. D	32. C	42. B
3. A	13. B	23. C	33. A	43. C
4. B	14. D	24. B	34. B	44. A
5. C	15. B	25. C	35. B	45. D
6. C	16. C	26. D	36. D	46. C
7. C	17. B	27. B	37. C	47. A
8. B	18. B	28. B	38. B	48. B
9. D	19. B	29. A	39. B	49. D
10. C	20. B	30. A	40. B	50. C

TEST 2

DIRECTIONS: Each question or incomplete statement is followed by several suggested answers or completions. Select the one that BEST answers the question or completes the statement. *PRINT THE LETTER OF THE CORRECT ANSWER IN THE SPACE AT THE RIGHT.*

1. If a 1 mfd capacitor and a 2 mfd capacitor are connected in series across a 120-volt source, the potential difference across the 1 mfd capacitor is _____ volts. 1.___
   A. 30 B. 40 C. 60 D. 80

2. If a battery having an EMF of 6 volts and internal resistance of 0.6 ohms is supplying a current of 0.5 amperes, the terminal voltage is _____ volts. 2.___
   A. 5.4 B. 5.5 C. 5.6 D. 5.7

3. A capacitor C and inductor L are connected in series across an AC source. 3.___
   As the value of C is increased, the current in the circuit
   A. increases B. decreases
   C. remains constant D. may increase or decrease

4. As the frequency of a generator is decreased, the impedance of a circuit 4.___
   A. increases
   B. decreases
   C. may increase or decrease
   D. does not change

5. It is desired to determine the direction of electron flow in a vertical conductor carrying direct current. This may be done with the aid of a compass placed 5.___
   A. to the right of the wire
   B. to the left of the wire
   C. behind the wire
   D. in any of the above listed positions

6. If the current flow in a circular coil consisting of a single turn of wire of 1 cm radius is 2 abamperes, it will produce a magnetic field intensity at the center of the coil equal, in oersteds, to 6.___
   A. π B. 2π C. 4π D. 8π

7. Ampere's Law is concerned with 7.___
   A. the force on a wire carrying a current in a magnetic field
   B. electrochemical equivalents
   C. rms values
   D. unit magnetic poles

8. A 60-watt, 120-volt and a 40-watt, 120-volt lamp are joined in series and connected to a 120-volt line. The current flow in the circuit, in amperes, is 8.____

A. more than 0.5
B. between 0.5 and 0.3
C. 0.2
D. less than 0.2

9. The function of the grid in a three element vacuum tube is to 9.____

A. aid electron flow at reduced cathode temperatures
B. reduce the loss of heat from the cathode
C. prevent secondary emission of electrons from the plate
D. control the electron flow to the plate

10. Three capacitors of 4 mfd, 10 mfd, and 20 mfd are connected in parallel. The equivalent capacitance of this group equals, in mfd, 10.____

A. 2.5 B. 17 C. 34 D. 800

11. Two 60-watt, 120-volt heaters are connected in series on a 120-volt DC line. The power consumption is now X times as great as it would be if they were connected in parallel. Assuming no change in resistance, X would be 11.____

A. 1/4 B. 1/2 C. 2 D. 4

12. The current in an alternating current circuit is equal to the voltage divided by the 12.____

A. impedance
B. capacitance
C. reluctance
D. inductance

13. The core of a transformer is laminated largely for the purpose of 13.____

A. reducing eddy currents
B. aiding in heat dissipation
C. increasing self-inductance
D. increasing impedance

14. Kirchhoff's First Law is really a restatement of 14.____

A. Lenz' Law
B. Ohm's Law
C. Faraday's Law of Electrolysis
D. Law of Conservation of Energy

15. A condenser designed for use across a 220-volt AC line should have a peak inverse voltage rating of AT LEAST ______ volts. 15.____

A. 110 B. 220 C. 250 D. 325

16. The voltage between the cathode and target of an x-ray tube is V volts. If e is the charge on the electron in e.s.u., then Ve has the dimensions of 16.____

A. work B. current C. force D. momentum

17. A transformer with 50 turns on the primary and 100 turns on the secondary is connected to a 6-volt battery. The voltage on the secondary will be 17.___
   A. zero
   B. equal to that on the primary
   C. twice the voltage on the primary
   D. half the voltage on the primary

18. When a 60 cycle source of e.m.f. is connected between the plate and the cathode of a diode, the current between the cathode and plate will 18.___
   A. reverse 60 times per second
   B. reverse 120 times per second
   C. flow continuously
   D. flow intermittently in the same direction

19. Inside a dry cell that is delivering current, 19.___
   A. electrons flow from + to -
   B. there is no resistance
   C. there is no current
   D. electrons flow from - to +

20. In a circuit containing an alternating source and a coil, increasing both the frequency of the source and the inductance of the coil (without changing the e.m.f.) will result in 20.___
   A. increased current
   B. decreased current
   C. same current
   D. increased or decreased current, depending on which factor is increased more

21. If the plate current of a triode electronic tube increases 10 milliamperes when the plate voltage is increased from 60 to 80 volts or when the grid potential changes 2 volts, the amplification factor of the tube is 21.___
   A. 5 B. 10 C. 20 D. 40

22. A series circuit with a capacitive reactance of 20 ohms, an inductive reactance of 50 ohms, and a resistance of 40 ohms has an overall impedance, in ohms, of 22.___
   A. 10 B. 50 C. 70 D. 110

23. Waves of 3 cm in length radiated by an electronic oscillator would have a frequency, in cycles/sec, of 23.___
   A. 10^6 B. 10^8 C. 10^{10} D. 10^{12}

24. If a student, in finding the resistance of a lamp on a 120-volt line, places the voltmeter in series with the lamp and the ammeter in parallel with the lamp, the 24.___
   A. ammeter will burn out B. voltmeter will burn out
   C. lamp will burn out D. lamp may not light

25. If a 120-volt, 60-cycle source is connected to a circuit containing resistance, inductive reactance, and capacitive reactance, each of 16 ohms, the current in the circuit is _____ amperes. 25.____

A. 5 B. 7.5 C. 15 D. 30

26. If a nail is attracted to an electromagnet carrying direct current, and the current is quickly reversed, the nail will 26.____

A. fall to the ground
B. be repelled by the electromagnet
C. heat up
D. still be attracted

27. The volt is NOT a unit of 27.____

A. charge per unit time
B. e.m.f.
C. work per unit charge
D. potential difference

28. The energy used to carry unit charge around an electrical circuit is measured by the 28.____

A. current
B. potential difference
C. power
D. resistance of the circuit

29. Two 1 mfd capacitors are connected in parallel and the combination is then charged. 29.____
If the capacitance, charge, and potential of each capacitor are C, Q, and V, respectively, the corresponding three values for the combination is which one of the following?

A. C/2, 2Q, V
B. 2C, 2Q, V
C. 2C, Q, V
D. 2C, 2Q, V/2

30. The resonant frequency f of an alternating e.m.f. in a circuit containing in series an inductance L, a capacitance C, and a resistance R, is given by the formula f = 30.____

A. $2\pi\sqrt{LC}$ B. $\frac{1}{2}\pi\sqrt{LC}$ C. $\frac{1}{2\pi\sqrt{LC}}$ D. $\frac{R}{2\pi\sqrt{LC}}$

31. In a series resonant circuit, the circuit impedance is ALWAYS 31.____

A. equal to the inductive reactance
B. greater than the capacitive reactance
C. equal to the capacitive reactance
D. equal to the resistance

32. Two resistances of 30 ohms and 20 ohms, respectively, are joined in parallel in an electric circuit. 32.____
The equivalent resistance, in ohms, of this parallel pair is

A. 12 B. 25 C. 50 D. 600

33. A solenoid has an inductance of 0.32 henry. Its reactance, in ohms, to an alternating current having a frequency of 1000 cycles/sec is 33.___
   A. 2010 B. 3125 C. 3200 D. 3310

34. If, with an impressed voltage of 240 volts and a current of 13 amp, a shunt-wound DC motor delivers 4 horsepower, its efficiency, in percent, is 34.___
   A. 80 B. 86 C. 90 D. 96

35. If the maximum value of an alternating voltage is 110, its value, in volts, at the 30° phase is 35.___
   A. 55 B. 78 C. 155 D. 220

36. A coil of 2.5 henries inductance would resonate with a 1 microfarad capacitance at a frequency, in cycles/sec, of 36.___
   A. 100 B. 1,000 C. 10,000 D. 100,000

37. When an alternating emf is supplied to a circuit consisting, in addition to the source, of wiring a 40 watt lamp and an 80 mf capacitor, the voltage in the circuit 37.___
   A. leads to current
   B. lags behind the current
   C. is in phase with the current
   D. becomes and remains zero after a few minutes

38. Of the following, the one which is incorporated in a DC generator but NOT in an AC generator is the 38.___
   A. slip rings B. commutator
   C. series field D. permanent magnets

39. When a capacitor C, an inductance L, and a resistance R are joined in series and an alternating emf is supplied to the circuit, the resonant frequency of this circuit can be decreased by doing which one of the following? 39.___
   A. Increasing L B. Decreasing R
   C. Decreasing C D. Decreasing L

40. Three 6 mfd capacitors are connected in parallel across a 120-volt AC line. The equivalent capacitance, in mfd, of this circuit is 40.___
   A. 2 B. 6 C. 18 D. 20

41. Assume that two capacitors, one of 3 microfarad capacitance and the other of 6 microfarad capacitance, are connected in series and charged to a difference of potential of 120 volts. The potential difference, in volts, across the 3 microfarad capacitor is 41.___
   A. 40 B. 80 C. 180 D. 360

42. Three resistors are connected to form the sides of a triangle ABC. The resistance of side AB is 40 ohms, of side BC 60 ohms, and of side CA 100 ohms. The effective resistance between points A and B, in ohms, is 42.____

A. 32 B. 50 C. 64 D. 200

43. When a current of 2 amperes flows through a conductor of 2 ohms resistance for 3 seconds, the heat produced, in joules, is 43.____

A. 12 B. 24 C. 36 D. 72

44. A length of wire, diameter 2 mils, has a resistance of 6 ohms. The same length of wire of the same material having a diameter of 4 mils has a resistance, in ohms, of 44.____

A. 1.5 B. 3 C. 12 D. 24

45. The generalization that the algebraic sum of the currents at a junction in a circuit equals zero was postulated by 45.____

A. Ohm B. Kirchhoff C. Onnes D. Seebeck

46. It is desired to charge an electroscope negatively by induction. One of the steps that must be performed is to 46.____

A. use a negatively charged rod
B. remove positive charges
C. remove electrons
D. ground the electroscope

47. A series AC circuit contains an inductance L, a capacitance C, and a resistor R. The impedance of this circuit equals 47.____

A. $R^2 + X_L + X_C$ B. $\sqrt{R^2 + (X_L - X_C)^2}$

C. $R^2 + \sqrt{X_L - X_C}$ D. $R^2 - X_L - X_C^2$

48. In a selenium rectifier, current flow practically ceases when the 48.____

A. selenium becomes negative
B. selenium becomes positive
C. accompanying alloy becomes negative
D. applied voltage exceeds the critical value

49. An alternating current generator having 4 poles rotates at 60 revolutions per second. The frequency of the current produced, in cycles per second, is 49.____

A. 60 B. 15 C. 120 D. 240

50. If an AC circuit contains resistance only, then current 50.___
   A. and voltage are in phase
   B. lags by 45°
   C. leads by 90°
   D. lags by 45° and voltage leads by 45°

KEY (CORRECT ANSWERS)

1. D	11. A	21. B	31. D	41. B
2. D	12. A	22. B	32. A	42. A
3. D	13. A	23. C	33. A	43. B
4. C	14. D	24. D	34. D	44. A
5. D	15. D	25. B	35. A	45. B
6. C	16. A	26. D	36. A	46. D
7. A	17. A	27. A	37. B	47. B
8. C	18. D	28. B	38. B	48. A
9. D	19. A	29. B	39. A	49. C
10. C	20. B	30. C	40. C	50. A

EXAMINATION SECTION

TEST 1

DIRECTIONS: Each question or incomplete statement is followed by several suggested answers or completions. Select the one that BEST answers the question or completes the statement. *PRINT THE LETTER OF THE CORRECT ANSWER IN THE SPACE AT THE RIGHT.*

1. A normally operating NPN common emitter transistor has the GREATEST voltage drop between 1.____
   A. emitter and collector B. base and collector
   C. base and emitter D. plate and grid

2. When a superheterodyne receiver with an IF of 455 Khz is tuned to a station of 1010 Khz, the frequency of the local oscillator should be _____ Khz. 2.____
   A. 455 B. 1010 C. 555 D. 1465

3. The DC voltage at the rectifier output is zero. The AC input is 125 volts. This indicates 3.____
   A. a shorted choke
   B. capacitor is wrong value
   C. defective rectifier tube
   D. open output filter capacitor

4. A radio receiver has a loud hum which is NOT affected by the setting of the volume control. The trouble is *probably* caused by 4.____
   A. an open volume control
   B. a gassy output tube
   C. improper IF alignment
   D. an open filter capacitor

5. A speaker designed to reproduce high audio frequencies is called 5.____
   A. woofer B. magnetic speaker
   C. PM speaker D. tweeter

6. If two .02 MFd. 400 volt capacitors are connected in series, they are better than one .01 MFd. 600 volt unit because the two capacitors together will 6.____
   A. have more capacity
   B. have a higher coulomb rating
   C. cost less
   D. have a higher voltage rating

7. Powdered iron cores are *generally* used in 7.____
   A. power transformers
   B. non-inductive resistors
   C. dynamic speakers
   D. IF transformers

8. A resistor having a value of less than 100 ohms is in series with a selenium rectifier to 8.___
   A. improve the filtering action
   B. eliminate hum distortion
   C. prevent current surges from damaging the rectifier
   D. prevent damage to a series filament circuit

9. Hum troubles, when caused by the power supply, are *generally* due to the 9.___
   A. power transformer B. filter capacitor
   C. filter resistor D. fuse

10. The ripple frequency of a half wave voltage doubler is ____ cps. 10.___
   A. 60 B. 129 C. 240 D. 30

11. Pre-amplifiers are needed with a magnetic pickup because 11.___
   A. the pick-up output voltage is too weak
   B. the more tubes you use, the more things can go wrong
   C. it sounds impressive
   D. the pick-up output voltage is too strong

12. A transformerless power supply has 12.___
   A. a voltage doubler and series heaters
   B. a voltage doubler and parallel heaters
   C. a full-wave rectifier and parallel heaters
   D. one side of the power line connected to B+

13. After detection, the carrier frequency is eliminated by a 13.___
   A. tube B. transistor C. resistor D. capacitor

14. Inductance is provided by a 14.___
   A. resistor B. tube C. coil D. capacitor

15. Frequency band width of an FM wave depends upon 15.___
   A. ratio detector
   B. discriminator
   C. RF carrier
   D. strength of the impressed audio voltage

16. The ____ tube CANNOT be used as an amplifier. 16.___
   A. diode B. triode C. tetrode D. pentode

17. The reading of resistance between base and emitter in a forward direction should be 17.___
   A. high B. low
   C. very high D. none of the above

18. A resistor is color coded red-red-silver. Its value is ____ ohms. 18.___
   A. 22 B. .22 C. 220 D. 2.2

19. If we know the direction of electron flow through a resistor, we 19.____
   A. know the amount of voltage across the resistor
   B. are sure the resistor is in good condition
   C. know the value of resistance
   D. know the polarity of the voltage across the resistor

20. To see a sine wave on an oscilloscope screen, the signal is *generally* injected into the ____ circuit. 20.____
   A. horizontal B. high voltage
   C. brightness D. vertical

21. It is important, when coupling a transistor output to a speaker, to match the 21.____
   A. impedance B. voltage C. current D. size

22. A line isolation transformer is used to 22.____
   A. *increase* sensitivity
   B. protect a hot chassis from ground
   C. get more volume
   D. *decrease* hum

23. Positive feedback is used in 23.____
   A. audio amplifiers B. oscillators
   C. amplifiers D. mixers

24. Transistors are damaged by 24.____
   A. heat B. vibration
   C. high heater voltage D. too little negative bias

25. A heat sink 25.____
   A. heats up the cathodes
   B. removes heat from a power transistor
   C. shields the power supply
   D. insulates the chassis

KEY (CORRECT ANSWERS)

1. B
2. D
3. C
4. D
5. D

6. D
7. D
8. C
9. B
10. A

11. A
12. A
13. D
14. C
15. D

16. A
17. B
18. B
19. D
20. D

21. A
22. B
23. B
24. A
25. B

TEST 2

DIRECTIONS: Each question or incomplete statement is followed by several suggested answers or completions. Select the one that BEST answers the question or completes the statement. *PRINT THE LETTER OF THE CORRECT ANSWER IN THE SPACE AT THE RIGHT.*

1. The frequency range of the FM band is 1.___
   A. 60 to 75 Mhz
   B. 88 to 108 Mhz
   C. 550 to 1650 Khz
   D. 455 to 456 Khz

2. What type of solid state device is the BEST replacement for a voltage regulator tube? 2.___
   A. Silicon controlled rectifier
   B. Alloy junction diode
   C. Zener diode
   D. NPN transistor

3. The MAJORITY of current carriers in an N-channel FET are 3.___
   A. protons
   B. holes
   C. electrons
   D. ions

4. A grounded emitter transistor amplifier is *most nearly* like a 4.___
   A. grounded grid amplifier
   B. grounded cathode amplifier
   C. power amplifier
   D. triode tube amplifier

5. The base of a transistor is *similar* to a vacuum tube's 5.___
   A. plate
   B. screen
   C. cathode
   D. grid

6. The name of the stage that obtains audio from a modulated RF signal is called 6.___
   A. audio amplifier
   B. RF amplifier
   C. detector
   D. oscillator

7. The point on the tube characteristic curve at which current stops flowing is called 7.___
   A. saturation
   B. cut-off
   C. linear portion
   D. curved portion

8. The band width in the IF amplifier of an AM broadcast receiver is 8.___
   A. 15 Khz
   B. 5 Khz
   C. 3 Mhz
   D. 1 Mhz

9. A power transistor is ALWAYS mounted on a piece of metal so that the metal can act as the 9.___
   A. heat sink
   B. base bias
   C. emitter bias
   D. spark bias

10. The length of an FM antenna is determined by 10.____
   A. frequency received B. material used
   C. IF input D. the audio band

11. 1000 milliamperes is equivalent to ____ amperes. 11.____
   A. .001 B. .1 C. 1 D. 10

12. To widen the trace on an oscilloscope, adjust the ____ control. 12.____
   A. vertical amplifier B. horizontal amplifier
   C. vertical positioning D. horizontal positioning

13. A rectifier will 13.____
   A. pass current in both directions
   B. change DC to AC
   C. pass current in one direction only
   D. carry high current

14. The process by which a transistor or vacuum tube *increases* the strength of a signal is known as 14.____
   A. regeneration B. degeneration
   C. amplification D. oscillation

15. The phase inverter circuit supplies the push-pull circuit with two signals 15.____
   A. in phase B. 90° out of phase
   C. 270° out of phase D. 180° out of phase

16. 180 PF is equal to ____ MFD. 16.____
   A. .00000018 B. .00018 C. 1800 D. 180000

17. The element of a transistor is the 17.____
   A. base collector grid B. plate grid screen
   C. base plate emitter D. emitter base collector

18. A resistance check of a silicon diode in good condition should read 18.____
   A. *low* resistance in both directions
   B. *high* resistance in one direction, low resistance in the other
   C. *high* resistance in both directions
   D. zero in both directions

19. The property of a coil which opposes current changes is called 19.____
   A. inductance B. capacitance
   C. resistance D. conductance

20. A push-pull circuit is used to 20.____
   A. double the frequency range that can be handled
   B. increase the harmonic content
   C. cancel out even harmonics, reduce hum, and increase power output
   D. increase total harmonic content and increase power output

21. A limiter tube in a FM receiver is operated 21.___
   A. with the plate at zero B+
   B. without a screen by-pass capacitor
   C. with low plate and screen voltages
   D. with large cathode resistors

22. The current flowing in the emitter circuit of a transistor consists of ____ current. 22.___
   A. collector B. base
   C. base and collector D. collector minus the base

23. Two 100 ohm, 10 watt resistors are connected in parallel. The net resistance and power rating will be ____ ohms at ____ watts. 23.___
   A. 200; 5 B. 50; 10 C. 100; 20 D. 50; 20

24. If the emitter-base forward bias of a transistor *increases*, the 24.___
   A. emitter current remains constant
   B. collector current decreases
   C. base current approaches cutoff
   D. emitter current increases

25. The terminals of a junction FET are known as the 25.___
   A. emitter, base, and collector
   B. anode, cathode, and gate
   C. grid, cathode, and plate
   D. source, drain, and gate

KEY (CORRECT ANSWERS)

1. B
2. C
3. C
4. B
5. D

6. C
7. B
8. B
9. A
10. A

11. C
12. B
13. C
14. C
15. D

16. B
17. D
18. B
19. A
20. C

21. C
22. C
23. D
24. D
25. D

TEST 3

DIRECTIONS: Select the lettered word or phrase in Column II which matches the numbered item in Column I. Write this letter next to the corresponding number at the right.

COLUMN I

1. Inductance
2. Parallel
3. Zener diode
4. Reactance
5. Surge
6. Kilo
7. Farad
8. Source
9. Series
10. Attenuate
11. Cathode
12. Capacitor
13. Binary
14. Ampere
15. Current
16. Potentiometer
17. EMF
18. Anode
19. Amplification
20. Dielectric

COLUMN II

A. Unit of measure of current
B. Device which stores electrons
C. Electron flow
D. Insulator
E. Unit of measure of capacitors
F. Coil
G. Variable resistor with 3 contacts
H. Circuit with multiple paths
J. Regulated voltage supply
K. Device which supplies electrical power
L. Voltage
M. Metric prefix meaning one thousand
N. Increasing the strength of a signal
O. Positively charged electrode toward which electrons flow
P. To decrease in strength
R. Number of systems using zeros and ones
S. Opposition to AC by a coil or capacitor
T. Element which emits electrons
U. Circuit in which current has one path
V. Sudden increase in voltage or current
W. Unit of measure of inductance
X. In series
Y. Provides protection in case of arcing
Z. Oscillator

1.____
2.____
3.____
4.____
5.____
6.____
7.____
8.____
9.____
10.____
11.____
12.____
13.____
14.____
15.____
16.____
17.____
18.____
19.____
20.____

KEY (CORRECT ANSWERS)

1. F
2. H
3. J
4. S
5. V

6. M
7. E
8. K
9. U
10. P

11. T
12. B
13. R
14. A
15. C

16. G
17. L
18. O
19. N
20. D

EXAMINATION SECTION
TEST 1

DIRECTIONS: Each question or incomplete statement is followed by several suggested answers or completions. Select the one that BEST answers the question or completes the statement. *PRINT THE LETTER OF THE CORRECT ANSWER IN THE SPACE AT THE RIGHT.*

1. 1.___

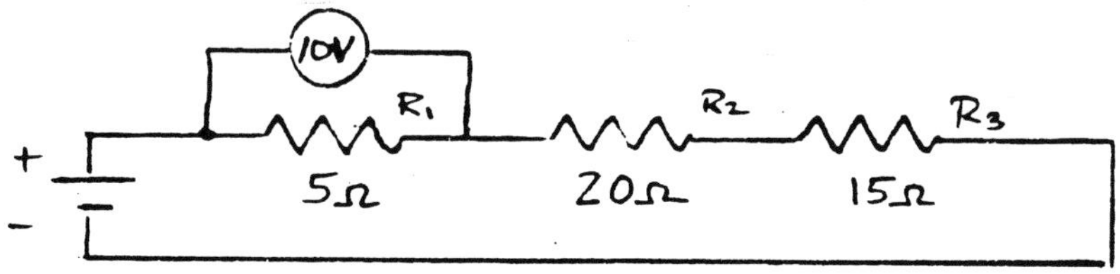

What is the voltage across R_2 in the circuit diagram shown above?
A. 2.5V B. 10V C. 40V D. 80V

2. What is the TOTAL resistance in the diagram shown at the right? 2.___
A. 3Ω
B. 4Ω
C. 6Ω
D. 18Ω

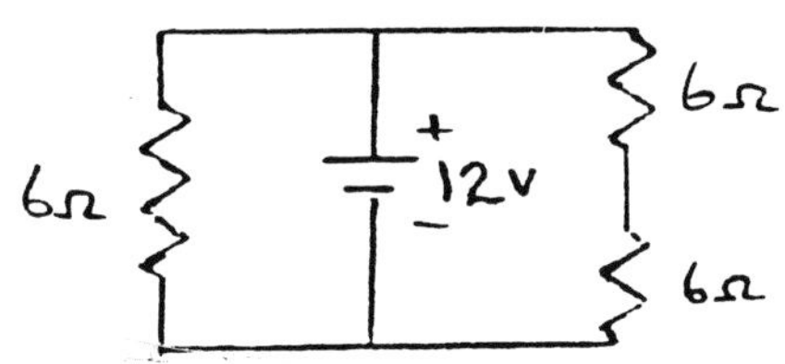

3. Any two-terminal network containing voltage and current sources and resistors, no matter how interconnected, may be represented as a single voltage source in series with a single resistance is a statement of 3.___
A. Thevenin's Theorem B. Norton's Theorem
C. Lenz's Law D. Faraday's Law

4. If the cross-sectional area of a conductor is tripled and its length is tripled, the resistance of the conductor will 4.___
A. be cut in half
B. be doubled
C. be four times as large
D. remain the same

5. 5.___

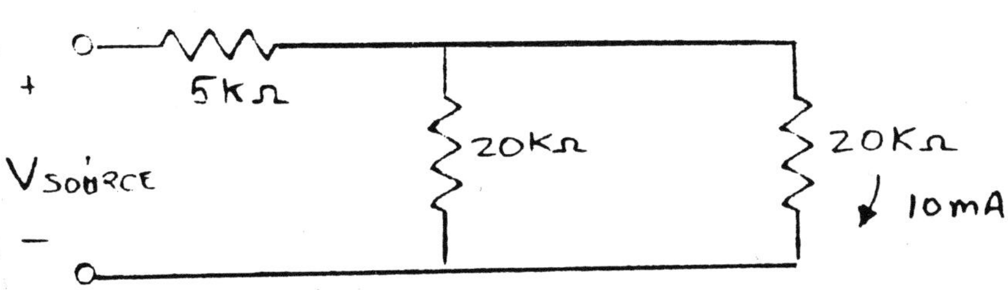

What is the source voltage in the circuit diagram shown above?
A. 50V B. 200V C. 250V D. 300V

6. Which electrical component stores energy in its electric field? 6.____
   A. Capacitor B. Inductor
   C. Transistor D. Resistor

7. The core of a transformer is laminated to reduce 7.____
   A. copper losses B. weight
   C. eddy currents D. cost

8. What is the voltage induced in a 5 henry coil which has a current change from 0 to 2 amperes in 0.2 seconds? 8.____
   A. 5V B. 50V C. 200V D. 500V

9. A glowing lamp and an air core coil are connected in series across a 120V, 60Hz supply. 9.____
   If an iron core is slowly slipped into the coil, the brightness of the lamp will
   A. become dimmer
   B. become brighter
   C. remain the same
   D. become brighter then extinguish

10. 10.____

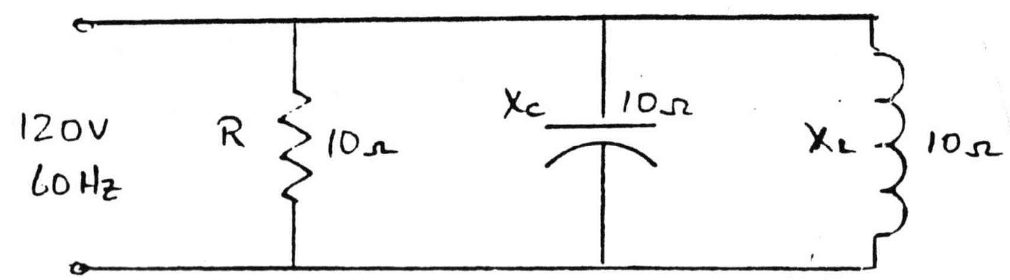

What is the impedance of the circuit diagram shown above?
   A. 10Ω B. 14.14Ω C. 30Ω D. Infinite

11. If two identical capacitors are placed in series, the TOTAL capacitive reactance will 11.____
   A. double B. triple
   C. remain the same D. be halved

12. A fuse is identified by 12.____
   A. current *only*
   B. voltage *only*
   C. current and voltage *only*
   D. current, voltage, and physical size

13. The ideal resistance of a voltmeter is 13.____
   A. zero B. 500Ω
   C. very large D. cannot be determined

14. The field intensity in ampere turns per meter produced by a coil of 1000 turns, length 0.04 meters, and a current of 800 milliamperes is 14.____
   A. 0.000008 B. 50 C. 20,000 D. 31,250

15. A sinusoidal wave for has a period of 2 milliseconds. Its frequency is ____ Hz. 15.____
A. 5 B. 60 C. 0.5k D. 500,000

16. A 12 pole, synchronous motor is operated from a 60 hertz supply. 16.____
The motor rpm is
A. 600 B. 1200 C. 1800 D. 2400

17. The fourth wire in a wye connected three-phase line is known as the 17.____
A. neutral
B. phase wire
C. line wire
D. fourth wire

18. In a balanced three-phase wye connection, the phase voltage is equal to 18.____
A. E_{phase}
B. E_{line}
C. $\sqrt{3}\ E_{line}$
D. $E_{line}/\sqrt{3}$

19. A balanced three-phase delta has an impedance of 200Ω per phase and a line voltage of 400V. 19.____
The APPROXIMATE line current is
A. 1.15A B. 2A C. 3.46A D. 6A

20. The purpose of the suppressor grid in a vacuum pentode is to 20.____
A. reduce grid-to-plate capacitance
B. reduce secondary emission
C. provide an AC by-pass
D. block DC

21. A zener diode is normally used 21.____
A. in forward bias
B. in reverse bias
C. to convert AC to DC
D. to amplify voltage

22. An amplifier which produces 80 watts at midband will produce at out-off _____ watts. 22.____
A. 20 B. 40 C. 56.56 D. 160

23. G_m is a transistor tube parameter known as the 23.____
A. transconductance
B. amplification factor
C. plate resistance
D. standoff ratio

24. A 120V RMS sinusoidal wave when displayed on an oscilloscope would have a peak-to-peak voltage of APPROXIMATELY 24.____
A. 120V B. 240V C. 170V D. 340V

25. The FET amplifier producing a 180 degree voltage phase shift between input and output is a 25.____
A. common source
B. common drain
C. common gate
D. source follower

KEY (CORRECT ANSWERS)

1. C
2. B
3. A
4. D
5. D

6. A
7. C
8. B
9. A
10. A

11. D
12. D
13. C
14. C
15. C

16. A
17. A
18. D
19. B
20. B

21. B
22. C
23. A
24. D
25. A

TEST 2

DIRECTIONS: Each question or incomplete statement is followed by several suggested answers or completions. Select the one that BEST answers the question or completes the statement. *PRINT THE LETTER OF THE CORRECT ANSWER IN THE SPACE AT THE RIGHT.*

1. The output frequency of full-wave bridge amplifier is 400 Hz. 1.____
The input frequency is _____ Hz.
A. 60 B. 200 C. 400 D. 800

2. The introduction of negative feedback in an amplifier causes 2.____
A. increased voltage gain B. decreased distortion
C. increased distortion D. oscillation

3. The average output voltage of an unfiltered half-wave rectifier which is connected to a 120V AC supply is APPROXIMATELY 3.____
A. 54V B. 76V C. 108V D. 170V

4. A full-wave rectifier delivers 2.0 amperes DC to a resistive load. 4.____
The AVERAGE load current is
A. 1.0A B. 1.4A C. 2A D. 2.8A

5. If two identical amplifiers each with a voltage gain of 50 are operated in cascade, when the input voltage is 4 millivolts, the output voltage will be 5.____
A. 200 mV B. 400 mV C. 5,000 mV D. 10V

6. The bandwidth of a circuit whose resonant frequency is 200 kHz and Q is 40 is _____ kHz. 6.____
A. 2 B. 5 C. 50 D. 8,000

7. A bipolar transistor with an alpha of 0.99 has a beta of 7.____
A. 0.2 B. 0.5 C. 49 D. 99

8. An amplifier whose output current flows 180° for each 360° input cycle is said to be operating in class 8.____
A. A B. B C. C D. D

9. What should be the turns ratio of a transformer if it is used to match a 2000Ω load to a 5Ω load? 9.____
_____:1.
A. 160,000 B. 400 C. 20 D. 10

10. A fundamental parameter of a unijunction transistor is its 10.____
A. eddy currents B. power gain
C. bandwidth D. standoff ratio

11. The type of computer memory which can be erased by ultra- violet light is known by 11.___
   A. ROM
   B. static RAM
   C. dynamic RAM
   D. EPROM

12. The logic equivalent to two inverters feeding into an OR gate is a(n) 12.___
   A. NOR gate
   B. exclusive OR gate
   C. NAND gate
   D. latch

13. *The complement of the sum is equal to the product of the complements* is part of 13.___
   A. Lenz's Law
   B. DeMorgan's Law
   C. Marshall's Theorem
   D. Kirchhoff's Law

Questions 14-19.

DIRECTIONS: Questions 14 through 19 are to be answered on the basis of the circuit shown below. 1.5A is flowing through the 60 ohm resistor.

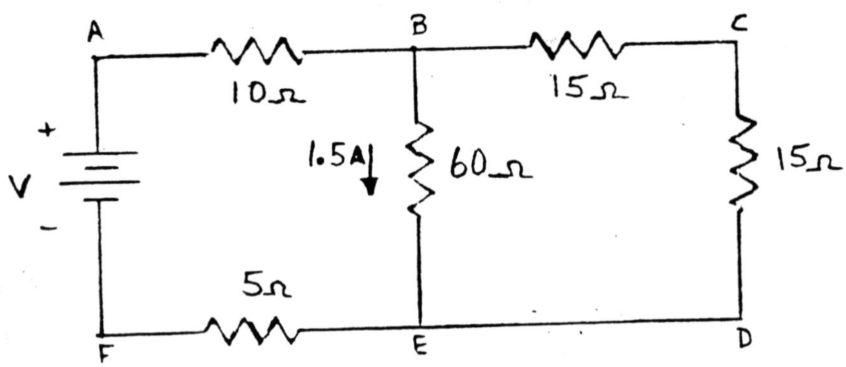

14. Determine R_T, the total resistance opposing battery, V 14.___

15. Determine V_{BE} 15.___

16. Determine I_{BC} 16.___

17. Determine I_{AB} 17.___

18. Determine V_{EF} 18.___

19. Determine V, the battery 19.___

Questions 20-25.

DIRECTIONS: Questions 20 through 25 are to be answered on the basis of the circuit shown below.

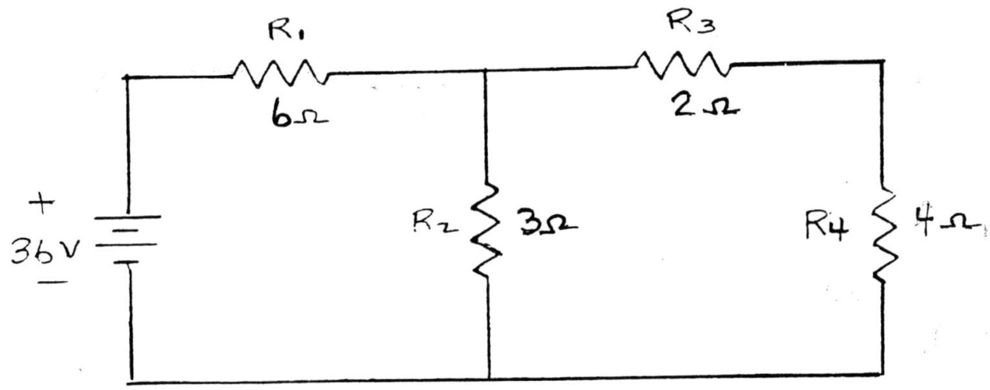

20. Determine the Thevenin voltage seen by R_4. 20.____

21. Determine the Thevenin resistance seen by R_4. 21.____

22. Draw the Thevenin equivalent circuit seen by R_4. 22.____

23. Use the results of the above question to find the current through R_4. 23.____

24. Find the Norton Current seen by R_4. 24.____

25. Find the Norton Resistance seen by R_4. 25.____

KEY (CORRECT ANSWERS)

1. B
2. B
3. A
4. C
5. D

6. D
7. D
8. B
9. C
10. D

11. D
12. A
13. B
14. 35Ω
15. 90V

16. 3A
17. 4.5A
18. 22.5V
19. 157.5V
20. 6V

21. 4Ω
22. 

23. -1.6A
24. 3A
25. 4Ω

TEST 3

DIRECTIONS: Each question or incomplete statement is followed by several suggested answers or completions. Select the one that BEST answers the question or completes the statement. *PRINT THE LETTER OF THE CORRECT ANSWER IN THE SPACE AT THE RIGHT.*

Questions 1-7.

DIRECTIONS: Questions 1 through 7 are to be answered on the basis of the circuit shown below.

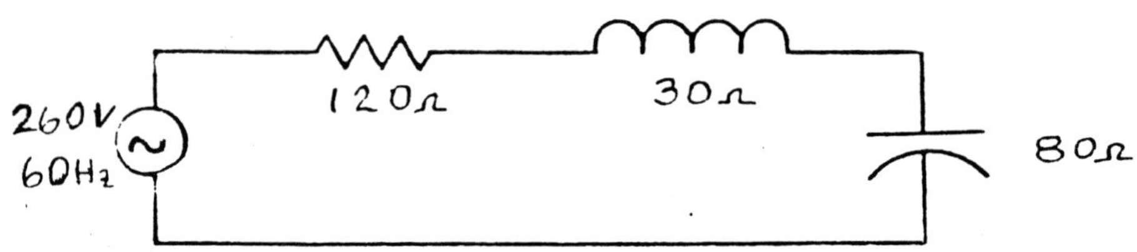

1. Determine the total impedance, z. 1.___
2. Find the current through the resistor. 2.___
3. Find the voltage across the inductor. 3.___
4. Determine the circuit power factor. 4.___
5. Determine the real power. 5.___
6. Find L, the inductance. 6.___
7. Determine how much inductive reactance must be added in series with the inductor to make the power factor unity. 7.___

Questions 8-14.

DIRECTIONS: Questions 8 through 14 are to be answered on the basis of the circuit shown below.

A 6:1 step-down transformer connects a 120V supply to a half-wave rectifier in series with a 300 ohm load. Assume the diode is ideal.

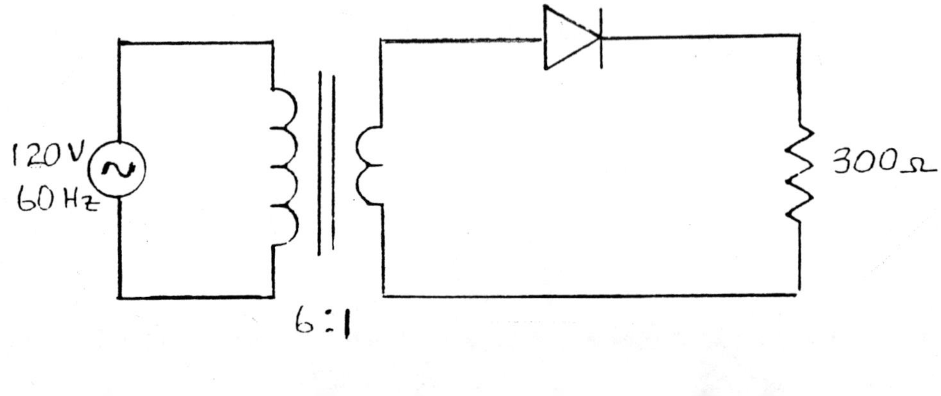

8. Determine the secondary RMS voltage. 8.____

9. Determine the peak secondary voltage. 9.____

10. Determine the DC output voltage. 10.____

11. Find the diode P.I.V. 11.____

Assume a very large capacitor is placed in parallel with the 300Ω; then answer Questions 12 through 14.

12. Determine the approximate DC output voltage. 12.____

13. Find the diode P.I.V. 13.____

14. Determine the ripple frequency. 14.____

Questions 15-19.

DIRECTIONS: Questions 15 through 19 are to be answered on the basis of the circuit shown below.

R_1 is 40k ohms, R_2 is 20k ohms, and C is 50 microfarads. Prior to the switch closing, the capacitor is uncharged.

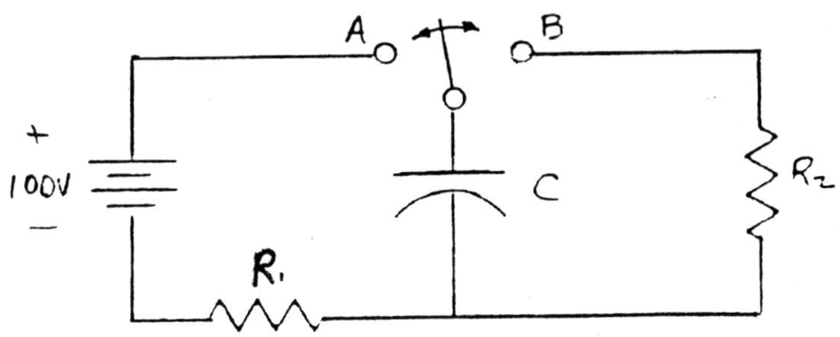

At t = 0, the switch goes to position A.
At t = 15 sec., the switch goes to position B and remains there.

15. In position A, what is the time constant? 15.____

16. After two time constants in position A, what is the capacitor voltage? 16.____

17. What is the current through R_1 after one time constant in position A? 17.____

18. What is the capacitor voltage at t = 14 sec.? 18.____

19. What is Vcap at t = 16 sec.? 19.____

Questions 20-25.

DIRECTIONS: Questions 20 through 25 are to be answered on the basis of the following information.

A DC shunt motor has a rated line voltage of 120V and a line current of 21 amperes. The armature resistance is 0.1 ohm, the field resistance is 120 ohms, and the rotational losses amount to 150 watts.

20. Determine the field current. 20.____

21. Determine the armature current. 21.____

22. Determine the total copper losses. 22.____

23. Determine the total losses. 23.____

24. Find the horsepower output. 24.____

25. Find the overall efficiency. 25.____

KEY (CORRECT ANSWERS)

1. 130Ω
2. 2A
3. 60V
4. .927
5. 478 watt

6. 79.6 MHY
7. X_L = 50Ω added
8. 20V RMS
9. 28.28 V_p
10. 12.74 V DC

11. 28.28 V
12. 28.28 V
13. 56.56 V
14. 60 Hz
15. 2 sec.

16. 86.5 V
17. 1.58A
18. 99.90 V
19. 99.94 V
20. 1A

21. 20A
22. 40W
23. 310W
24. 3 H.P.
25. 87.7%

TEST 4

DIRECTIONS: Each question or incomplete statement is followed by several suggested answers or completions. Select the one that BEST answers the question or completes the statement. *PRINT THE LETTER OF THE CORRECT ANSWER IN THE SPACE AT THE RIGHT.*

Questions 1-6.

DIRECTIONS: Questions 1 through 6 are to be answered on the basis of the diagram shown below.

A full wave bridge rectifier with a simple capacitor filter produces the wave form below when directly connected to a 120V, 60 Hertz supply. The load resistance is 3400 ohms.

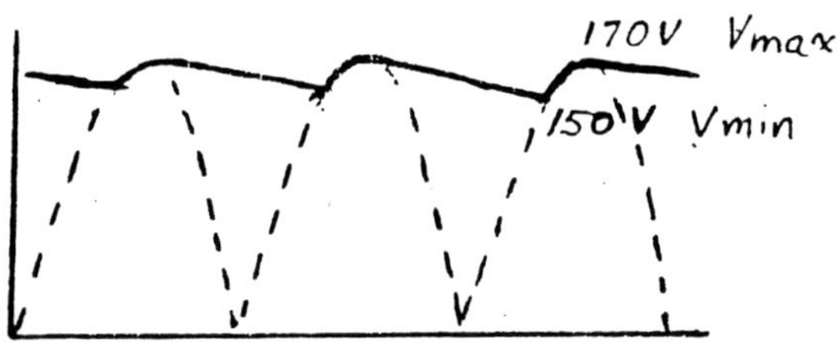

1. Determine the DC delivered to the load. 1.____
2. Determine the RMS ripple voltage. 2.____
3. Find the ripple factor. 3.____
4. Determine the ripple frequency. 4.____
5. Determine the minimum value for C. 5.____
6. Find the diode conduction angle. 6.____

Questions 7-13.

DIRECTIONS: Complete the following base conversions.

7. 27_{10} = ______$_2$ 7.____
8. 1101 0111_2 = ______$_{10}$ 8.____
9. 1011 1001_2 = ______$_{16}$ 9.____
10. $2A_{16}$ = ______$_{10}$ 10.____

11. 34_{10} = ______$_{16}$ 11.___

12. $B5_{16}$ = ______$_{2}$ 12.___

13. 42_{10} = ______$_{8}$ 13.___

Questions 14-18.

DIRECTIONS: Questions 14 through 18 are to be answered on the basis of the diagram shown below.

A wye connected 120V per phase alternator is supplying power to a balanced delta connected load of 104 ohms per arm at a 90% power factor.

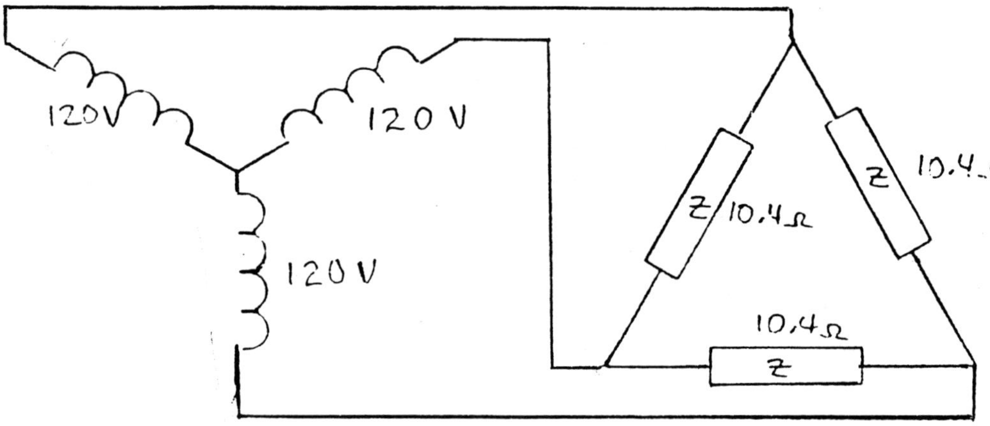

14. Determine the alternator line voltage, E_L. 14.___

15. Determine the load phase voltage, E_{pL}. 15.___

16. The load phase current, I_{pL}. 16.___

17. The alternator line current, I_L. 17.___

18. The alternator phase current, I_{pa}. 18.___

Questions 19-25.

DIRECTIONS: In the space at the right, print the name of the logic gate for the parts in Questions 19 through 25.

19. 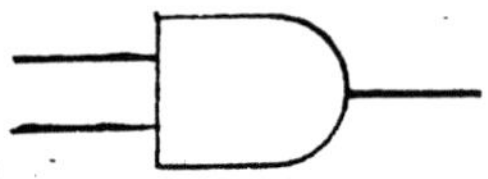 19.___

20. 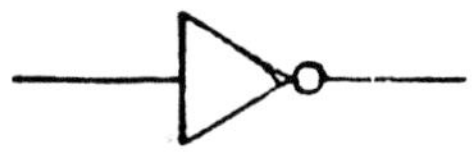 20.____

21. 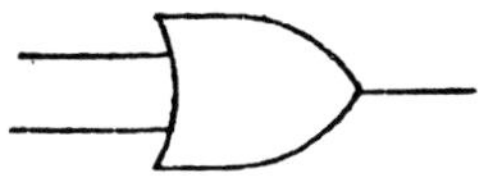 21.____

22.  22.____

23. 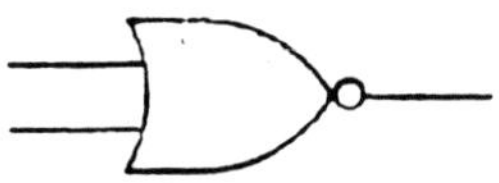 23.____

24. 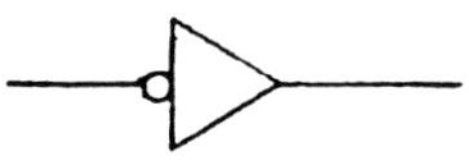 24.____

25. 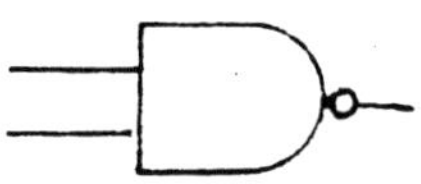 25.____

KEY (CORRECT ANSWERS)

1. 160V
2. 7.09V RMS
3. .125
4. 120 Hz
5. 14 Mfd

6. 43°
7. 1011
8. 215
9. 185
10. 42

11. 22
12. 10110101
13. 52
14. 208V AC
15. 120

16. 20A
17. 20A
18. 34A
19. AND gate
20. inverter

21. OR gate
22. exclusive OR
23. NOR gate
24. inverter
25. NAND gate

EXAMINATION SECTION

TEST 1

DIRECTIONS: Each question or incomplete statement is followed by several suggested answers or completions. Select the one that BEST answers the question or completes the statement. *PRINT THE LETTER OF THE CORRECT ANSWER IN THE SPACE AT THE RIGHT.*

1. The majority of simple electronics circuits are based upon a logic 1.___
   A. A and a logic B state
   B. *on* and a logic *off* state
   C. 0 and a logic 1 state
   D. *go* and *no-go* state

2. What voltage level exists in the 7400-series of IC chips? 2.___
   A. -15 volts, ground, and +5 volts
   B. Ground, +5 volts, and +12 volts
   C. -12 volts, ground, and +5 volts
   D. +5 volts and ground

3. A clock pulse is a transition 3.___
   A. from a logic 0 to a logic 1
   B. from a logic 1 to a logic 0
   C. from a logic 0 to a logic 1 and back to logic 0
   D. both A and B are correct

4. The symbols shown to the right are those for a 2-input ____ gate and 2-input ____ gate. 4.___

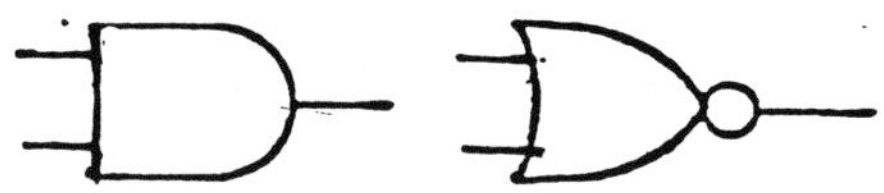

   A. AND; NOR
   B. EXCLUSIVE-OR; NAND
   C. OR; NOR
   D. AND; NAND

5. For which type of gate is the following definition CORRECT? 5.___
   A binary circuit with two or more inputs and single output, in which the output is a logic 1 only when all inputs are logic 1, and the output is logic 0 if any one of the inputs is logic 0.
   A. AND gate B. NAND gate C. OR gate D. NOR gate

6. If the two inputs to a 2-input AND, NAND, OR, and NOR gate are both at logic 1, the outputs from the four gates are 6.___
   A. 1,1,0 and 0 B. 1,0,0 and 1
   C. 0,1,1 and 0 D. 1,0,1 and 0

7. The term *to enable* a gate is the opposite of which of the following terms? 7.___
   A. To close a gate B. To disable a gate
   C. To block a gate D. All of the above

8. A gate and a switch differ in which of the following ways? 8.___
   A. A gate is essentially a one-way device, whereas a switch is a two-way device.
   B. When a gate is closed, a signal can pass; when a switch is closed, a signal cannot pass.
   C. When a gate is open, a signal can pass; when a switch is closed, a signal cannot pass.
   D. A gate and a switch are identical in function.

9. A gating input of logic 0 to a 2-input NOR gate will allow what type of digital information to appear at the output of the gate? 9.___
   A. Logic 1 state B. Logic 0 state
   C. Unchanged input data D. Inverted input data

10. A gating input of logic 0 to a 2-input NAND gate will allow what type of digital information to appear at the output of the gate? 10.___
   A. Logic 1 state B. Logic 0 state
   C. Unchanged input data D. Inverted input data

11. The 7400, 7402, 7408, and 7432 IC chips each contain 11.___
   A. *only* a single gate B. four gates
   C. three gates D. two gates

12. A common-anode 7 segment display has the following inputs to its segments: A,D,E=1 and B,C,F,G=0. 12.___
What does the display read out?
   A. Nine B. Four C. Six D. Three
   E. None of these

13. A common cathode display has the following inputs to its segments: C,F.=0 and A,B,D,E,G=1. 13.___
What does the display indicate?
   A. Nine B. Four C. Two D. Five

14. What is the difference between a 7490 and a 7493 counter? 14.___
   A. There is no difference.
   B. The 7490 is a decade counter, whereas the 7493 is a binary counter.
   C. The 7490 is a binary counter, whereas the 7493 is a decade counter.
   D. They have different pin configurations

15. The truth table to the right applies for the _____ gate. 15.___
   A. AND
   B. OR
   C. NAND
   D. NOR
   E. None of these

IN AB	OUT X
00	1
01	1
10	1
11	1

16. In the truth tables for a 2-input AND, NAND, OR, and NOR gate, the unique output state is, respectively, 16.___
   A. logic 1, logic 1, logic 0, logic 1
   B. logic 1, logic 0, logic 0, logic 1
   C. logic 0, logic 1, logic 1, logic 1
   D. logic 1, logic 1, logic 1, logic 1

17. When one inverts the output from an AND gate, he converts it into a(n) ____ gate. 17.____
    A. EXCLUSIVE-OR B. NAND
    C. NOR D. AND-OR-INVERT

18. If the two inputs to a 2-input AND gate and a 2-input NOR gate are left unconnected, the outputs from these two gates are, respectively, 18.____
    A. logic 1 and logic 0 B. logic 0 and logic 1
    C. logic 0 and logic 0 D. logic 1 and logic 1

19. The 7400, 7420, and 7430 IC chips have gates that have, respectively, the following numbers of inputs: 19.____
    A. two, three, five, and ten
    B. one, two, three, four
    C. two, three, four, eight
    D. two, three, four, six

20. The 7400, 7420, 7430 IC chips are all ____ gates. 20.____
    A. AND B. NAND
    C. both AND and NAND D. NOR

21. A four decade counter can count from 21.____
    A. 0001 to 10000 B. 0001 to 9999
    C. 0000 to 9999 D. 0000 to 100000

22. In a positive clock pulse, the transition from logic 1 to logic 0 occurs on the ____ edge. 22.____
    A. positive trailing B. positive leading
    B. negative trailing D. negative leading

23. The 74121 IC chip is a 23.____
    A. counter B. flip-flop
    C. monostable multivibrator D. programmable timer

24. When one debounces a SPDT switch, he 24.____
    A. makes sure that the switching action occurs quickly
    B. makes sure that the output from the switch can be controlled to produce a single clock pulse at a time
    C. turns the SPDT into, basically, a monostable multivibrator
    D. none of the above

25. When R=1 megohm and C=1 microfarad, the RC time constant has a value of 25.____
    A. 1 second B. .000001 seconds
    C. .001 seconds D. .000001 hz

KEY (CORRECT ANSWERS)

1. C
2. D
3. C
4. A
5. A

6. D
7. D
8. A
9. D
10. A

11. B
12. E
13. C
14. B
15. E

16. B
17. B
18. A
19. C
20. B

21. C
22. C
23. C
24. B
25. A

TEST 2

DIRECTIONS: Each question or incomplete statement is followed by several suggested answers or completions. Select the one that BEST answers the question or completes the statement. *PRINT THE LETTER OF THE CORRECT ANSWER IN THE SPACE AT THE RIGHT.*

1. In HEXADECIMAL notation, the binary number DCBA = 1100 represents either _____ or the letter _____. 1.____
   A. eleven; B B. eleven; A
   C. twelve; C D. thirteen; D

2. The binary number, 11111, represents which of the following decimal numbers? 2.____
   A. 32 B. 16 C. 15 D. 31

3. A twelve-bit binary number can encode ____ decimal numbers. 3.____
   A. two thousand and forty-eight
   B. four thousand and ninety-six
   C. five hundred and twelve
   D. two hundred and fifty-six

4. The binary number, 1111, appears on the seven-segment LED display as a 4.____
   A. blank display B. decimal 15
   C. the letter F D. 0

5. The LARGEST binary number that can exist in binary-coded decimal (BCD) is 5.____
   A. 1000 B. 1111 C. 1001 D. 1010

6. The quantity, 0111 = DCBA, is in binary-coded decimal equal to decimal 6.____
   A. 7 B. 5 C. 9 D. 6

7. In which of the following choices are all of the IC chips either decoders or decoder-drivers? 7.____
   A. 7447, 7451, and 74150 B. 7448, 7451, and 74160
   C. 7442, 7447, and 74154 D. 7442, 7448, and 74150

8. Which of the following IC chips is a 4-line-to-10-line decoder? 8.____
   A. 7447 B. 7451 C. 74150 D. 7442

9. In a decade sequencer, one requires a 9.____
   A. 74150 chip and a decade counter such as the 7490
   B. 74150 chip and a binary counter such as the 7493
   C. 7451 chip and a decoder such as the 7442
   D. 7442 chip and a 7490 chip

10. A demultiplexer is similar to a 10.__
   A. shift register
   B. decoder
   C. device that can select one of a number of inputs and pass the logic level on to the output
   D. data selector

11. When connected in the proper way, the 74150 and 74154 chips can serve as a 11.__
   A. sequencer
   B. simultaneous decoder/driver
   C. multiplexer/demultiplexer circuit
   D. programmable sequencer

12. The current that passes through a light-emitting diode (LED) should generally NOT exceed 12.__
   A. one ampere B. one milliampere
   C. 30 to 50 milliamperes D. 300 milliamperes

13. When the anode of a LED is connected to +5 volts and the cathode is connected to ground, the LED will 13.__
   A. remain unlit
   B. become lit, although only slightly
   C. light up immediately, but it may become unlit owing to the lack of a current-limiting resistor
   D. burn out

14. One can construct a simple logic probe from a 14.__
   A. light-emitting diode, capacitor, and battery
   B. LED, transistor, and battery
   C. LED and resistor
   D. LED and capacitor

15. When a logic probe is constructed from a LED and other components, a transistor is *usually* employed to 15.__
   A. make the LED light a bit more brighter
   B. conserve power
   C. *decrease* the current required to light the lamp monitor circuit
   D. *decrease* the voltage across the LED

16. A typical J-K flip-flop can have the following inputs: 16.__
   A. Strobe, enable, count, clear, and present
   B. Clock, preset, J, K, and clear
   C. Clock, count, J, K, and clear
   D. Clock, J, K, clear, preset, Q, and $\overline{Q}$

17. A flip-flop is a 17.__
   A. three state device
   B. two state device
   C. one state device
   D. either a one state or a two state depending upon the logic state appearing at the strobe input

18. With the aid of a single gate, a 74126 gate with three-state output can be converted into a 18.____
   A. monostable multivibrator
   B. bi-stable memory element
   C. tri-stable memory element
   D. none of the above

19. When J = 0 and K = 1 in a J-K flip-flop, 19.____
   A. Q can go to or stay at logic 1, but cannot go to logic 0
   B. Q can go to or stay at logic 0, but cannot go to logic 1
   C. the flip-flop toggles
   D. Q remains at its logic state; the clock has no effect

20. When present at the inputs or outputs of logic devices, the small circle o represents 20.____
   A. that a positive clock pulse may be required to enable the device
   B. that a positive leading edge may be required to enable the device
   C. inversion
   D. a shorthand form for an AND gate

21. The preset and clear inputs to a flip-flop 21.____
   A. do not take precedence to the J-K inputs, but do take precedence to the clock input
   B. do not take precedence to the clock input, but do take precedence to the J-K inputs
   C. take precedence over all other inputs
   D. none of the above

22. A typical read-only memory (ROM) IC chip has 22.____
   A. memory cell select inputs, memory enable input, read/write select input, data inputs, and data outputs
   B. data inputs, data outputs, memory enable input, and memory cell select inputs
   C. memory enable input, memory cell select inputs, and data outputs
   D. data inputs, data outputs, read/write select inputs, memory cell select input, and clock input

23. A typical random access memory (RAM) IC chip has 23.____
   A. memory cell select inputs, memory enable input, read/write select input, data inputs, and data outputs
   B. data inputs, data outputs, and memory cell select inputs
   C. data outputs, memory cell select inputs, and memory enable input
   D. data inputs, data outputs, read/write select input, memory cell select input, and clock input

24. Shift registers can 24.___
    A. convert serial data into parallel data
    B. convert parallel data into serial data
    C. store both serial and parallel data
    D. do all of the above, depending, of course, upon the nature of the specific shift register used

25. The 74194 shift register can 25.___
    A. parallel load
    B. shift right
    C. shift left
    D. all of the above

KEY (CORRECT ANSWERS)

1. C
2. D
3. B
4. C
5. B

6. A
7. C
8. D
9. D
10. C

11. C
12. C
13. C
14. C
15. C

16. B
17. B
18. D
19. B
20. C

21. C
22. C
23. D
24. D
25. D

EXAMINATION SECTION

TEST 1

DIRECTIONS: Each question or incomplete statement is followed by several suggested answers or completions. Select the one that BEST answers the question or completes the statement. *PRINT THE LETTER OF THE CORRECT ANSWER IN THE SPACE AT THE RIGHT.*

1. Which of the following is the equivalent NAND-NAND network for x = a+c? 1.___

A.

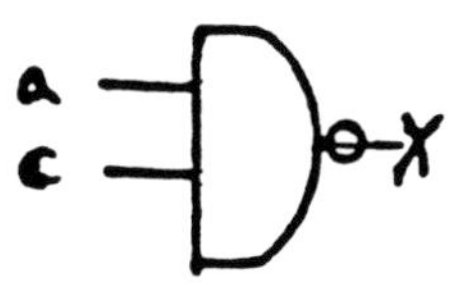

B.

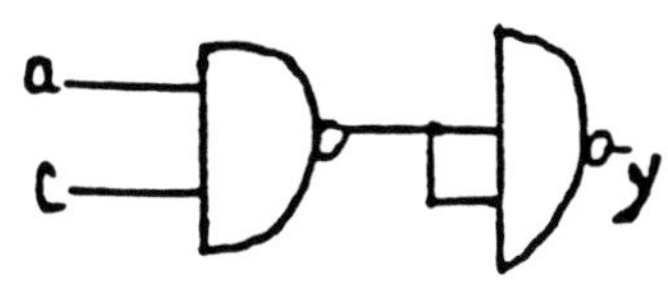

C.

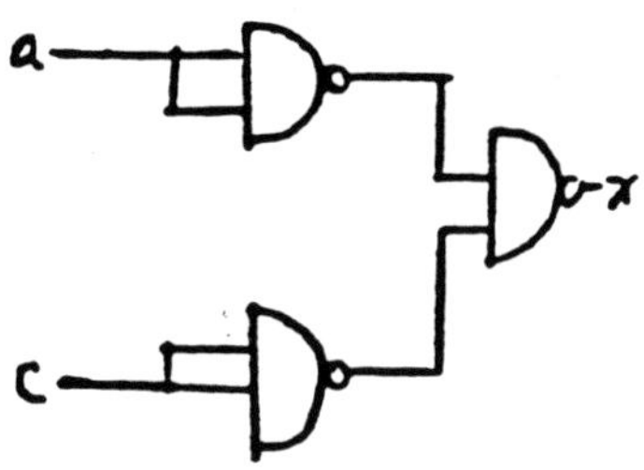

D.

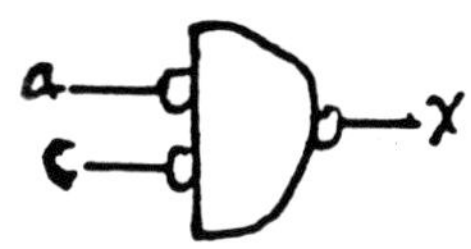

2. 1101_2 divided by 10_2 equals 2.___
   A. 11_2 B. 110.1_2 C. 101_2 D. 10_2

3. 1101_2 multiplied by 110_2 equals 3.___
   A. decimal 78
   B. decimal 64
   C. decimal 8
   D. octal 14

4. 1101_2 minus 110_2 equals 4.___
   A. 111_2
   B. 100_2
   C. decimal 8
   D. octal 14

5. The BCD equivalent of decimal 46 is 5.___
   A. 10101110 B. 111010 C. 101110 D. 1000110

6. In boolean algebra, a and b is equal to c is written 6.___
   A. a+b = c B. ab+c C. a=b=c D. a/b=c

7. Inspect the truth table shown at the right. From which logic function was it derived? 7.___
   A. and
   B. or
   C. nand
   D. exclusive or

a	b	x
0	0	0
0	1	0
1	0	0
1	1	1

8. Propagation delay in TTL devices refers to 8.___
   A. the noise immunity of the TTL device
   B. time between input change and output change
   C. how many TTL devices can be connected to output of the device
   D. how long it takes to connect these devices together

9. Choose the CORRECT output signal. 9.___

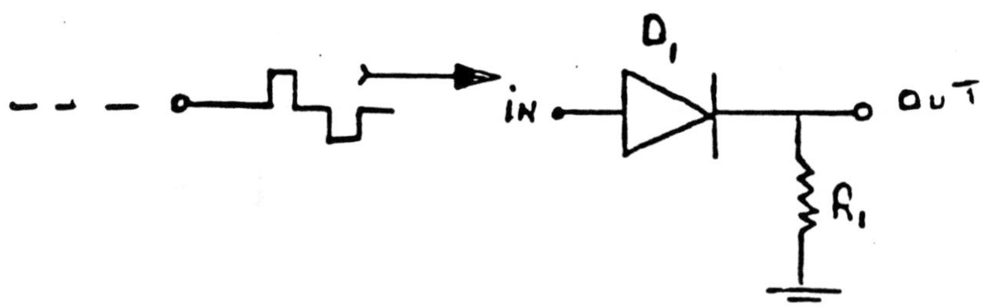

A.  B.  C.  D. 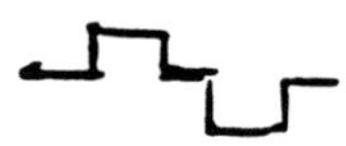

10. Analog electronic techniques vs. digital electronic techniques. 10.___
The digital techniques allow
   A. multiplexing many signals on a common carrier
   B. greater accuracy of signal measurement
   C. long distance signal transmission with less distortion and signal loss
   D. all of the above

11. The binary number 1010 is equal to the decimal 11.___
   A. 4   B. 5   C. 10   D. 17

12. The octal number 166 is equal to the decimal number 12.___
   A. 83   B. 120   C. 63   D. 118

13. There are _____ different digits in the octal number system. 13.___
   A. 2   B. 8   C. 10   D. 7

14. The advantage of using integrated circuits in logic design is 14.___
   A. small size   B. low price
   C. reliability   D. all of the above

15. If a logic zero is placed on the *J* input of J-K flip flop, 15.___
   A. it can never be set by the clock
   B. it can never be reset by the clock
   C. the set output of the f/f will become a one
   D. the f/f will topple on every clock pulse

16. The Nand gate is the backbone of the 7400 TTL Series because 16.___
   A. it is the least expensive gate in the line
   B. other logic elements can be constructed using only Nand gates
   C. it is readily available
   D. all of the above

17. Which type of multivibrator circuit can change a non-pulse type input into a perfect pulse output? 17.___
   A. Stable B. *One-shot*
   C. Schmitt trigger D. Bistable

18. Which type of multivibrator can be used to change the width of the input pulse? 18.___
   A. Stable B. *One-shot*
   C. Ujt oscillator D. Bistable

19. Which type of multivibrator is used to construct flip flops? 19.___
   A. Stable B. Mono-stable
   C. Schmitt trigger D. Bistable

20. A one khz clock pulse is applied to the *T* input of an RST flip flop. The flip flop is initially reset. What is the state of the flip flop after 5 clock pulses? 20.___
   A. No change (reset) B. Set output = 0
   C. Set output = 1 D. Indeterminate

KEY (CORRECT ANSWERS)

1. B	11. C
2. B	12. D
3. A	13. B
4. A	14. D
5. C	15. B
6. A	16. A
7. A	17. C
8. B	18. B
9. A	19. D
10. A	20. C

DIRECTIONS: Each question or incomplete statement is followed by several suggested answers or completions. Select the one that BEST answers the question or completes the statement. *PRINT THE LETTER OF THE CORRECT ANSWER IN THE SPACE AT THE RIGHT.*

1. The following counter is initially cleared (reset). What is the count after 5 clock pulses? 1.___
   A. a=0, b=0, c=1 B. a=1, b=1, c=1
   C. a=1, b=0, c=1 D. a=1, b=1, c=0

2. Which flip flop type has only one input and will toggle on every input pulse? 2.___
   A. D-type B. RS C. J-K D. T-type

3. Which flip flop type has two inputs, one of which is the clock? 3.___
   A. D-type B. RS C. J-K D. T-type

4. The output of a certain TTL device has very low impedance. This means that the device will have 4.___
   A. very fast switching speed
   B. very slow switching speed
   C. a reduced fan-out
   D. less output sensitivity

5. What voltage level exists in the 7400-series of IC chips? 5.___
   A. -15 volts, ground, and +5 volts
   B. Ground, +5 volts, and +12 volts
   C. -12 volts, ground, and +5 volts
   D. +5 volts and ground

6. For which type of gate is the following definition CORRECT? 6.___
   A binary circuit with two or more inputs and single ouput, in which the output is a logic 1 only when all inputs are logic 1, and the output is logic 0 if any one of the inputs is logic 0.
   ______ gate.
   A. AND B. NAND C. OR D. NOR

7. A gating input of logic 0 to a 2-input NOR gate will allow what type of digital information to appear at the output of the gate? 7.___
   A. A logic 1 state B. A logic 0 state
   C. Unchanged input data D. Inverted input data

8. A gating input of logic 0 to a 2-input NAND gate will allow what type of digital information to appear at the output of the gate? 8.___
   A. A logic 1 state B. A logic 0 state
   C. Unchanged input data D. Inverted input data

9. A common-anode 7 segment display has the following inputs to its segments: A, D, E=1 and B, C, F, G=0. What does the display read out? 9.____
   A. Nine B. Four C. Six D. Three

10. A common cathode display has the following inputs to its segments: C, F=0 and A, B, D, E, G=1. What does the display indicate? 10.____
   A. Nine B. Four C. Two D. Five

11. When one inverts the output from an AND gate, he converts it into a ______ gate. 11.____
   A. exclusive-OR
   B. NAND
   C. NOR
   D. AND-OR-invert

12. A four decade counter can count from ______ to ______. 12.____
   A. 0001; 10000
   B. 0001; 9999
   C. 0000; 9999
   D. 0000; 100000

13. When one debounces a SPDT switch, he 13.____
   A. makes sure that the switching action occurs quickly
   B. makes sure that the output from the switch can be controlled to produce a single clock pulse at a time
   C. turns the SPDT into, basically, a monostable multivibrator
   D. none of the above

14. The binary number, 11111, represents which of the following decimal numbers? 14.____
   A. 32 B. 16 C. 15 D. 31

15. The binary number, 1111, appears on the seven-segment LED display as a 15.____
   A. blank display
   B. decimal 15
   C. the letter F
   D. 0

16. The largest binary number that can exist in binary coded decimal (BCD) is 16.____
   A. 1000 B. 1111 C. 1001 D. 1010

17. A demultiplexer is similar to a 17.____
   A. shift register
   B. decoder
   C. a device that can select one of a number of inputs and pass the logic level on to the output
   D. data selector

18. When a logic probe is constructed from a LED and other components, a transistor is USUALLY employed to 18.____
   A. make the LED light a bit more brightly
   B. conserve power
   C. decrease the current required to light the lamp monitor circuit
   D. decrease the voltage across the LED

19. A typical J-K flip flop can have the following inputs: 19.___
   A. Strobe, enable, count, clear, and preset
   B. Clock, preset, J, K, and clear
   C. Clock, count, J, K, and clear
   D. Clock, J, K, clear, preset, Q and Q

20. A preset and clear inputs to a flip flop 20.___
   A. don't take precedence to the J-K inputs, but do take precedence to the clock input
   B. don't take precedence to the clock input, but do take precedence to the J-K inputs
   C. take precedence over all other inputs
   D. none of the above

KEY (CORRECT ANSWERS)

1. C
2. D
3. A
4. D
5. D

6. A
7. A
8. A
9. B
10. C

11. B
12. C
13. B
14. D
15. C

16. B
17. C
18. C
19. B
20. C

EXAMINATION SECTION
TEST 1

DIRECTIONS: Each question or incomplete statement is followed by several suggested answers or completions. Select the one that BEST answers the question or completes the statement. *PRINT THE LETTER OF THE CORRECT ANSWER IN THE SPACE AT THE RIGHT.*

1. A chip that is classified as *non-hermetic* will be encased in a package made from 1.___
   A. metal B. plastic C. glass D. ceramic

2. Of all the chips mounted on a computer's print board, the chip that typically breaks down LEAST often is the 2.___
   A. MPU B. RAM C. ROM D. ALU

3. The horizontal frequency range of a typical oscilloscope is 3.___
   A. 40 Hz B. 100 kHz C. 750 kHz D. 25 MHz

4. Once the computer's power supply has been identified as the source of a problem, a(n) _____ should be used to check it out. 4.___
   A. VOM B. current tracer
   C. capacitor D. frequency tracer

5. What computer component is represented by the schematic diagram symbol shown at the right? 5.___

   A. Variable resistor
   B. Transistor
   C. Diode
   D. Bridge rectifier

6. Video signals are typically traced with a(n) 6.___
   A. continuity tester B. VOM
   C. ordinary oscilloscope D. logic pulser

7. On a 6800 processor, the interrupt request line (IRQ) signal is located at pin number 7.___
   A. 4 B. D5 C. 40 D. A15

8. Which of the following devices should NEVER be attached to a computer? 8.___
   A. The ohmmeter of a VOM
   B. A multitrace oscilloscope
   C. A logic pulser
   D. An ordinary service oscilloscope

9. A soldering iron used for replacing chips should be NO higher than _____ watts. 9.___
   A. 15 B. 30 C. 60 D. 75

10. A technician uses a current tracer to test the output of a TTL chip.
If the chip has a normal high output, the current flow should measure about 10.___
A. .5 amps B. 40 microamps
C. 600 microamps D. 1.6 milliamps

11. The purpose of a march pattern is to diagnose the computer's 11.___
A. RAM B. MPU
C. ROM D. I/O circuits

12. Which of the following color codes, appearing on a diode, would indicate the HIGHEST resistance value (in ohms)? 12.___
A. Red-red-red B. Yellow-violet-orange
C. Orange-orange-green D. Red-violet-gold

13. On a Dual In-line Package (DIP) chip, the pin number should be counted 13.___
A. clockwise from the date code
B. clockwise from the keyway notch
C. counterclockwise from the date code
D. counterclockwise from the keyway notch

14. The purpose of a PIA chip is to assist in 14.___
A. storage
B. synchronizing clock pulses
C. digital/analog conversion
D. data addressing

15. What computer component is represented by the schematic diagram symbol shown at the right? 15.___

A. Toggle
B. Fixed resistor
C. Trimpot
D. Diode

16. If a technician has performed a logic probe of the supply voltage of a bus line, the NEXT place to test is the 16.___
A. VCC and ground pins
B. MPU chip
C. print board ground planes
D. address and data pins

17. If a monitor is dead, the technician's FIRST step should be to check the 17.___
A. green amps B. ac line fuse
C. sync processors D. inverters

18. Which of the following test instruments is used LEAST often by computer technicians? 18.___
A. Voltmeter B. Continuity tester
C. Current tracer D. Logic probe

19. The program counter of a 68000 16-bit processer is a(n) _____-bit register. 19.____
   A. 8   B. 16   C. 32   D. 64

20. If a video display appears rippled or wavy, the MOST likely cause is a 20.____
   A. faulty PIA
   B. defective monitor power supply
   C. blown ac line fuse
   D. faulty VDG

21. The term *fanout* is used to describe 21.____
   A. the MPU's pulse distribution pattern
   B. the architecture of units surrounding the arithmetic logic unit
   C. the display on a VOM voltmeter
   D. how well a chip is able to drive a number of parallel loads

22. A way to INCREASE the resolution of a video monitor display is to 22.____
   A. increase the vertical sweep frequency
   B. selecting a lower bandwidth
   C. increase the horizontal sync signal coming from the video generator
   D. decrease the horizontal sweep frequency

23. If a TTL chip is examined with a VOM, a high would be any voltage that reads from 23.____
   A. -2.5V to +2.5V   B. 0V to +2.5V
   C. +2.5V to +5V   D. +3V to +10V

24. When a monitor loses its horizontal sweep, the MOST likely symptom would be 24.____
   A. *ghosting* images
   B. a bright vertical line on the screen
   C. a loss in vertical hold
   D. a folded appearance at the bottom of the screen

25. What computer component is represented by the schematic diagram symbol shown at the right? 25.____

   A. Connection
   B. Capacitor
   C. Ground
   D. Voltage regulator

KEY (CORRECT ANSWERS)

1. B	6. C	11. A	16. D	21. D
2. A	7. A	12. C	17. B	22. C
3. B	8. A	13. D	18. C	23. C
4. A	9. B	14. C	19. C	24. B
5. A	10. B	15. D	20. B	25. C

TEST 2

DIRECTIONS: Each question or incomplete statement is followed by several suggested answers or completions. Select the one that BEST answers the question or completes the statement. *PRINT THE LETTER OF THE CORRECT ANSWER IN THE SPACE AT THE RIGHT.*

1. To show up shorts and opens in a circuit, a technician would MOST likely use a(n) 1.___
   A. multitrace oscilloscope B. voltmeter
   C. continuity tester D. logic probe

2. On a 6800 processor, the reset signal is located at pin number 2.___
   A. 6 B. A12 C. 40 D. D8

3. An instruction can throw one or more _____, which in turn cause changes in different registers. 3.___
   A. stacks B. flags C. nodes D. strings

4. After a logic probe is connected, its LED reads as follows 4.___

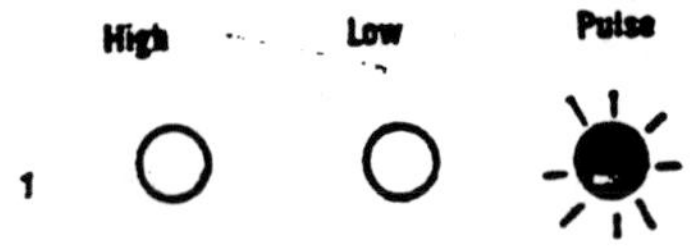

Which of the following graphic depictions of the test signal would be a match for this reading?

A. 

B. 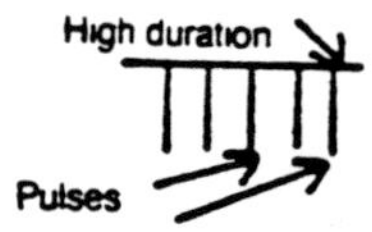

C. 

D. 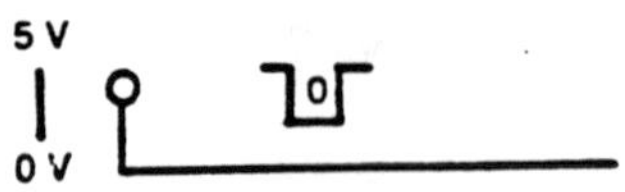

5. A CRC run through a computer's ROM indicates a defect, but the chip proves to be good. When the indicated chip is replaced, the identical trouble is experienced. The MOST likely problem is a(n) 5.___
   A. open input circuit
   B. defective MPU
   C. short-circuited data bus
   D. faulty CRC

6. Each of the following is a difference between DIP and SMD (Surface-Mounted Device) chips EXCEPT 6.____
   A. side or sides of print board used
   B. dc/ac capacity
   C. method of mounting on print board
   D. spacing of legs

7. If a computer starts up, but then displays nothing but junk, each of the following is a possible cause EXCEPT 7.____
   A. defective CPU
   B. improperly seated boards
   C. intermittent clock failure
   D. defective video memory

8. On an 8088 processor, the read output is located at pin number 8.____
   A. 29 B. D6 C. 32 D. 40

9. A video display appears folded over on the bottom, with a whitish haze appearing. 9.____
   The MOST likely cause for this is
   A. too much vertical sweep
   B. defective monitor power supply
   C. loss of horizontal sweep
   D. PIA chip defect

10. What computer component is represented by the schematic diagram symbol shown at the right? 10.____

   A. Crystal
   B. Inductor
   C. Fuse
   D. Integrated circuit

11. How many control lines are there from the PIA chip to the Video Display Generator? 11.____
   A. 2 B. 5 C. 10 D. 17

12. Which of the following instruments is used MOST often during the examination of digital circuits? 12.____
   A. Logic probe B. Voltmeter
   C. Current tracer D. Frequency counter

13. When a computer stops dead, the technician's FIRST step should be to check the 13.____
   A. power supply B. MPU
   C. RAM D. data bus lines

14. How many address-type registers are included in a 68000 16-bit processor? 14.____
   A. 2 B. 7 C. 18 D. 32

Questions 15-17.

DIRECTIONS: Questions 15 through 17 refer to the figure below, a cross-sectional diagram of a field-effective transistor (FET). Place the letter that corresponds to each diagrammed signal in the space at the right.

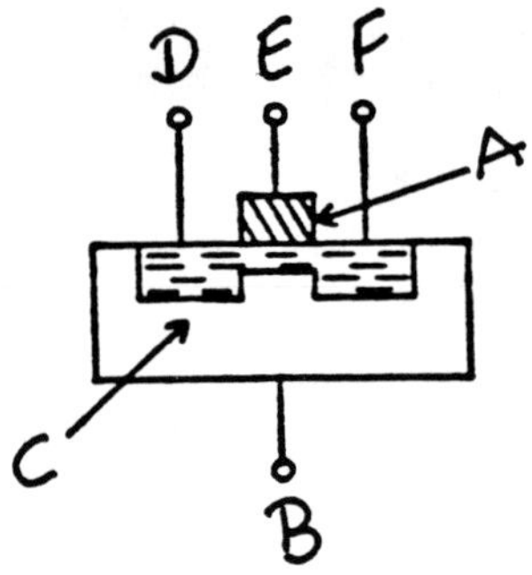

15. Gate 15.___

16. Drain 16.___

17. Source 17.___

18. To read the digital voltages of a circuit, a technician should use a(n) 18.___
   A. VOM
   B. logic probe
   C. ohmmeter
   D. regular duty oscilloscope

19. What computer component is represented by the schematic diagram symbol shown at the right? 19.___

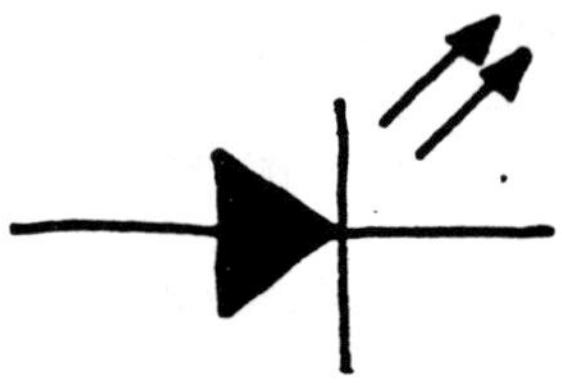

   A. Capacitor
   B. Bridge rectifier
   C. LED
   D. Transformer

20. The BEST way to examine a SAM chip that takes over the address bus is by 20.___
   A. a board check with a current tracer
   B. a pin-by-pin check with a continuity tester
   C. replacement of the chip
   D. a pin-by-pin check with a logic probe

21. The MAIN *disadvantage* associated with DIP chips is that they 21.___
   A. need both sides of the print board to be attached
   B. require four-sided leg mounting
   C. have pins that protrude out the bottom of the chip
   D. have legs spaced inconveniently close together

22. Which of the following would be LEAST likely to respond in static tests with a Substitution Machine? 22.___
A. Regular RAM
B. ROM
C. I/O chips
D. Dynamic RAM

23. If a CMOS chip is examined with a VOM, a *low* voltage reading would be any voltage that reads lower than 23.___
A. -2.5V
B. 0.8
C. 1.8V
D. 4.2V

24. Which of the following color codes, appearing on a diode, would indicate the LOWEST resistance value (ohms)? 24.___
A. Blue-gray-red
B. Red-violet-gold
C. Orange-white-yellow
D. Blue-blue-green

25. Each of the following is a likely cause for an open circuit EXCEPT 25.___
A. cold solder joint
B. splashed solder
C. faulty IC socket
D. broken connector wire

KEY (CORRECT ANSWERS)

1. C
2. C
3. B
4. C
5. C

6. B
7. C
8. C
9. A
10. A

11. B
12. A
13. A
14. B
15. E

16. F
17. D
18. A
19. C
20. D

21. A
22. D
23. C
24. B
25. B

EXAMINATION SECTION

TEST 1

DIRECTIONS: Each question or incomplete statement is followed by several suggested answers or completions. Select the one that BEST answers the question or completes the statement. *PRINT THE LETTER OF THE CORRECT ANSWER IN THE SPACE AT THE RIGHT.*

1. Files are used to store 1.____
   A. data
   B. programs
   C. operating systems
   D. source programs
   E. all of the above

2. MOST hard disks hold ____ bytes. 2.____
   A. 1-100 trillion
   B. 1-100 billion
   C. 1-100 million
   D. 1-1000
   E. less than 1000

3. MOST floppy disks can store ____ bytes. 3.____
   A. 1-100 trillion
   B. 1-100 billion
   C. 1-100 million
   D. .1-1 million
   E. less than 1000

4. A master file stores 4.____
   A. data about particular events
   B. relatively permanent data
   C. source copies of programs
   D. copies of other files
   E. data extracted from another file and held for a short term

5. A transaction file stores 5.____
   A. data about particular events
   B. relatively permanent data
   C. source copies of programs
   D. copies of other files
   E. data extracted from another file and held for a short term

6. A program file stores 6.____
   A. data about particular events
   B. relatively permanent data
   C. source copies of programs
   D. copies of other files
   E. data extracted from another file and held for a short term

7. A backup file stores 7.____
   A. data about particular events
   B. relatively permanent data
   C. source copies of programs
   D. copies of other files
   E. data extracted from another file and held for a short term

8. Which of the following is NOT a type of storage method? 8.___
   A. EBCDIC
   B. Packed hexadecimal
   C. Packed decimal
   D. True binary
   E. ASCII

9. Which of the following is the storage method COMMONLY used by IBM? 9.___
   A. EBCDIC
   B. Packed hexadecimal
   C. Packed decimal
   D. True binary
   E. ASCII

10. Which of the following is a very efficient numerical storage method? 10.___
   A. EBCDIC
   B. Packed hexadecimal
   C. Packed decimal
   D. True binary
   E. ASCII

11. Which of the following is the type of storage method COMMONLY found on home or personal computers, as well as many mini-computers? 11.___
   A. EBCDIC
   B. Packed hexadecimal
   C. Packed decimal
   D. True binary
   E. ASCII

12. In the EBCDIC storage method, 12.___
   A. numbers follow letters
   B. letters follow numbers
   C. numbers are intermixed with letters
   D. numbers cannot be stored
   E. letters cannot be stored

13. In the ASCII storage method, 13.___
   A. numbers follow letters
   B. letters follow numbers
   C. numbers are intermixed with letters
   D. numbers cannot be stored
   E. letters cannot be stored

14. The collating sequence refers to 14.___
   A. the order of the letters in the alphabet
   B. the order of the digits 0 through 9
   C. the order of manufacturers of computers
   D. the order of numbers and letters relative to each other
   E. none of the above

15. A two-letter state abbreviation takes how many bytes of computer memory when stored in ASCII? 15.___
   A. 0
   B. 1
   C. 2
   D. 5
   E. None of the above

16. A two-letter state abbreviation takes how many bytes of computer memory when stored in EBCDIC? 16.___
   A. 0
   B. 1
   C. 2
   D. 5
   E. None of the above

17. An alternative to storing numeric data in EBCDIC is to store it in 17.___
    A. ZIP format   B. true trinary   C. true hexabinary
    D. true binary   E. all of the above

18. Packed decimal means each decimal digit is stored in 18.___
    A. one nibble   B. one byte   C. ASCII format
    D. EBCDIC format   E. all of the above

19. Generally speaking, alphanumeric data should be stored in 19.___
    A. nibbles   B. straight binary   C. EBCDIC
    D. packed decimal   E. all of the above

20. Which statement below BEST describes a capability associated with virtual storage? 20.___
    A. It is possible to program as if more core is available than exists in the system
    B. All computers are now automatically compatible
    C. Only tapes and disks can be used for storage
    D. The programmer can write efficient programs while completely ignoring the nature of the computer system being used
    E. None of the above

KEY (CORRECT ANSWERS)

1.	A	11.	E
2.	C	12.	A
3.	D	13.	B
4.	B	14.	D
5.	A	15.	C
6.	C	16.	C
7.	D	17.	D
8.	B	18.	A
9.	A	19.	C
10.	D	20.	A

TEST 2

DIRECTIONS: Each question or incomplete statement is followed by several suggested answers or completions. Select the one that BEST answers the question or completes the statement. *PRINT THE LETTER OF THE CORRECT ANSWER IN THE SPACE AT THE RIGHT.*

1. Which of the following is NOT a common tape density? 1.___
   A. 800
   B. 1600
   C. 2400
   D. 6250
   E. All are common densities

2. Which of the following is a common tape length? 2.___
   A. 800
   B. 1600
   C. 2400
   D. 6250
   E. All are common densities

3. Blocking refers to the 3.___
   A. number of physical records in a logical record
   B. number of bytes in a record
   C. number of bytes per inch of tape
   D. number of logical records in a physical record
   E. the space between physical records

4. The inter block gap refers to the 4.___
   A. number of physical records in a logical record
   B. number of bytes in a record
   C. number of bytes per inch of tape
   D. number of logical records in a physical record
   E. the space between physical records

5. Density refers to the 5.___
   A. number of physical records in a logical record
   B. number of bytes in a record
   C. number of bytes per inch of tape
   D. number of logical records in a physical record
   E. the space between physical records

6. The record length refers to the 6.___
   A. number of physical records in a logical record
   B. number of bytes in a record
   C. number of bytes per inch of tape
   D. number of logical records in a physical record
   E. the space between physical records

7. Disks are BEST used in situations where 7.___
   A. we need to store data inexpensively
   B. we need to store historical data
   C. we want to process data sequentially
   D. we need to store data for on-line applications
   E. All of the above

8. Tape is BEST used in situations where 8.___
   A. we need to store data inexpensively
   B. we need to store historical data
   C. we want to process data sequentially
   D. we do not need to store data for on-line applications
   E. All of the above

9. Which of the following is NOT a direct access method? 9.___
   A. VSAM
   B. Sequential
   C. KSAM
   D. ISAM
   E. All of the above are direct access methods

10. A hasing algorithm calculates a records location in a file using a(n) 10.___
   A. record address
   B. social security number
   C. key field like a social security number
   D. using the binary search strategy
   E. all of the above

11. The soundex algorithm converts 11.___
   A. numeric keys to disk addresses
   B. disk addresses to numeric keys
   C. disk addresses to alphanumeric keys
   D. alphanumeric keys to disk addresses
   E. disk addresses to collisions

12. A collision occurs if two records have the 12.___
   A. same record key
   B. hash to the same disk address
   C. same length
   D. same blocking factor
   E. same density

13. In designing a tape file, an analyst should consider which of the following factors? 13.___
   A. Record fields
   B. Sequential order of records
   C. Estimate the number of records in the file
   D. Calculate the record length
   E. All of the above

14. In designing a tape file, an analyst should consider which of the following factors? 14.___
   A. Order of fields in a record
   B. Placement of fields in a record
   C. An expansion area for future use
   D. Data storage method, EBCDIC or packed decimal
   E. All of the above

15. A record count tallies the number of 15.___
   A. records in the file
   B. number of fields in each record
   C. number of bytes in a record
   D. files in the database
   E. All of the above

16. The term backup means 16.___
   A. copying each record to a new record
   B. copying a file to tape
   C. copying a disk to memory
   D. deleting a file from disk
   E. deleting a file from a tape

17. The interblock gap is typically ____ inch(es). 17.___
   A. .05 B. .005 C. .5 D. 5 E. 50

18. In writing the schema, the analyst defines 18.___
   A. data sets
   B. data elements
   C. data type, numeric or alphanumeric
   D. slave data sets
   E. All of the above

19. Which of the following is a typical data manipulation language command? 19.___
   A. QUERY B. LOCK C. DML
   D. DDL E. None of the above

20. Which of the following statements concerning index files and backup programs is TRUE? 20.___
   A. Index files may not be backed up.
   B. All appropriate index files are automatically backed up whenever a database file backup is created.
   C. Index files are often not backed up because they are so easily rebuilt.
   D. Index files must be backed up whenever their database files are backed up.
   E. None of the above

KEY (CORRECT ANSWERS)

1. B	11. D
2. C	12. B
3. D	13. E
4. E	14. E
5. C	15. A
6. B	16. B
7. D	17. C
8. E	18. E
9. B	19. B
10. C	20. C

EXAMINATION SECTION

TEST 1

DIRECTIONS: Each question or incomplete statement is followed by several suggested answers or completions. Select the one that BEST answers the question or completes the statement. *PRINT THE LETTER OF THE CORRECT ANSWER IN THE SPACE AT THE RIGHT.*

1. The maintenance and use of computer hardware is assigned to the ____ group in the data processing department. 1.____
   A. systems B. programming C. database
   D. analysis E. operations

2. The equipment which makes up a computer system is called 2.____
   A. hardcopy B. software
   C. CPU D. peripheral devices
   E. hardware

3. The entering of data into a computer system is the responsibility of the 3.____
   A. programmers B. analysts
   C. data control clerk D. data entry clerk
   E. computer operator

4. The MOST common input media is(are) the 4.____
   A. keypunch cards B. magnetic tape
   C. magnetic disk D. compact disk
   E. keyboard

5. The coding, testing, and debugging of computer software is the duty of the 5.____
   A. programmer B. analyst
   C. operator D. data control clerk
   E. data entry clerk

6. ____ is the term associated with the off-line preparation of data later submitted for processing with other data that has been prepared off-line. 6.____
   A. Timesharing B. Batch processing
   C. Interactive processing D. Real-time processing
   E. Aggregate processing

7. Multiple users share a single computer's resources in 7.____
   A. batch processing B. interactive processing
   C. microprocessing D. timesharing
   E. all of the above

8. The ____ field is the LARGEST consumer of computer products and services. 8.____
   A. science B. business C. health
   D. military E. education

9. The ____ is considered to be an end-user of computer services. 9.____
   A. accountant
   B. programmer
   C. operator
   D. analyst
   E. database administrator

10. In the organization of data, a group of related fields that pertains to a single data entity is called a 10.____
   A. character
   B. record
   C. file
   D. database
   E. key field

11. A ____ field uniquely identifies a record in a file. 11.____
   A. descriptive
   B. indicative
   C. normal key
   D. key
   E. none of the above

12. A collection of records within the SAME classification is called a 12.____
   A. character
   B. database
   C. field
   D. record
   E. none of the above

13. The CORRECT hierarchy of data is 13.____
   A. character, field, file, record
   B. character, file, record, field
   C. field, record, file, database
   D. character, field, record, file, database
   E. database, file, record, character, field

14. Which of the following data items located on a record would be BEST suited as a key field? 14.____
   A. Hours worked
   B. Rate of pay
   C. Social security number
   D. First name
   E. Last name

15. An example of a logic function is determining that 15.____
   A. 10 is 3 more than 7
   B. 12 plus 6 is 18
   C. 11 is an odd number
   D. 10 is not equal to 20
   E. all of the above

16. Which of the following is an example of a logic function performed by the computer? 16.____
   A. Determining that 60 is an even number
   B. Knowing that 120 is 12 dozen
   C. Arranging a group of numbers in ascending sequence
   D. Determining that 15 is 8 more than 7
   E. Determining that 12 minus 7 is 5

17. Which of the following is an arithmetic process performed by the computer? 17.____
   A. Arranging a group of numbers in descending order
   B. Arranging a group of numbers in ascending order
   C. Determining that 5 plus 4 is 9
   D. Knowing that 6 is an even number
   E. All of the above

18. An example of numeric data is 18.___
 A. 3
 B. THREE
 C. $300.45
 D. three hundred dollars forty-five cents
 E. none of the above

19. A coding technique used by banks to process checks is 19.___
 A. OCR B. UPC C. MICR
 D. Hollerith E. SQL

20. The duplicate copy of an existing file is a(an)___ file. 20.___
 A. duplicate B. grandfather C. backup
 D. archive E. save

21. ____ code is used for the internal storage of data in the computer. 21.___
 A. ASCII B. EBCDIC C. Binary
 D. Packed decimal E. All of the above

22. An example of a peripheral device is the 22.___
 A. CPU B. ALU C. control unit
 D. disk drive E. memory

23. The part of the computer that performs logical and arithmetic functions is the 23.___
 A. memory B. control unit C. disk
 D. ALU E. CPU

24. Of the following, the secondary storage media that offers the FASTEST data storage and retrieval is the 24.___
 A. computer memory B. magnetic tape
 C. magnetic disk D. keypunch cards
 E. paper tape

25. An example of an input/output device is the 25.___
 A. magnetic tape B. magnetic disk
 C. CPU D. disk drive
 E. all of the above

KEY (CORRECT ANSWERS)

1. E	11. D
2. E	12. E
3. D	13. D
4. C	14. C
5. A	15. D
6. B	16. C
7. D	17. C
8. B	18. A
9. A	19. C
10. B	20. C

21. E
22. D
23. D
24. C
25. D

TEST 2

DIRECTIONS: Each question or incomplete statement is followed by several suggested answers or completions. Select the one that BEST answers the question or completes the statement. *PRINT THE LETTER OF THE CORRECT ANSWER IN THE SPACE AT THE RIGHT.*

1. The computerized retrieval of microfilm is referred to as 1.____
   A. OCR B. COM C. microfax
   D. MICR E. CAD/CAM

2. The ____ is a very high speed non-impact printer. 2.____
   A. daisy wheel B. chain printer
   C. laser printer D. plotter
   E. dot-matrix printer

3. A computer which inputs data from physical measurements such as heat, motion, or touch is called a ____ computer. 3.____
   A. digital B. analog C. mainframe
   D. micro E. mini

4. A ____ computer performs general purpose, multiple concurrent operations with many users. 4.____
   A. micro B. mini C. mainframe
   D. timesharing E. All of the above

5. Some smaller computers are comparable to larger computers in performance, but are considered to be *task-oriented*. They may handle multiple users, but they primarily work on a single type of application such as accounting, billing, or inventory. 5.____
   This paragraph BEST describes ____ computers.
   A. mainframe B. micro C. mini
   D. super E. general purpose

6. The MOST widely used computer language is 6.____
   A. Cobol B. Pascal C. RPG
   D. Fortran E. Basic

7. The software which converts a high-level language like Fortran into the machine language which is directly understood by the computer is called a(n) 7.____
   A. preprocessor B. multiplexor
   C. source originator D. compiler
   E. all of the above

8. ____ is NOT a *high-level* computer language. 8.____
   A. Cobol B. Basic C. RPG
   D. PL/1 E. Assembler

9. ____ is used PRIMARILY in the scientific and engineering community. 9.___
   A. Cobol B. Basic C. Machine language
   D. PL/1 E. Fortran

10. An example of a computer language used effectively for business applications is 10.___
   A. Cobol B. RPG C. Basic
   D. PL/1 E. all of the above

11. Which of the following terms is CLOSELY associated with the term *structured programming*? 11.___
   A. HIPO B. Top-down design
   C. Hierarchy chart D. Modular programming
   E. All of the above

12. A storage capacity of 640K is APPROXIMATELY _____ bytes. 12.___
   A. 640,000,000 B. 64,000 C. 640,000
   D. 6,400,000 E. 640

13. A computer disk is divided into concentric circles called 13.___
   A. tracks B. cylinders C. rings
   D. sectors E. segments

14. In order to ensure the accuracy of data stored on tape or disk, the computer will append a ____ to each byte of data. 14.___
   A. check digit B. parity bit C. check bit
   D. modula 11 bit E. validation bit

15. Direct access is a feature of the 15.___
   A. magnetic tape B. keypunch cards
   C. magnetic disk D. compiler
   E. none of the above

16. Magnetic tape is BEST suited for situations where 16.___
   A. data is accessed sequentially
   B. VSAM is the data access method
   C. ISAM is the data access method
   D. data is accessed using direct access
   E. KSAM is the data access method

17. A term used to describe the algorithmic process of converting a key field into a storage location is 17.___
   A. hashing B. ISAM C. VSAM
   D. indexing E. dynamic storage

18. The software which controls the general operating procedures of the computer is the 18.___
   A. applications program B. source program
   C. object program D. operating system
   E. systems program

19. The computer language used to DIRECTLY communicate with the computer's operating software is 19.___
   A. RPG B. DOS C. JCL
   D. VSAM E. none of the above

20. A common pathway on which all data travels to and from the CPU to peripheral devices is the 20.____
  A. channel B. line C. bus
  D. modem E. multiplexor

21. The ____ converts computer signals to be transferred over telephone lines and vice versa. 21.____
  A. coaxial cable B. multiplexor
  C. modem D. controller
  E. digital teleprocessor

22. *Baud* is a measurement of 22.____
  A. storage capacity B. CPU performance
  C. disk speed D. printer speed
  E. transmission speed

23. ______ is a type of network COMMONLY used on micro-computers. 23.____
  A. Narrowband B. LAN C. Star
  D. Ring D. Hub

24. A type of computer that will handle the preliminary processing of data BEFORE it is sent to the mainframe computer is called a 24.____
  A. pre-processor B. front-end processor
  C. back-end processor D. minicomputer
  E. slave computer

25. The _____ would NOT be a member of the data processing staff in an organization. 25.____
  A. analyst B. programmer C. auditor
  D. operator E. data entry clerk

KEY (CORRECT ANSWERS)

1. B
2. C
3. B
4. C
5. C

6. A
7. D
8. E
9. E
10. E

11. E
12. C
13. B
14. B
15. C

16. A
17. A
18. D
19. C
20. C

21. C
22. E
23. B
24. B
25. C

Basic Mathematics
EXAMINATION SECTION

DIRECTIONS: Each question or incomplete statement is followed by several suggested answers or completions. Select the one that BEST answers the question or completes the statement. *PRINT THE LETTER OF THE CORRECT ANSWER IN THE SPACE AT THE RIGHT.*

1. $\begin{array}{r} 534 \\ 18 \\ +1291 \\ \hline \end{array}$ 1.____

   A. 1733 B. 1743 C. 1833 D. 1843 E. 1853

2. (17×23) - 16 + 20 = 2.____
   A. 459 B. 427 C. 411 D. 395 E. 355

3. 3/7 + 5/11 = 3.____
   A. 33/35 B. 4/9 C. 8/18 D. 68/77 E. 15/77

4. 4832 ÷ 6 = 4.____
   A. 905 1/3 B. 805 1/3 C. 95 1/3 D. 95 E. 85 1/3

5. 62.3 - 4.9 = 5.____
   A. 5.74 B. 7.4 C. 57.4 D. 58.4 E. 67.4

6. 3/5 × 4/9 = 6.____
   A. 4/15 B. 7/45 C. 27/20 D. 12/14 E. 15/4

7. 14/16 - 5/16 = 7.____
   A. 8/16 B. 9/16 C. 11/16 D. 8 E. 9

8. 5.03 + 2.7 + 40 = 8.____
   A. .570 B. 4.773 C. 5.70 D. 11.73 E. 47.73

9. 5.37 × 21.4 = 9.____
   A. 11491.8 B. 1149.18 C. 114.918 D. 11.4918 E. 1.14918

10. 5 1/4 + 2 7/8 = 10.____
    A. $8\frac{1}{4}$ B. 8 1/8 C. 7 2/3 D. $7\frac{1}{4}$ E. 7 1/8

11. -14 + 5 = 11.____
    A. -19 B. -9 C. 9 D. 19 E. 70

12. 2/7 of 28 = 12.____
    A. 98 B. 16 C. 14 D. 8 E. 4

13. 2/5 = 13.____
    A. .10 B. .20 C. .25 D. .40 E. .52

14. 20% of ______ is 38. 14.____
    A. 7.6 B. 19 C. 76 D. 190 E. 760

15. $\frac{8.4}{400} =$ 15. ____

A. .0021 B. .021 C. .21 D. 2.1 E. 21

16. $\frac{4}{5} = \frac{?}{60}$ 16. ____

A. 240 B. 48 C. 20 D. 15 E. 12

17. What is the area of the rectangle shown at the right? 17. ____

A. 47 mm^2
B. 94 mm^2
C. 240 mm^2
D. 480 mm^2
E. 960 mm^2

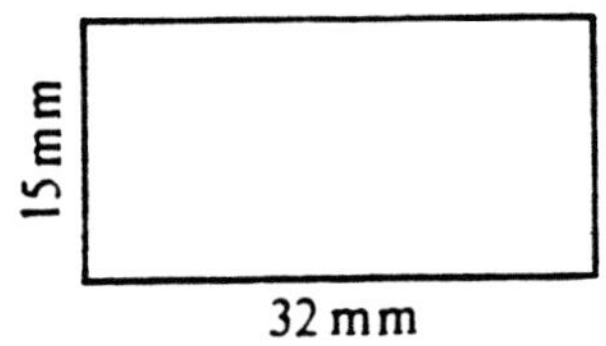

18. What number does □ represent in this equation? 18. ____

25 - □ - □ - □ - □ = 13

A. 13 B. 12 C. 7 D. 4 E. 3

19. Approximate lengths are given in the right triangles shown at the right. 19. ____
What does length x equal?

A. 48
B. 39
C. 37
D. 35
E. 32

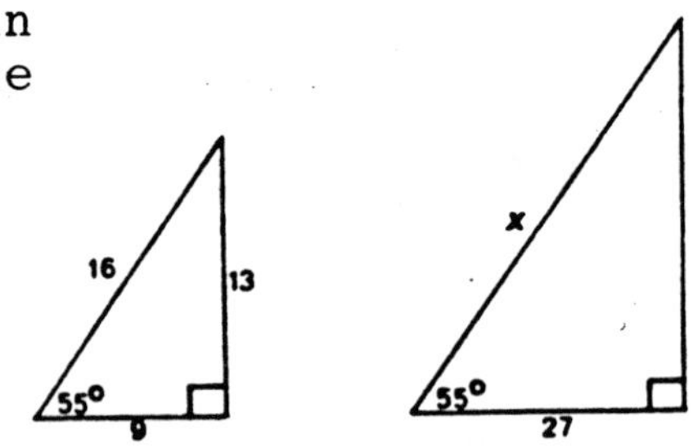

20. What is the perimeter of the triangle shown at the right? 20. ____

A. 10 × 15 × 17
B. 10 + 15 + 17
C. $\frac{1}{2}$ × 10 × 15
D. $\frac{1}{2}$ × 10 × 17
E. $\frac{1}{2}$(10 + 15 + 17)

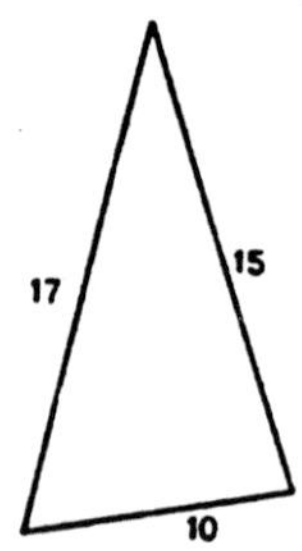

21. Which of the following expressions will give the same answer as 45 × 9? 21. ____

A. 5×3^3
B. (4×9) + (5×9)
C. (40+9) × 5
D. (45×3) + (45×3)
E. (45×10) - (45×1)

22. Find the average of 19, 21, 21, 22, and 27. 22.____
A. 23 B. 22 C. 21 D. 20 E. 19

23. In the triangle at the right, how many degrees is ∠T? 23.____
A. 75°
B. 85°
C. 95°
D. 115°
E. 180°

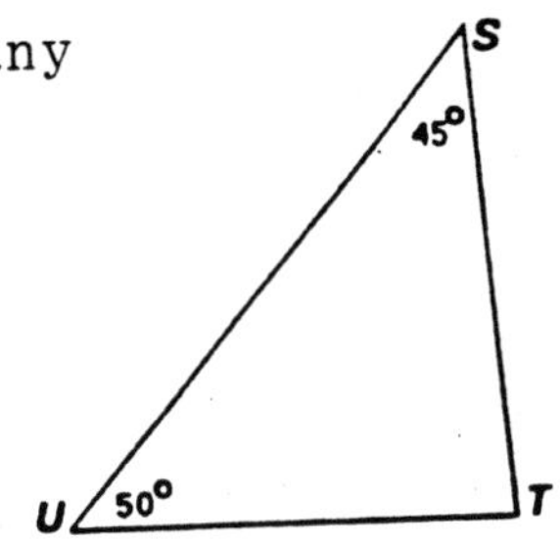

24. 24.____

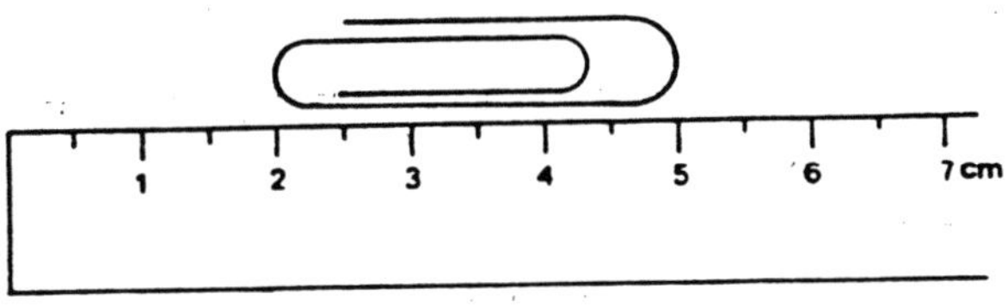

About how long is the paper clip?
A. 5 cm B. 4 cm C. 3 cm D. 2 cm E. 1 cm

25. Five stores sell the same size cans of tomato soup. Their prices are listed below. 25.____
Which sells the soup for the LOWEST price per can?
______ cans for ______.
A. 6; 99¢ B. 6; 90¢ C. 5; 93¢ D. 3; 56¢ E. 3; 50¢

26. Rock star Peter Giles receives $1.97 royalty on each of his albums that is sold. 14,127 albums are sold. 26.____
Estimate how much Peter Giles will receive.
A. $7,000 B. $14,000 C. $20,000 D. $26,000 E. $28,000

27. An amplifier is advertised for 20% off the list price of $430. 27.____
What is the sale price?
A. $516 B. $454 C. $354 D. $344 E. $215

28. If 9 dozen eggs cost $3.60, what do 25 dozen eggs cost? 28.____
A. $90.00 B. $10.00 C. $9.00 D. $2.54 E. $.40

29. The distance between New York and San Antonio is 1860 miles. 29.____
If a jet averages 465 miles per hour, how many hours will it take to travel the distance?
A. 9 B. 5 C. 4 D. 3 E. 2

30. In a high school homeroom of 32 students, 24 are girls. 30.____
What percent are girls?
A. 3/4% B. 24% C. 25% D. 75% E. 80%

31. Which problem could give the answer shown on the calculator? 31. ____

A. 2 + .3
B. 2 × 3/10
C. 2 × 1/3
D. 33333 + .2
E. 7 ÷ 3

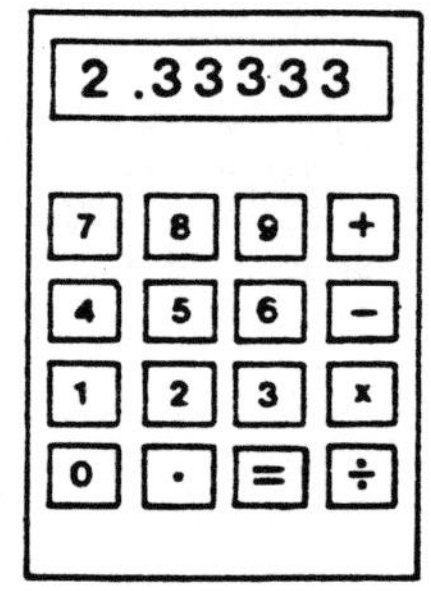

32. According to the table at the right, how much will it cost in a typical week for the 3 members of the Wright family to eat at home? Mr. Wright is 56 years old; Mrs. Wright, 52; and their son Harry, 17. 32. ____

A. $125
B. $52
C. $49
D. $42
E. $40

Cost of Eating at Home
(one week)

Age	Male	Female
6-11 yrs.	$14	$14
12-19 yrs.	$19	$16
20-54 yrs.	$20	$16
55 and up	$14	$14

33. According to the table shown in the previous question, how much does it cost in a typical four-week month to feed a 12-year-old girl? 33. ____

A. $4 B. $16 C. $48 D. $64 E. $78

34. Reverend Wilhite jogs for $1\frac{1}{2}$ hours each day, 6 days a week. If he burns 800 calories per hour of jogging, how many calories does he burn in a week? 34. ____

A. 4800 B. 5600 C. 7200 D. 8400 E. 9000

35. Ground meat costs 90¢ per pound. How much does the meat on the scale cost? 35. ____

A. $1.80
B. $1.60
C. $1.54
D. $1.44
E. $.90

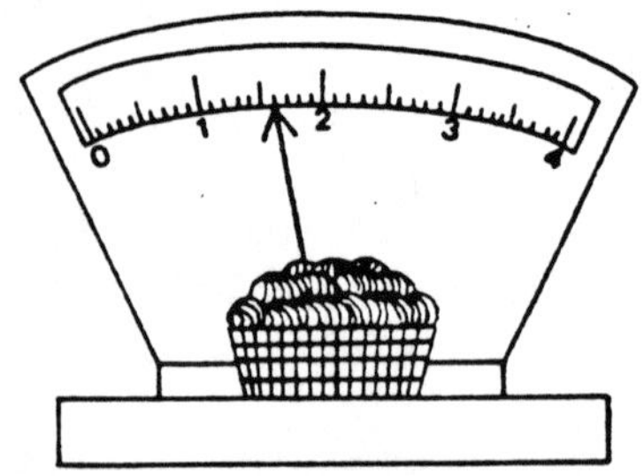

36. According to the graph at the right, about when did the weekly wages for a minimum wage worker go over $100? 36. ____

A. 1965
B. 1970
C. 1975
D. 1979
E. 1980

37. According to the bar graph at the right, what is the approximate height of the Crystal Beach Comet? 37.____

A. 40 ft.
B. 90 ft.
C. 92 ft.
D. 94 ft.
E. 98 ft.

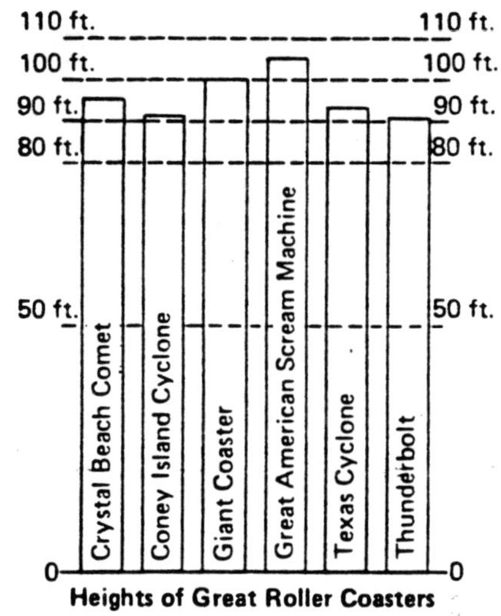

38. According to the bar graph shown in the previous question, what is the difference in height between the tallest and shortest roller coasters? 38.____

_____ feet.

A. 5 B. 10 C. 15 D. 20 E. 50

39. How much change will you receive from a $10 bill when you buy 4 grapefruits at 90¢ each and 3 apples at 40¢ each? 39.____

A. $6.20 B. $5.20 C. $4.80 D. $4.20 E. $4.00

40. A medical supplier packages medicine in boxes. The cost of packaging is computed with the flow chart at the right. 40.____

What is the cost of packaging medicine in a box that is 30 cm long, 20 cm wide, and 20 cm high?

A. $.20
B. $.24
C. $2.00
D. $2.40
E. $3.00

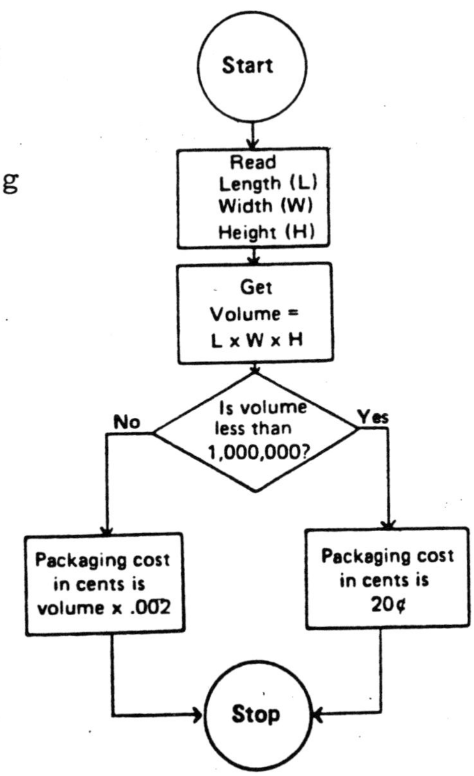

KEY (CORRECT ANSWERS)

1. D	11. B	21. E	31. E
2. D	12. D	22. B	32. C
3. D	13. D	23. B	33. D
4. B	14. D	24. C	34. C
5. C	15. B	25. B	35. D
6. A	16. B	26. E	36. C
7. B	17. D	27. D	37. D
8. E	18. E	28. B	38. C
9. C	19. A	29. C	39. B
10. B	20. B	30. D	40. A

SOLUTIONS TO PROBLEMS

1. $534 + 18 + 1291 = 1843$

2. $(17 \times 23) - 16 + 20 = 391 - 16 + 20 = 395$

3. $\frac{3}{7} + \frac{5}{11} = \frac{33}{77} + \frac{35}{77} = \frac{68}{77}$

4. $4832 \div 6 = 805\ \frac{1}{3}$

5. $62.3 - 4.9 = 57.4$

6. $\frac{3}{5} \times \frac{4}{9} = \frac{12}{45} = \frac{4}{15}$

7. $\frac{14}{16} - \frac{5}{16} = \frac{9}{16}$

8. $5.03 + 2.7 + 40 = 47.73$

9. $5.37 \times 21.4 = 114.918$

10. $5\frac{1}{4} + 2\frac{7}{8} = 7\ \frac{9}{8} = 8\ \frac{1}{8}$

11. $-14 + 5 = -9$

12. $\frac{2}{7}$ of $28 = (\frac{2}{7})(\frac{28}{1}) = 8$

13. $\frac{2}{5} = .40$ as a decimal

14. Let x = missing number. Then, $.20x = 38$. Solving, $x = 190$

15. $\frac{8.4}{400} = .021$

16. Let x = missing number. Then, $\frac{4}{5} = \frac{x}{60}$. $5x = 240$, so $x = 48$

17. Area $= (15)(32) = 480\text{mm}^2$

18. Let $x = \square$. Then, $25 - 4x = 13$. So, $-4x = -12$. Solving, $x = 3$

19. $\frac{9}{27} = \frac{16}{x}$. Then, 9x = 432. Solving, x = 48.

20. Perimeter = 17 + 10 + 15 = 42

21. 45 × 9 = 405 = (45×10) - (45×1)

22. 19 + 21 + 21 + 22 + 27 = 110. Then, 110 ÷ 5 = 22

23. ∠T = 180° - 50° - 45° = 85°

24. The paper clip's length is about 5 - 2 = 3 cm.

25. For A: price per can = $\frac{.99}{6}$ = .165;

For B: price per can = $\frac{.90}{6}$ = .15

For C: price per can = $\frac{.93}{5}$ = .186

For D: price per can = $\frac{.56}{3}$ = $.18\overline{6}$

For E: price per can = $\frac{.50}{3}$ = $.1\overline{6}$

Lowest price is for B.

26. \$1.97 ≈ \$2.00. Then, (\$2.00)(14,127) = \$28,254 ≈ \$28,000.

27. Sale price = (\$430)(.80) = \$344

28. Let x = cost. Then, $\frac{9}{360} = \frac{25}{x}$. 9x = \$90, so x = \$10.00

29. $\frac{1860}{465}$ = 4 hours

30. $\frac{24}{32}$ = 75%

31. $\frac{7}{3} = 2.\overline{3}$ = 2.33333 on this calculator shown.

32. Total cost = \$14 + \$16 + \$19 = \$49

33. Cost = (\$16)(4) = \$64

34. (800)($1\frac{1}{2}$)(6) = 7200 calories

35. (.90)(1.6) = \$1.44

36. Around 1975, the minimum weekly wages exceeded \$100.

37. The Crystal Beach Comet's height is about 94 ft.

38. Tallest $\approx$ 105 ft. and the shortest $\approx$ 90 ft.
Difference = 15 ft.

39. \$10 - (4)(.90) - (3)(.40) = \$5.20 change

40. (30)(20)(20) = 12,000 cm^3. Since 12,000 < 1,000,000, the price is 20 cents.

EXAMINATION SECTION

DIRECTIONS: Each question or incomplete statement is followed by several suggested answers or completions. Select the one that BEST answers the question or completes the statement. *PRINT THE LETTER OF THE CORRECT ANSWER IN THE SPACE AT THE RIGHT.*

1. $2/3 \times 12$ equals 1.___
   A. 4 B. 6 C. 8
   D. 18 E. None of the above

2. 2.___
   83.97
   1.78
   14.36
   9.03

   The sum of the above column is
   A. 99.13 B. 99.24 C. 109.14
   D. 109.23 E. 109.24

3. The value of x in the equation $5x = 75$ is 3.___
   A. 13 B. 15 C. 70
   D. 80 E. none of the above

4. $65.13 \div .13$ equals 4.___
   A. .501 B. 5.01 C. 50.1
   D. 501 E. none of the above

5. The sum of 6 feet 8 inches and 3 feet 4 inches is 5.___
   A. 2 ft. 2 in. B. 9 ft. C. 10 ft.
   D. 10 ft. 12 in. E. none of the above

6. $3/4 - 1/2 + 1/8$ equals 6.___
   A. 3/10 B. 3/8 C. 5/8
   D. 1 3/8 E. none of the above

7. 4 5/16 - 2 3/8 equals 7.___
   A. 1 15/16 B. 2 1/16 C. 2 1/4
   D. 2 15/16 E. none of the above

8. $(-12) + (-3)$ equals 8.___
   A. -9 B. +15 C. +9
   D. -15 E. none of the above

9. The ratio of the lengths of two lines is 5 to 3. The length of the shorter line is 30 inches. The length of the longer line is _____ inches. 9.___
   A. 18 B. 48 C. 50
   D. 150 E. none of the above

10. .025 written as a common fraction is 10.___
A. 25/10 B. 25/100 C. 25/1000
D. 25/10,000 E. none of the above

11. In the proportion $\frac{5}{2} = \frac{9}{x}$, the value of x is 11.___
A. 1.8 B. 3.6 C. 22.5
D. 36 E. none of the above

12. 33 1/3 percent of 3 equals 12.___
A. 1 B. 10 C. 100/3
D. 100 E. none of the above

13. $\sqrt{225}$ equals 13.___
A. 15 B. 20.5 C. 25
D. 112.5 E. none of the above

14. On the portion of the scale shown at the right, the reading to which the arrow points is _____ units. 14.___

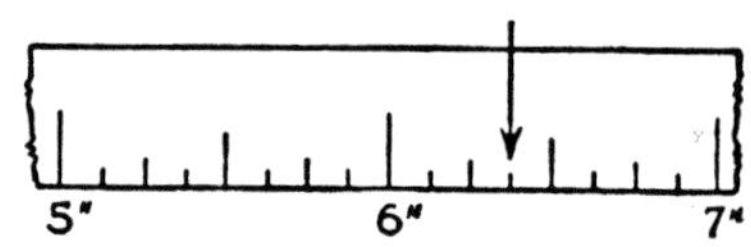

A. 6 3/16
B. 6 3/5
C. 6 3/4
D. 7 5/8
E. none of the above

15. If $\frac{4x}{5} - 6 = 10$, then x equals 15.___
A. 15 1/5 B. 5 C. 4
D. 3 1/5 E. none of the above

16. The difference between 8 hours 0 minutes 6 seconds and 6 hours 4 minutes 15 seconds is _____ hr. _____ min. _____ seconds. 16.___
A. 0; 54; 51 B. 1; 54; 51 C. 2; 4; 9
D. 2; 54; 45 E. none of the above

17. The scores made by nine pupils on a science test are: 2, 4, 6, 6, 8, 10, 12, 14, 19. 17.___
The MEAN score is
A. 6 B. 8 C. 9
D. 81 E. none of the above

18. A certain cost formula is represented graphically in the figure at the right. 18.___
From the graph, when n = 7, the value of C is about

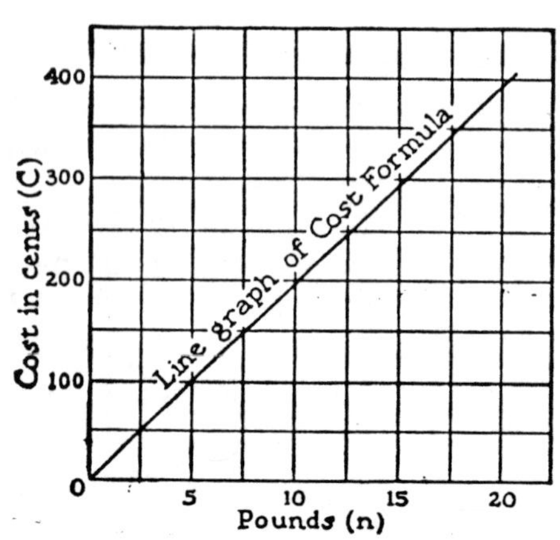

A. 140
B. 120
C. 110
D. 102
E. none of the above

19. A simplified form of the expression $A = \frac{1}{2}bh + \frac{1}{2}ah$ is 19.____

A. $A = \frac{1}{2}h(b+a)$ B. $A = bh + ah$ C. $A = abh$

D. $\frac{A}{\frac{1}{2}bh} = \frac{1}{2}ah$ E. none of the above

20. The ratio of 6 inches to 3 feet is 20.____

A. 6/1 B. 2/1 C. 1/2

D. 1/18 E. none of the above

21. The value of s in the equation $3s = 12 - s$ is 21.____

A. 6 B. 4 C. 3 2/3

D. 3 E. none of the above

22. 16 2/3 percent of what number is 30? 22.____

A. 5 B. 18 C. 160

D. 180 E. none of the above

23. The line graph shown at the right represents the temperature readings in Albany, New York, at two-hour intervals from 4 A.M. to 10 P.M. on a certain day in February. The APPROXIMATE change in temperature between 7 A.M. and 9 A.M. is _____ degrees. 23.____

A. 3.5
B. 3.0
C. 2.5
D. 2.0
E. none of the above

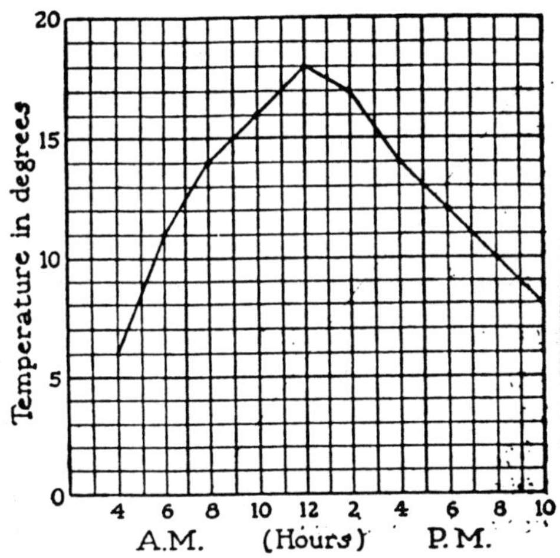

Questions 24-25.

DIRECTIONS: Questions 24 and 25 are to be answered on the basis of the following figure and information.

In the figure below, a square whose side is b is cut from a square whose side is a.

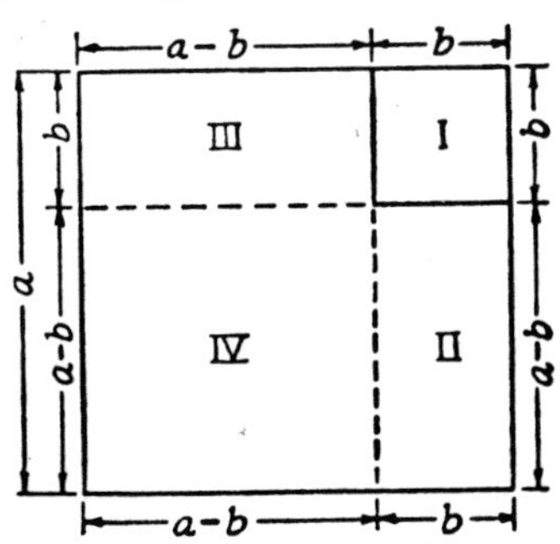

24. The sum of the perimeters of Section I and Section III can be represented by 24.___
A. b^2 B. 4a - 2b C. 2a + 3b
D. a(a-b) E. none of the above

25. The sum of the areas of Section II and Section IV can be represented by 25.___
A. b^2 B. 4a - 2b C. 2a + 3b
D. a(a-b) E. none of the above

26. The temperature reading (F) on the Fahrenheit scale equals 32 more than 9/5 of the Centigrade reading (C). This rule when translated into symbols is expressed by 26.___
A. F = 9/5C + 32 B. F = 9/5(C + 32) C. F = 9/5 + 32C
D. F + 32 = 9/5C E. none of the above

27. In the equation 6x - 114 = .3x, the value of x is 27.___
A. 38 B. 20 C. 12 2/3
D. 2 E. none of the above

28. What percent of 42 is 84? 28.___
A. ½% B. 2% C. 50%
D. 200% E. none of the above

29. The CORRECT name of the solid figure at the right is 29.___
A. semicircle
B. circle
C. sphere
D. cone
E. cylinder

30. Which of these fractions has the LARGEST value? 30.___
A. 1/2 B. 5/9 C. 7/12
D. 2/3 E. 3/4

31. The formula for the area of a circle is A = 31.___
A. πr^2 B. $2/3\pi r^2$ C. $2\pi r$
D. bh E. none of the above

32. The CORRECT name of the figure at the right is 32.___
A. pentagon
B. hexagon
C. rectangle
D. trapezoid
E. square

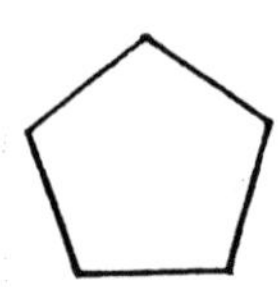

33. The figure at the right is a 33.___
A. rectangle
B. square
C. pentagon
D. trapezoid
E. parallelogram

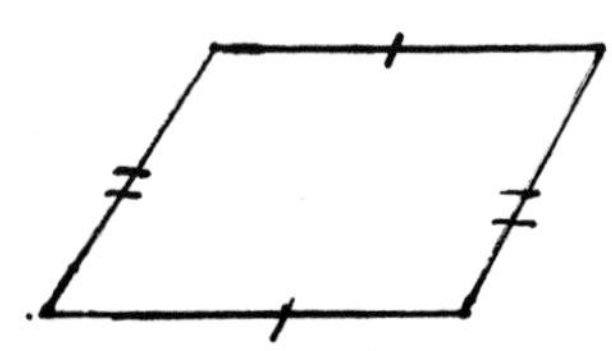

34. If x = -18, y = 3, and z = -2, then x - y + z equals 34.____
A. 3 B. -3 C. -23 D. -52 E. -56

35. The number 335,560 rounded off to the nearest thousand is 35.____
A. 335,000 B. 335,500 C. 336,000
D. 340,000 E. none of the above

36. In the triangle ABC at the right, the sum of the angles is _____ degrees. 36.____
A. 360
B. 180
C. 90
D. 45
E. none of the above

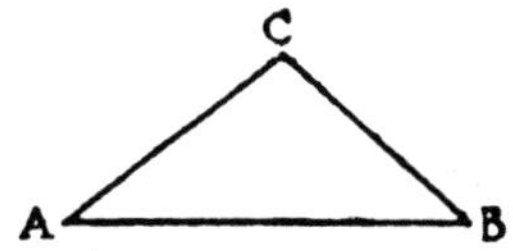

37. According to the map shown at the right, the APPROXIMATE distance between the southern point of New York City and Albany is _____ miles. 37.____
A. 50
B. 75
C. 130
D. 180
E. 200

38. If 6 is added to a certain number n, the result is 1. An equation which expresses this relationship is 38.____
A. n + 6 = 1 B. n - 1 = 6 C. 6 - n = 1
D. n + 1 = 6 E. none of the above

39. In the expression $2n^3$, the 3 is called a(n) 39.____
A. coefficient B. factor C. exponent
D. multiplicand E. none of the above

40. The number of inches in n feet is represented by 40.____
A. 12n B. 3n C. n/3
D. n/12 E. none of the above

41. The simple interest on $600 for 3 months at 4 percent per year is represented by 600 × .04 × 41.____
A. 1/4 B. 1/3 C. 3
D. 4 E. none of the above

42. The circle graph at the right indicates how a family's annual budget of $3000 was planned. 42._

Food	40 percent
Shelter	25 percent
Clothes	15 percent
Operating Expenses	10 percent
Insurance & Savings	10 percent

The part of the circle representing Shelter is _____ degrees.

A. 25
B. 45
C. 90
D. 250
E. none of the above

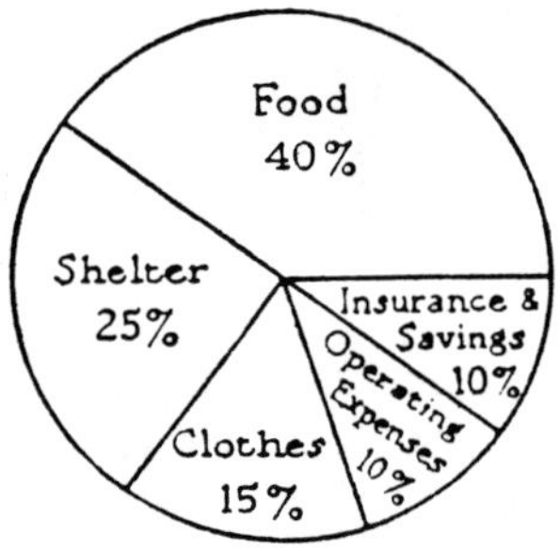

43. In the parallelogram ABCD at the right, each small square represents 4 square inches. 43._
The area of the right triangle AED represents _____ square inches.

A. 3
B. 12
C. 24
D. 48
E. none of the above

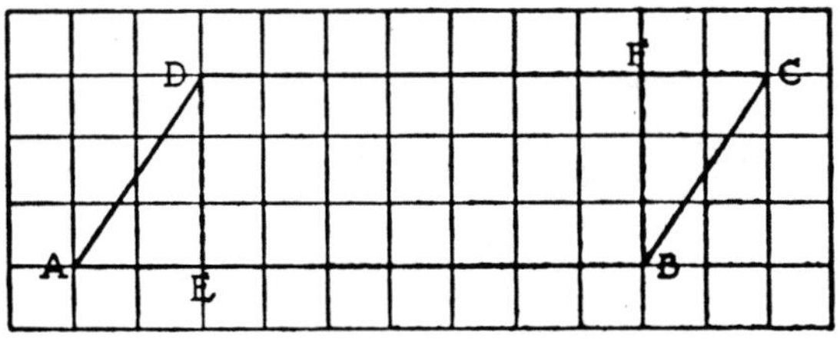

44. A surveyor measured angle x with a transit. (See figure at the right.) 44._
Angle x is called

A. the angle of depression of B from A
B. an obtuse angle
C. the supplement of angle y
D. the angle of elevation of B from A
E. none of the above

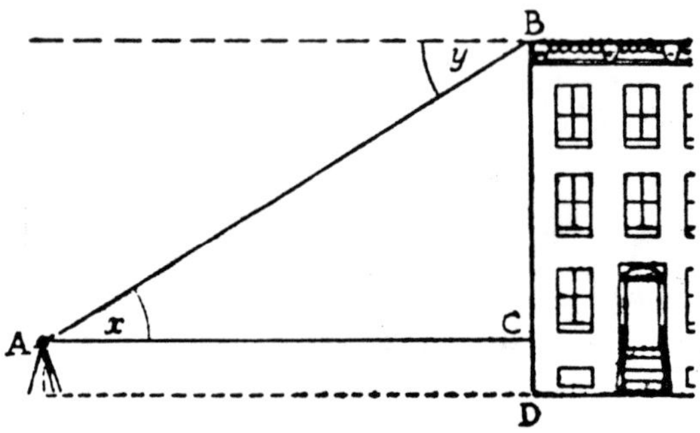

45. In the figure at the right, AOB is a straight line. 45._
An equation showing the relationship between u and v is

A. $u = \frac{1}{2}v$
B. $u = 180 - v$
C. $u + v = 90$
D. $v = 3u$
E. none of the above

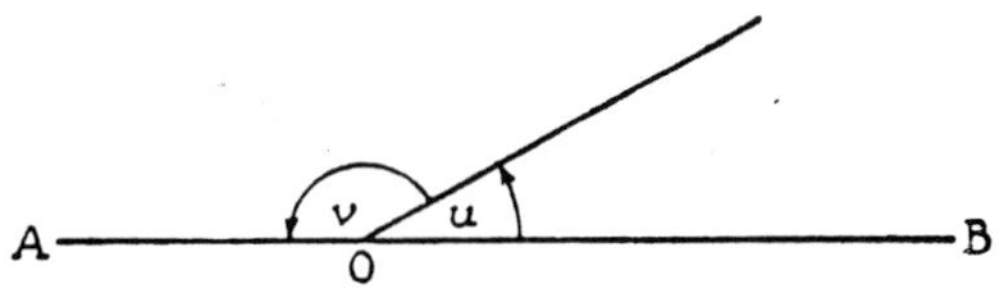

46. If x = 4 when y = 6 and x varies directly as y, then when y = 15, x equals 46.____
A. 20 B. 10 C. 1 3/5
D. 1 1/3 E. none of the above

47. A discount of 15 percent from a marked price produces a net price which is _____ of the marked price. 47.____
A. .15% B. .85% C. 15% D. 85% E. 115%

48. When the formula A = P + Prt is solved for t, t equals 48.____
A. A - P - Pr B. $\frac{A - Pr}{P}$ C. $\frac{A - P}{1 + r}$

D. $\frac{A - P}{Pr}$ E. none of the above

49. The Greek letter π 49.____
A. was assigned the value 3.1416 by the International Court of Law
B. was given an arbitrary value of 22/7 by a famous mathematician
C. was discovered to be exactly 3.142
D. when multiplied by the radius of a circle equals the area
E. is used as a symbol for the ratio of the circumference of a circle to its diameter

50. If the base and altitude of a triangle are doubled, the area 50.____
A. remains constant B. is multiplied by 4
C. is doubled D. is divided by 4
E. is none of the above

51. Each side of the equilateral triangle in the figure at the right is s inches long. The length of an altitude of the triangle is represented as 51.____

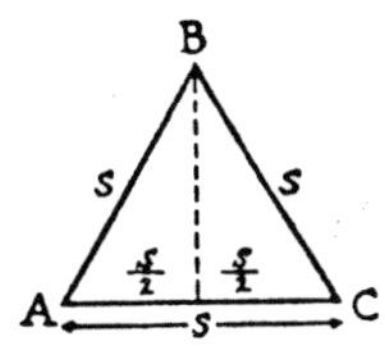

A. s in.
B. $s\sqrt{2}$ in.
C. $s\sqrt{3}$ in.
D. $\frac{s\sqrt{3}}{2}$ in.

E. none of the above

52. The length of a meter is about _____ inches. 52.____
A. 1 B. 6 C. 12 D. 40 E. 100

53. A point which lies on the straight-line graph of the equation 2x - 3y = 12 is 53.____
A. (3,-2) B. (2,-3) C. (-4,0)
D. (0,6) E. none of the above

54. If the two parallel lines AB and CD in the figure at the right are cut by a third line, EF, then the FALSE statement is 54.____

A. $\angle r + \angle s = \angle s + \angle y$
B. $\angle y + \angle w = \angle t + \angle s$
C. $\angle u + \angle w = \angle s + \angle x$
D. $\angle r + \angle x = \angle t + \angle w$
E. $\angle s + \angle u = \angle r + \angle t$

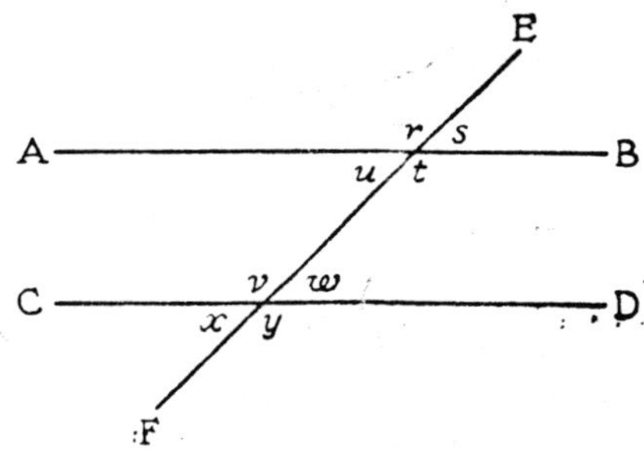

55. The product of n^4 and n^2 equals 55.____

A. $2n^8$ B. $2n^6$ C. n^8
D. n^2 E. none of the above

56. The volume of the rectangular solid shown at the right is 56.____

A. 12 cu. in.
B. 44 sq. in.
C. 48 cu. in.
D. 88 sq. in.
E. none of the above

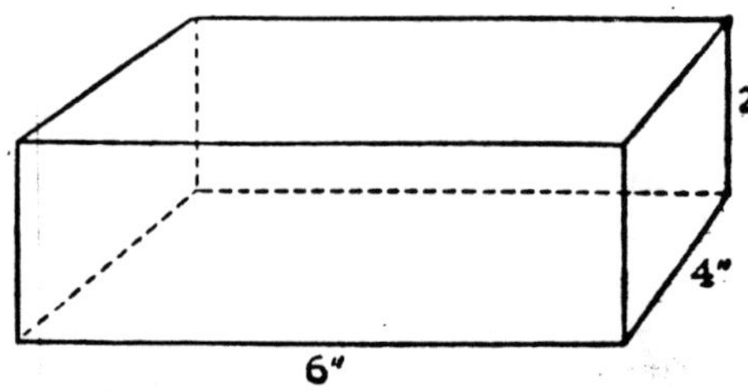

57. Baseball bats listed at twenty-one dollars per dozen are sold to schools at a discount of 20 percent. How much do they cost the schools per dozen? 57.____

A. \$4.20 B. \$16.80 C. \$20.80
D. \$25.20 E. None of the above

58. Last year a Chicago merchant's total business amounted to \$30,000. For the goods sold, he paid \$12,000, for rent he paid \$2500, for clerk services \$4742, and for other expenses \$1058. His average monthly net profit was 58.____

A. \$676.67 B. \$891.67 C. \$2500.00
D. \$9700.00 E. none of the above

59. If the marked price of an article is \$100 and the first discount is 10 percent and the second discount 2 percent, the sale price is 59.____

A. \$78.20 B. \$88.00 C. \$88.20
D. \$88.80 E. none of the above

60. Mr. Smith agreed to pay an automobile agency a commission of 18 percent of the selling price of his car. If the selling price was \$1250, Mr. Smith would receive 60.____

A. \$225.00 B. \$1025.00 C. \$1227.50
D. \$1475.00 E. none of the above

61. Mr. Browne receives \$30.45 per year on an investment of \$870. At this rate, if his total investment was \$1500, his annual interest would be 61.____

A. \$52.50 B. \$62.50 C. \$625.00
D. \$655.45 E. none of the above

62. The Ephrata National Bank discounted a 60-day note for \$3500 at $3\frac{1}{2}$ percent per year. The proceeds of the note were 62.____

A. \$3377.50 B. \$3479.58 C. \$3520.42
D. \$3622.50 E. none of the above

63. The normal weight of an adult can be found by using the formula w = 5.5(20+d), where w represents the weight in pounds and d the number of inches one's height exceeds 5 feet. By this formula, the normal weight of an adult who is 5'6" tall is _____ pounds. 63.____

A. 134 B. 140.25 C. 140.8
D. 143.0 E. none of the above

64. In the figure at the right, triangles ACB and ADE are similar triangles. The length of side DE is _____ feet. 64.____

A. 30
B. 32
C. 48
D. 50
E. none of the above

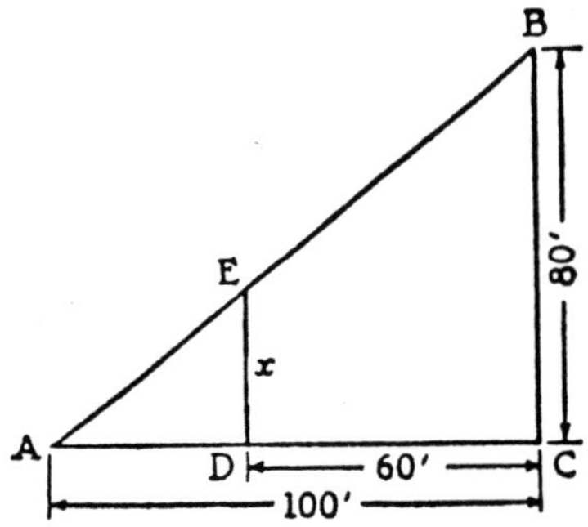

65. A square piece of tin shown in the figure at the right is used to make an open box. One-inch squares are cut from each corner of the piece of tin and the sides then turned up, to form a box containing 49 cubic inches. The length of a side of the original square piece of tin required to make this box is _____ inches. 65.____

A. 5
B. 7
C. 8
D. 9
E. none of the above

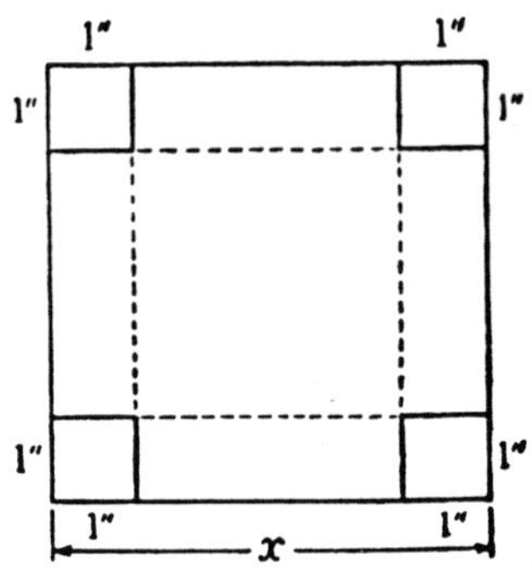

KEY (CORRECT ANSWERS)

1. C
2. C
3. B
4. D
5. C

6. B
7. A
8. D
9. C
10. C

11. B
12. A
13. A
14. E
15. E

16. E
17. C
18. A
19. A
20. E

21. D
22. D
23. C
24. E
25. D

26. A
27. B
28. D
29. E
30. E

31. A
32. A
33. E
34. C
35. C

36. B
37. C
38. A
39. C
40. A

41. A
42. C
43. B
44. D
45. B

46. B
47. D
48. D
49. E
50. B

51. D
52. D
53. A
54. E
55. E

56. C
57. B
58. E
59. C
60. B

61. A
62. B
63. D
64. B
65. D

SOLUTIONS TO PROBLEMS

1. $2/3 \times 12 = \frac{2}{3} \cdot \frac{12}{1} = \frac{24}{3} = 8$

2. Adding, we get 109.14

3. If $5x = 75$, $x = 75/5 = 15$

4. $65.13 \div .13 = 501$

5. 6 ft. 8 in. + 3 ft. 4 in. = 9 ft. 12 in. = 10 ft.

6. $3/4 - 1/2 + 1/8 = 6/8 - 4/8 + 1/8 = 3/8$

7. $4\frac{5}{16} - 2\frac{3}{8} = 3\frac{21}{16} - 2\frac{6}{16} = 1\frac{15}{16}$

8. $(-12) + (-3) = -15$

9. Let x = length of longer line. Then, $5:3 = x:30$.
Solving, $x = 50$

10. $.025 = 25/1000$ (Can also be reduced to 1/40)

11. Cross-multiplying, $5x = 18$. Thus, $x = 18/5 = 3.6$

12. $33\frac{1}{3}\%$ of $3 = (\frac{1}{3})(3) = 1$

13. $\sqrt{225} = 15$, since $15^2 = 225$

14. The arrow points to $6\frac{3}{8}$

15. $\frac{4x}{5} - 6 = 10$. Adding 6, $\frac{4x}{5} = 16$. Then, $x = 16 \div \frac{4}{5} = 20$

16. 8 hrs. 0 min. 6 sec. - 6 hrs. 4 min. 15 sec. can be written as
7 hrs. 59 min. 66 sec. - 6 hrs. 4 min. 15 sec. to get
1 hr. 55 min. 51 sec.

17. Mean = $(2 + 4 + 6 + 6 + 8 + 10 + 12 + 14 + 19) \div 9 = 9$

18. When $n = 0$, $c = 0$. When $n = 5$, $c = 100$. Thus, $c = 20n$.
Finally, for $n = 7$, $c = (20)(7) = 140$

19. $A = \frac{1}{2}bh + \frac{1}{2}ah = \frac{1}{2}h(b+a)$

20. 6 inches : 3 feet = 6 inches : 36 inches = 1/6

21. Add 5 to both sides to get $4s = 12$, so $s = 3$

22. $16\frac{2}{3}\%$ of x is 30. Then, $\frac{1}{6}x = 30$, so $x = 180$

23. At 7:00 AM the temperature was 12.5°, while at 9:00 AM the temperature was 15°. The change was 2.5 degrees.

24. Perimeter of Section I is 4b and the perimeter of Section III is $2b + 2a - 2b = 2a$. The sum of the perimeters is $4b + 2a$.

25. Area of Section II is $b(a-b) = ab - b^2$ and the area of Section IV is $(a-b)^2 = a^2 - 2ab + b^2$. The sum of the areas is $a^2 - ab = a(a-b)$.

26. Direct translation of words to symbols yields $F = \frac{9}{5}C + 32$

27. Subtract 6x to get $-114 = -5.7x$. Solving, $x = 20$

28. (84/42)(100)% = 200%

29. The figure is a cylinder.

30. Converting each choice to a decimal, we get .5, $.\overline{5}$, $.58\overline{3}$, $.\overline{6}$, .75. The largest is .75 corresponding to 3/4.

31. For a circle, $A = \pi r^2$.

32. A five-sided enclosed figure with straight sides is called a pentagon.

33. A quadrilateral with opposite sides parallel is called a parallelogram. Rectangles and squares are parallelograms with 90° angles.

34. $x - y + z = -18 - 3 - 2 = -23$

35. Since the digit in the hundreds place is 5 or greater, the answer is 336,000.

36. The sum of the angles of any triangle is 180°.

37. The scale difference is about 2 inches, and since 50 miles corresponds to 3/4 inch, the actual distance is about $(50)(2 \div \frac{3}{4}) = 133\frac{1}{3}$ mi. Closest answer given is 130 mi.

38. 6 added to n means 6 + n. Thus, 6 + n = 1 or n + 6 = 1

39. 3 is an exponent for $2n^3$.

40. 12 inches in 1 foot means 12n inches in n feet.

41. 3 months = 1/4 year.

42. 25% of 360 degrees = 90 degrees

43. Area of $\triangle AED = (\frac{1}{2})(2)(3) = 3$ square units = 12 sq. inches

44. Angle x is the angle of elevation to B from A.

45. Since u + v = 180, we can also write u = 180 - v

46. 4/x = 6/15 Cross-multiplying, 6x = 60. Solving, x = 10

47. 100% - 15% = 85%

48. A = P + Prt becomes A - P = Prt. Dividing by Pr we get: t = (A-P)/Pr

49. π = ratio of circumference to diameter of a circle

50. Let B = base, H = altitude. Original area of triangle = $\frac{1}{2}BH$. If new base and altitude are 2B and 2H, new area = $\frac{1}{2}(2B)(2H) = 2BH$, which is 4 times the value of $\frac{1}{2}BH$.

51. Let x = altitude. Then, $x^2 + (\frac{S}{2})^2 = s^2$. This becomes $3/4s^2 = x^2$. Solving, $x = s\sqrt{3}/2$

52. 1 meter ≈ 39.37 inches ≈ 40 inches

53. Substituting (3,-2), 2(3) - 3(-2) = 12. The other points do not lie on 2x - 3y = 12.

54. The false statement is $\angle s + \angle u = \angle r + \angle t$ (It is only true that $\angle s = \angle u$ and $\angle r = \angle t$)

55. $n^4 \cdot n^2 = n^6$, since exponents are added in multiplication

56. Volume = (6)(4)(2) = 48 cu.in.

57. ($21)(.80) = $16.80

58. $30,000 - $12,000 - $2500 - $4742 - $1058 = $9700. The monthly amount is $9700 ÷ 12 = $808.33

59. ($100)(.90) = $90. Then, ($90)(.98) = $88.20

60. $1250 - (1250)(.18) = \$1025$

61. $\$30.45/\$870 = 3.5\%$ Then, 3.5% of $\$1500 = \$52.50$

62. $(.035)(60/360) = .0058\overline{3}$ = discount for 60 days.
The value of the note $= (1 - .0058\overline{3})(\$3500) = \$3479.58$

63. $W = 5.5(20 + 6) = (5.5)(26) = 143$

64. $x/80 = 40/100$. Solving, $x = 32$. Note that AD:AC = DE:BC

65. When folded, each new side is $\sqrt{49} = 7$ inches. Thus, the original length (unfolded) was $7 + 1 + 1 = 9$ inches.

EXAMINATION SECTION

DIRECTIONS: Each question or incomplete statement is followed by several suggested answers or completions. Select the one that BEST answers the question or completes the statement. *PRINT THE LETTER OF THE CORRECT ANSWER IN THE SPACE AT THE RIGHT.*

1. Which ordered pair of numbers (x,y) is the solution of the following system of equations? 1.____

$$3x - 2y = 5$$
$$2x + 2y = 10$$

A. (1,1) B. (1,2) C. (2,1) D. (2,3) E. (3,2)

2. A certain microcomputer's memory contains 16K (K=1,024) storage locations. 2.____
If a program being run uses 12,517 storage locations, how many storage locations are still available?
A. 3,767 B. 3,867 C. 4,867 D. 11,493 E. 16,384

3. $(3.5 + 0.3) - 4(0.82 + 1.08) =$ 3.____

A. -3.800 B. -0.380 C. 0.304 D. 1.700 E. 4.840

4. Which of the triangles shown below are congruent? 4.____

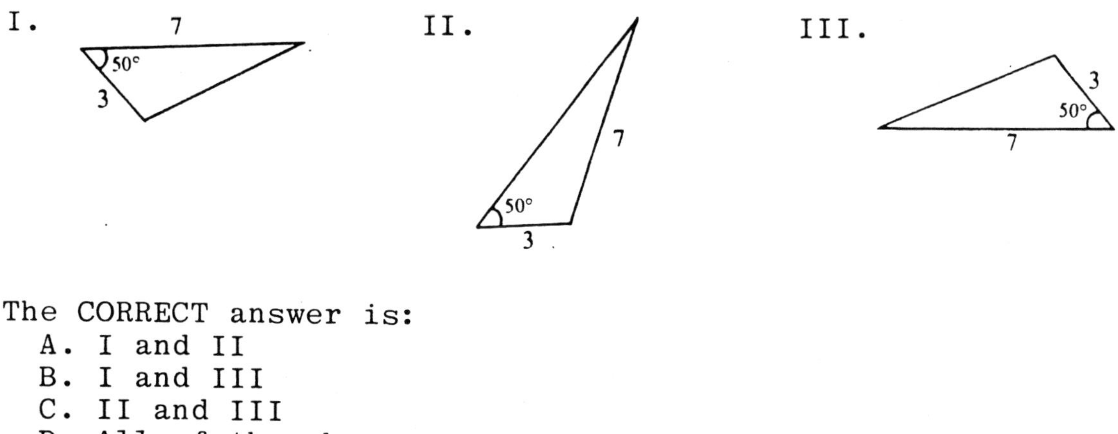

The CORRECT answer is:
A. I and II
B. I and III
C. II and III
D. All of the above
E. No triangle is congruent to any other triangle

5. A survey asked a sample of people to choose the better candidate in an upcoming election. Of the people surveyed, 20% said they would vote for Candidate A, 30% for Candidate B, and 50% said they were undecided. 5.____
If 1,000 people said they would vote for Candidate A, how many people said they would vote for Candidate B?
A. 300 B. 1,100 C. 1,500 D. 2,500 E. 5,000

6. Sheila's salary is $110 per day. Due to financial problems in her company, her employer has asked Sheila to take a 10% cut in pay. 6.____
How much will Sheila be earning per day if she takes the cut in pay?
A. $11 B. $99 C. $100 D. $109 E. $121

7. The 6 A.M. temperature one day last winter was -13°F. From 6 A.M. until 1 P.M., the temperature rose an average of 3°F per hour. 7.____
Which of the following expressions represents the temperature in °F at 1 P.M.?
A. 7(-13+3) B. -13-7(3) C. 7+3(-13)
D. -13+5(3) E. -13+7(3)

8. 8.____

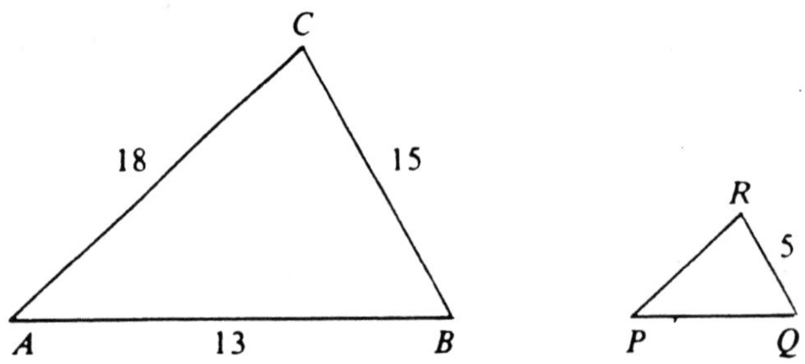

In the figure above, ΔABC is similar to ΔPQR, and the measure of $\angle A$ is equal to the measure of $\angle P$.
The length of PR is
A. 4 1/6 B. 4 1/3 C. 5 10/13 D. 6 E. 8

9. $|-5|+|6|+(-5)+6 =$ 9.____
A. -22 B. -10 C. 2 D. 10 E. 12

10. A bread recipe calls for $\frac{1}{2}$ cup of butter and $3\frac{1}{2}$ cups of flour. Using this recipe to make enough bread for a party, John will need $1\frac{1}{2}$ cups of butter. 10.____
How many cups of flour will he need?
A. $4\frac{1}{2}$ B. $5\frac{1}{2}$ C. $7\frac{1}{2}$ D. $9\frac{1}{2}$ E. $10\frac{1}{2}$

11. 11.____

Midland Stereo Supply House			
Item	Price each	Quantity ordered	Total for item(s)
Cassette tapes	$ 4.50	6	
Stereo headphones	$36.00	1	
Record protectors	$ 0.10	25	
		Subtotal	$
		Add 4% sales tax	+
		Shipping	+ 1.50
		Total	$

What would be the TOTAL cost of the order shown above?
A. $42.10 B. $65.50 C. $67.00 D. $69.62 E. $69.68

12. The distance, in miles, from an observer to the horizon is 1.35 times the square root of the observer's elevation, in feet. 12.____
If an observer's elevation is 16 feet, how many miles away is the horizon?
A. 5.4 B. 7.0 C. 10.8 D. 11.9 E. 48.6

13. If 3x - 2y = 6, then y equals which of these expressions? 13.____

A. $-\frac{3}{2}x - 3$ B. B. $\frac{3}{2}x + 6$ C. $\frac{3}{2}x - 3$

D. $\frac{2}{3}x + 3$ E. $3x - 3$

14. 14.____

Age	Number of Students
14	50
15	180
16	180
17	340
18	210
19	40
Total	1,000

The ages of the students attending City High School this year are listed in the table above.
If a student is picked at random from this school, what is the probability that he or she will be 18 or older?
A. 1/25 B. 1/4 C. 1/3 D. 1/2 E. 3/4

15. 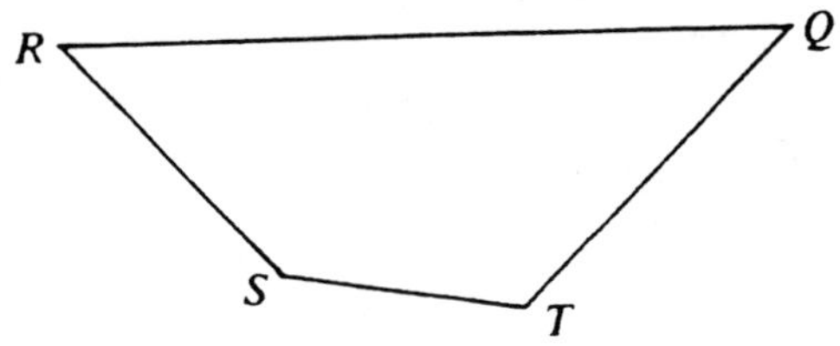 15.____

In quadrilateral QRST above, the measures of ∠Q, ∠R, and ∠S are 45°, 45°, and 140°, respectively.
The measure of ∠T is _____ degrees.
A. 120 B. 130 C. 135 D. 140 E. 220

16. If x = -3 and y = -2, then the GREATEST value is 16.____

A. $|x-y|$ B. $|x|-|y|$ C. $|x|-y$

D. $x-|y|$ E. $x-y$

17. 17.____

Name	Height in Inches
Adam	65
Barbara	64
Chris	69
Daniel	64
Ella	65

What is the average (arithmetic mean) height, in inches, of the 5 people whose heights are listed in the table above?
A. 65.0 B. 65.2 C. 65.4 D. 66.0 E. 66.5

18. The Jones family wants to buy a refrigerator that costs \$750. They agree to pay 15% of the cost initially and the balance in 5 equal monthly payments without interest. How much will each monthly payment be? 18.____
A. \$112.50 B. 127.50 C. \$129.50 D. \$147.00 E. \$150.00

19. What is the SMALLEST positive integer that gives a remainder of 1 when divided by any of the integers 12, 18, and 27? 19.____
A. 121 B. 109 C. 61 D. 55 E. 37

20. A serving of a certain cereal, with milk, provides 35% of the potassium required daily by the average adult. 20.____
A serving of this cereal with milk contains 112 milligrams of potassium.
How many milligrams of potassium does the average adult require each day?
A. 35 B. 39 C. 147 D. 320 E. 392

21. Three people share \$198 in the ratio 1:3:7. To the nearest dollar, how much is the LARGEST share? 21.____

A. \$18 B. \$28 C. \$54 D. \$126 E. \$134

22. Which of the following is a factorization of the polynomial $2x^2 + x - 10$? 22.____

A. $2(x^2+x-5)$ B. $(2x+2)(x-5)$ C. $(2x+5)(x-2)$
D. $(2x-5)(x+2)$ E. $(2x+10)(x-1)$

23. In the figure at the right, B, E, and C are collinear; A, D, and C are collinear; E is halfway between B and C; and $\overline{DE}$ and $\overline{AB}$ are each perpendicular to $\overline{BC}$. If $\overline{BE}$ is 40 units long and $\overline{AB}$ is 60 units long, how many units long is the perimeter of quadrilateral ABED? 23.____

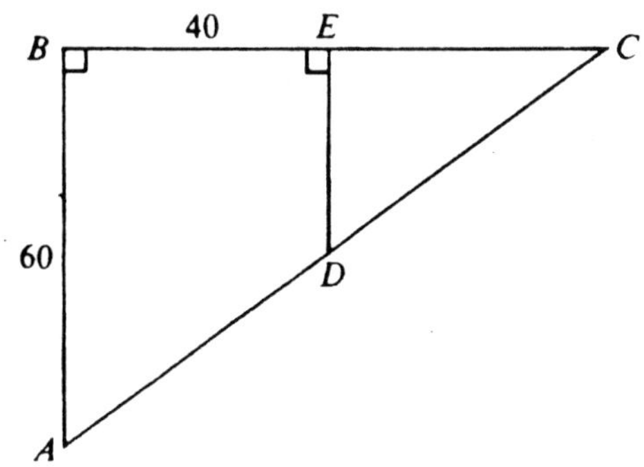

A. 100
B. 140
C. 180
D. 200
E. 220

24. The circle graph at the right represents the relative sizes of the sources of a tax dollar. The degree measure of the central angle of the sector labeled *Income* is _____ degrees. 24.____

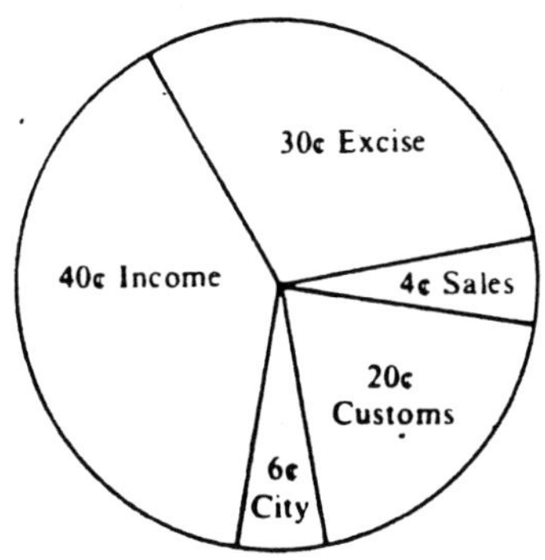

A. 40
B. 72
C. 100
D. 120
E. 144

25. $\sqrt{8} + \sqrt{16} + 3\sqrt{2} - \sqrt{3} =$ 25.____

A. $4 + 5\sqrt{2} - \sqrt{3}$ B. $11\sqrt{2} - \sqrt{3}$ C. $3\sqrt{26} - \sqrt{3}$
D. $15 - \sqrt{3}$ E. $3\sqrt{23}$

26. Two lines have the equations $2x+y = 4$ and $x-2y = 7$, respectively. At what (x,y) point do they intersect? 26.____

A. (3,-2) B. (6,-5) C. (5,-6) D. (-3,-2) E. (-2,3)

27.

x	0	2	4	6	8	10
y	4	7	10	13	16	19

27.____

Which of these equations expresses the relationship shown in the above table?

A. $y = 2x$ B. $y = x+4$ C. $y = x+9$
D. $2y = 3x+4$ E. $2y = 3x+8$

28. A life insurance policy costs $0.75 per month for each $1,000 worth of insurance. 28.____
At this rate, how much would someone have to pay in a year for $25,000 worth of this insurance?

A. $225.00 B. $187.50 C. $156.25
D. $75.00 E. $18.75

29. In the circle at the right, which has O as its center, $\overline{OA}$ and $\overline{AB}$ are each 4 units long. 29.____
If $\overline{OE}$ is perpendicular to $\overline{AB}$, how many units long is $\overline{OE}$?

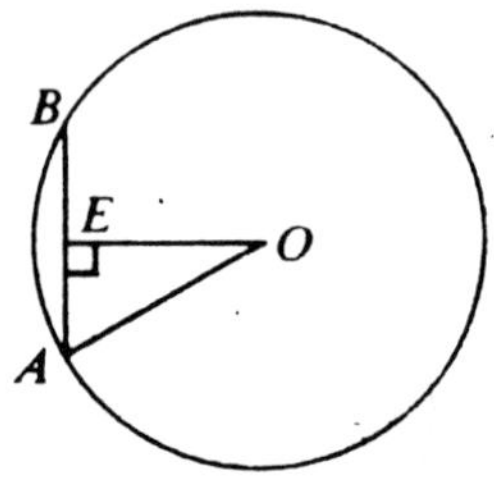

A. $\sqrt{3}$
B. 2
C. 3
D. $2\sqrt{3}$
E. 4

30. If the solutions of the equation $2x^2 - kx + 6 = 0$ are $x = 1$ and $x = 3$, then $k =$ 30.____

A. -4 B. 4 C. 7 D. 8 E. 10

31. If $f(x) = x + 3$ and $g(x) = 3 - x$, what is the value of $f[g(3)]$? 31.____

A. 6 B. 3 C. 0 D. -3 E. -6

32. Let x equal the numerator of a certain fraction. The denominator of that fraction is 2 more than the numerator. When 5 is added to both the numerator and the denominator, the resulting fraction equals 5/6. 32.____
Which of these equations determines the correct value of x, the numerator of the original fraction?

A. $\frac{x+5}{x+3} = \frac{5}{6}$ B. $\frac{x+3}{x+5} = \frac{5}{6}$ C. $\frac{x+5}{x+7} = \frac{5}{6}$

D. $\frac{x+5}{2x+5} = \frac{5}{6}$ E. $\frac{2x+5}{x+5} = \frac{5}{6}$

33. A man throwing darts at a dart board hit the board on 95% of the throws he made. He hit the board 114 times. Which equation determines the CORRECT value of x, the number of throws he made? 33.____

A. $(0.95)114 = x$ B. $0.95x = 114$ C. $114x = 95$

D. $\frac{x}{95} = 114$ E. $x = \frac{0.95}{114}$

34. Which equation determines the line that is parallel to the line with the equation $y = 3x+1$ and intersects the line with the equation $y = 6x-1$ at the y-axis? 34.____

$y =$

A. $3x-1$ B. $2x-1$ C. $\frac{1}{3}x - 1$ D. $\frac{1}{3}x + 1$ E. $\frac{1}{2}x - 1$

35. 35.____

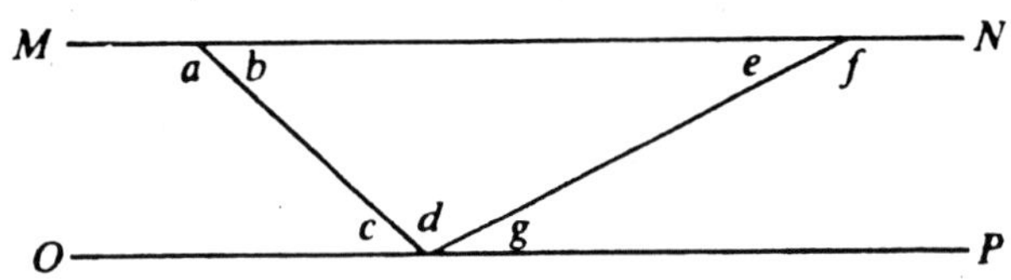

In the figure above, 2 line segments intersect $\overline{MN}$ and $\overline{OP}$, $\overline{MN}$ is parallel to $\overline{OP}$, and a, b, c, d, e, f, and g are the measures, in degrees, of the indicated angles. Which of these statements is NOT necessarily true?

A. $b = 180° - d - c$ B. $e = 180° - d - c$

C. $a = 180° - c$ D. $f = 180° - g$

E. $g = 180° - f$

36. If $x = -2$, $y = 3$, and $z = -5$, then the product xyz is how much GREATER than the sum $x + y + z$? 36.____

A. -34 B. -26 C. 20 D. 26 E. 34

37. If $n! = (n)(n-1)(n-2)\ldots(2)(1)$, then $5! =$ 37.____

A. 12 B. 15 C. 30 D. 120 E. 54,321

38. For all x, $(2x+3)^2 + 2(2x+4) - 2$ equals which of these expressions? 38.____

A. $4x^2 + 4x + 11$ B. $(4x+15)(x+1)$

C. $(2x+5)(2x+3)$ D. $(2x+5)(2x+2)$

E. $(2x+5)(2x-3)$

39. What is TRUE about the solutions of the equation $x^2 - 3x = -2$? 39.____

They are

A. real and unequal B. real and equal

C. real and negative D. irrational and negative

E. imaginary

40. If the retail price of a dinette set is $1\frac{1}{3}$ times the wholesale price, and the retail price of a dinette set is $200.00, what is its wholesale price? 40.____

A. $133.33 B. $150.00 C. $166.67
D. $266.67 E. $300.00

KEY (CORRECT ANSWERS)

1. E	11. D	21. D	31. B
2. B	12. A	22. C	32. C
3. A	13. C	23. C	33. B
4. B	14. B	24. E	34. A
5. C	15. B	25. A	35. A
6. B	16. C	26. A	36. E
7. E	17. C	27. E	37. D
8. D	18. B	28. A	38. C
9. E	19. B	29. D	39. A
10. E	20. D	30. D	40. B

SOLUTIONS TO PROBLEMS

1. Adding the equations, 5x = 15, so x = 3. Substituting this value of x into the first equation, 9 - 2y = 5. Then, y = 2.

2. (16)(1024) = 16,384. Then, 16,384 - 12,517 = 3867

3. Simplify to 3.8 - 4(1.9) = -3.8 or -3.800

4. Triangles I and III are congruent by SAS, which refers to two pairs of matching sides and an included equivalent angle.

5. The ratio of votes for Candidate A to votes for Candidate B is 2:3. Then, letting x = number of votes for Candidate B, 2:3 = 1000:x. Solving, x = 1500

6. Her pay after a 10% cut is ($110)(.90) = $99

7. Since the temperature rose 3° per hour for 7 hours, the temperature at 1:00 PM was -13° + 7(3°)

8. 18:PR = 15:5 Solving, PR = 6

9. $|-5| + |6| + (-5) + 6 = 5 + 6 - 5 + 6 = 12$

10. Let x = cups of flour. Then, $\frac{1}{2}:3\frac{1}{2} = 1\frac{1}{2}:x$ This becomes $\frac{1}{2}x = 5\frac{1}{4}$. Solving, $x = 10\frac{1}{2}$

11. Total for items before tax is 6($4.50) + 1($36.00) + 25($0.10) = $65.50. Then, $65.50 + (.04)($65.50) + $1.50 = $69.62

12. $(1.35)(\sqrt{16}) = (1.35)(4) = 5.4$

13. Subtract 3x from both sides of the equation to get -2y = - 3x + 6. Dividing by -2, $y = \frac{3}{2}x - 3$.

14. There are 250 students 18 or older out of 1000 students in the school. Probability of selecting a student 18 or older is $250/1000 = \frac{1}{4}$

15. ∠T = 360° - 45° - 45° - 140° = 130°. Note that the sum of the angles in any quadrilateral is 360°

16. Choice C has a value of $|-3| - (-2) = 3 + 2 = 5$, which exceeds the value of each of the other choices

17. Total of heights = 327. Average height is 327/5 = 65.4"

18. ($750)(.15) = $112.50. The balance is $637.50. If this is paid in 5 equal installments, each installment is $637.50/5 = $127.50

19. The number 109 when divided by any of 12, 18, or 27 gives a remainder of 1.

20. The amount of potassium required is $112 \div .35 = 320$ milligrams

21. Let x, 3x, 7x represent each person's share. Then, $x + 3x + 7x = 198$, so $x = \$18$. Then, largest share is $7(\$18) = \$1$

22. $2x^2 + x - 10 = (2x+5)(x-2)$, which can be checked by multiplicatior

23. Because $\triangle CED$ is similar to $\triangle CBA$, $ED = \frac{1}{2}(BA) = 30$. Now, $AC = \sqrt{80^2 + 60^2} = 100$, so $AD = \frac{1}{2}(100) = 50$. The perimeter of ABED is $40 + 60 + 50 + 30 = 180$

24. Total of all sources is 100 cents. The central angle for income is, in degrees, $(40/100)(360) = 144$.

25. $\sqrt{8} + \sqrt{16} + 3\sqrt{2} - \sqrt{3} = 2\sqrt{2} + 4 + 3\sqrt{2} - \sqrt{3} = 4 + 5\sqrt{2} - \sqrt{3}$

26. Double the 2nd equation to get $2x - 4y = 14$. Subtract the 1st equation to get $-5y = 10$, so $y = -2$. Substitute into the 1st equation to get $2x - 2 = 4$. Then, $x = 3$. The point of intersection is $(3,-2)$.

27. Determine the slope to be $(7-4)/(2-0) = \frac{3}{2}$. Then, $y = \frac{3}{2}x + k$, where k is a constant. Substituting the point (0,4), $4 = \frac{3}{2}(0) + k$, so $k = 4$. Thus, we get $y = \frac{3}{2}x + 4$ or equivalently $2y = 3x + 8$.

28. $(\$0.75)(25) = \$18.75$ per month $= \$225$ per year

29. $BE = \frac{1}{2}(4) = 2$. Then, $(OE)^2 + 2^2 = 4^2$. Solving, $OE = \sqrt{12} = 2\sqrt{3}$

30. Substituting $x = 1$, $2(1^2) - k(1) + 6 = 0$. Simplifying, $8 - k = 0$. So, $k = 8$. Note that if we used $x = 3$, $2(3^2) - k(3) + 6 = 0$. This would lead to $24 - 3k = 0$, and still $k = 8$.

31. $g(3) = 3 - 3 = 0$. Then, $f[g(3)] = f(0) = 0 + 3 = 3$

32. Since x = numerator, x + 2 = denominator. Adding 5 to each makes the new numerator x+5 and the new denominator x+7. Thus, $(x+5)/(x+7) = 5/6$

33. Let x = number of throws. Then, $.95x = 114$, since his 114 hits represent 95% of his throws

34. A line parallel to $y = 3x+1$ must be of the form $y = 3x + k$, where k is a constant. The equation $y = 6x-1$ crosses the y-axis at $(0,-1)$ and this point must lie on $y = 3x + k$. By substitution, $-1 = 3(0) + k$, $k = -1$. The resulting equation

35. The statement which is NOT necessarily true is $b = 180° - d - c$. We know that $c + d + g = 180°$, but b and g are not necessarily equal.

36. $xyz = 30$, $x + y + z = -4$. Then, $xyz - (x+y+z) = 34$

37. $5! = (5)(4)(3)(2)(1) = 120$

38. $(2x+3)^2 + 2(2x+4) - 2 = 4x^2 + 12x + 9 + 4x + 8 - 2 = 4x^2 + 16x + 15$. This last expression factors as $(2x+5)(2x+3)$.

39. Rewrite as $x^2 - 3x + 2 = 0$, which becomes $(x-2)(x-1) = 0$. The solutions are $x = 1$ and $x = 2$, which are real and unequal.

40. Let x = wholesale price. Then, $\$200 = 1\frac{1}{3}x$. Solving, $x = \$150$

READING COMPREHENSION
UNDERSTANDING AND INTERPRETING WRITTEN MATERIAL

EXAMINATION SECTION

TEST 1

DIRECTIONS: Each question or incomplete statement is followed by several suggested answers or completions. Select the one that BEST answers the question or completes the statement. *PRINT THE LETTER OF THE CORRECT ANSWER IN THE SPACE AT THE RIGHT.*

Questions 1-2.

DIRECTIONS: Questions 1 and 2 are to be answered SOLELY on the basis of the following paragraph.

When fixing an upper sash cord, you must also remove the lower sash. To do this, the parting strip between the sash must be removed. Now remove the cover from the weight box channel, cut off the cord as before, and pull it over the pulleys. Pull your new cord over the pulleys and down into the channel where it may be fastened to the weight. The cord for an upper sash is cut off 1" or 2" below the pulley with the weight resting on the floor of the pocket and the cord held taut. These measurements allow for slight stretching of the cord. When the cord is cut to length, it can be pulled up over the pulley and tied with a single common knot in the end to fit into the socket in the sash groove. If the knot protrudes beyond the face of the sash, tap it gently to flatten. In this way, it will not become frayed from constant rubbing against the groove.

1. When repairing the upper sash cord, the FIRST thing to do is to 1.____
   A. remove the lower sash
   B. cut the existing sash cord
   C. remove the parting strip
   D. measure the length of new cord necessary

2. According to the above paragraph, the rope may become frayed if the 2.____
   A. pulley is too small
   B. knot sticks out
   C. cord is too long
   D. weight is too heavy

Questions 3-4.

DIRECTIONS: Questions 3 and 4 are to be answered SOLELY on the basis of the following paragraph.

Repeated burning of the same area should be avoided. Burning should not be done on impervious, shallow, unstable, or highly erodible soils, or on steep slopes - especially in areas subject to heavy rains or rapid snowmelt. When existing vegetation is likely to be killed or seriously weakened by the fire, measures should be taken to assure prompt revegetation of the burned area. Burns should be limited to relatively small proportions of a watershed unit so that the stream channels will be able to carry any increased flows with a minimum of damage.

3. According to the above paragraph, planned burning should be limited to small areas of the watershed because 3.____
   A. the fire can be better controlled
   B. existing vegetation will be less likely to be killed
   C. plants will grow quicker in small areas
   D. there will be less likelihood of damaging floods

4. According to the above paragraph, burning USUALLY should be done on soils that 4.____
   A. readily absorb moisture
   B. have been burnt before
   C. exist as a thin layer over rock
   D. can be flooded by nearby streams

Questions 5-11.

DIRECTIONS: Questions 5 through 11 are to be answered SOLELY on the basis of the following paragraph.

FUSE INFORMATION

Badly bent or distorted fuse clips cannot be permitted. Sometimes, the distortion or bending is so slight that it escapes notice, yet it may be the cause for fuse failures through the heat that is developed by the poor contact. Occasionally, the proper spring tension of the fuse clips has been destroyed by overheating from loose wire connections to the clips. Proper contact surfaces must be maintained to avoid faulty operation of the fuse. Maintenance men should remove oxides that form on the copper and brass contacts, check the clip pressure, and make sure that contact surfaces are not deformed or bent in any way. When removing oxides, use a well-worn file and remove only the oxide film. Do not use sandpaper or emery cloth as hard particles may come off and become embedded in the contact surfaces. All wire connections to the fuse holders should be carefully inspected to see that they are tight.

5. Fuse failure because of poor clip contact or loose connections is due to the resulting 5.___
   A. excessive voltage B. increased current
   C. lowered resistance D. heating effect

6. Oxides should be removed from fuse contacts by using 6.___
   A. a dull file B. emery cloth
   C. fine sandpaper D. a sharp file

7. One result of loose wire connections at the terminal of a fuse clip is stated in the above paragraph to be 7.___
   A. loss of tension in the wire
   B. welding of the fuse to the clip
   C. distortion of the clip
   D. loss of tension of the clip

8. Simple reasoning will show that the oxide film referred to is undesirable CHIEFLY because it 8.___
   A. looks dull
   B. makes removal of the fuse difficult
   C. weakens the clips
   D. introduces undesirable resistance

9. Fuse clips that are bent very slightly 9.___
   A. should be replaced with new clips
   B. should be carefully filed
   C. may result in blowing of the fuse
   D. may prevent the fuse from blowing

10. From the fuse information paragraph, it would be reasonable to conclude that fuse clips 10.___
    A. are difficult to maintain
    B. must be given proper maintenance
    C. require more attention than other electrical equipment
    D. are unreliable

11. A safe practical way of checking the tightness of the wire connection to the fuse clips of a live 120-volt lighting circuit is to 11.___
    A. feel the connection with your hand to see if it is warm
    B. try tightening with an insulated screwdriver or socket wrench
    C. see if the circuit works
    D. measure the resistance with an ohmmeter

Questions 12-13.

DIRECTIONS: Questions 12 through 13 are to be answered SOLELY on the basis of the following paragraph.

For cast iron pipe lines, the middle ring or sleeve shall have *beveled* ends and shall be high quality cast iron. The middle ring shall have a minimum wall thickness of 3/8" for pipe up to 8", 7/16" for pipe 10" to 30", and ½" for pipe over 30", nominal diameter.

Minimum length of middle ring shall be 5" for pipe up to 10", 6" for pipe 10" to 30", and 10" for pipe 30" nominal diameter and larger. The middle ring shall not have a center pipe stop, unless otherwise specified.

12. As used in the above paragraph, the word *beveled* means MOST NEARLY 12.___
   A. straight B. slanted C. curved D. rounded

13. In accordance with the above paragraph, the middle ring of a 24" nominal diameter pipe would have a minimum wall thickness and length of ____ thick and ____ long. 13.___
   A. 3/8"; 5:
   B. 3/8"; 6"
   C. 7/16"; 6"
   D. ½"; 6"

Questions 14-17.

DIRECTIONS: Questions 14 through 17 are to be answered SOLELY on the basis of the following paragraph.

Operators spotting loads with long booms and working around men need the smooth, easy operation and positive control of uniform pressure swing clutches. There are no jerks or grabs with these large disc-type clutches because there is always even pressure over the entire clutch lining surface. In the conventional band-type swing clutch, the pressure varies between dead and live ends of the band. The uniform pressure swing clutch has excellent provision for heat dissipation. The driving elements, which are always rotating, have a great number of fins cast in them. This gives them an impeller or blower action for cooling, resulting in longer life and freedom from frequent adjustment.

14. According to the above paragraph, it may be said that conventional band-type swing clutches have 14.___
   A. even pressure on the clutch lining
   B. larger contact area
   C. smaller contact area
   D. uneven pressure on the clutch lining

15. According to the above paragraph, machines equipped with uniform pressure swing clutches will 15.___
   A. give better service under all conditions
   B. require no clutch adjustment
   C. give positive control of hoist
   D. provide better control of swing

16. According to the above paragraph, it may be said that the rotation of the driving elements of the uniform pressure swing clutch is ALWAYS 16.___
   A. continuous
   B. constant
   C. varying
   D. uncertain

17. According to the above paragraph, freedom from frequent adjustment is due to the 17.___
A. operator's smooth, easy operation
B. positive control of the clutch
C. cooling effect of the rotating fins
D. larger contact area of the bigger clutch

Questions 18-22.

DIRECTIONS: Questions 18 through 22 are to be answered SOLELY on the basis of the following paragraphs.

Exhaust valve clearance adjustment on diesel engines is very important for proper operation of the engine. Insufficient clearance between the exhaust valve stem and the rocker arm causes a loss of compression and, after a while, burning of the valves and valve seat inserts. On the other hand, too much valve clearance will result in noisy operation of the engine.

Exhaust valves that are maintained in good operating condition will result in efficient combustion in the engine. Valve seats must be true and unpitted, and valve stems must work smoothly within the valve guides. Long valve life will result from proper maintenance and operation of the engine.

Engine operating temperatures should be maintained between 160°F and 185°F. Low operating temperatures result in incomplete combustion and the deposit of fuel lacquers on valves.

18. According to the above paragraphs, too much valve clearance will cause the engine to operate 18.___
A. slowly B. noisily C. smoothly D. cold

19. On the basis of the information given in the above paragraphs, operating temperatures of a diesel engine should be between 19.___
A. 125°F and 130°F B. 140°F and 150°F
C. 160°F and 185°F D. 190°F and 205°F

20. According to the above paragraphs, the deposit of fuel lacquers on valves is caused by 20.___
A. high operating temperatures
B. insufficient valve clearance
C. low operating temperatures
D. efficient combustion

21. According to the above paragraphs, for efficient operation of the engine, valve seats must 21.___
A. have sufficient clearance
B. be true and unpitted
C. operate at low temperatures
D. be adjusted regularly

22. According to the above paragraphs, a loss of compression is due to insufficient clearance between the exhaust valve stem and the 22.___
A. rocker arm B. valve seat
C. valve seat inserts D. valve guides

Questions 23-25.

DIRECTIONS: Questions 23 through 25 are to be answered SOLELY on the basis of the following excerpt from A SPECIFICATION FOR ELECTRIC WORK FOR THE CITY.

Breakers shall be equipped with magnetic blowout coils...Handles of breakers shall be trip-free...Breakers shall be designed to carry 100% of trip rating continuously; to have inverse time delay tripping above 100% of trip rating...

23. According to the above paragraph, the breaker shall have provision for 23.___
A. resetting B. arc quenching
C. adjusting trip time D. adjusting trip rating

24. According to the above paragraph, the breaker 24.___
A. shall trip easily at exactly 100% of trip rating
B. shall trip instantly at a little more than 100% of trip rating
C. should be constructed so that it shall not be possible to prevent it from opening on overload or short circuit by holding the handle in the *ON* position
D. shall not trip prematurely at 100% of trip rating

25. According to the above paragraph, the breaker shall trip 25.___
A. instantaneously as soon as 100% of trip rating is reached
B. instantaneously as soon as 100% of trip rating is exceeded
C. more quickly the greater the current, once 100% of trip rating is exceeded
D. after a predetermined fixed time lapse, once 100% of trip rating is reached

TEST 2

Questions 1-4.

DIRECTIONS: Questions 1 through 4 are to be answered SOLELY on the basis of the following paragraph.

A low pressure hot water boiler shall include a relief valve or valves of a capacity such that with the heat generating equipment operating at maximum, the pressure cannot rise more than 20 percent above the maximum allowable working pressure (set pressure) if that is 30 p.s.i. gage or less, nor more than 10 percent if it is more than 30 p.s.i. gage. The difference between the set pressure and the pressure at which the valve is relieving is known as *over-pressure or accumulation*. If the steam relieving capacity in pounds per hour is calculated, it shall be determined by dividing by 1,000 the maximum BTU output at the boiler nozzle obtainable from the heat generating equipment, or by multiplying the square feet of heating surface by five.

1. In accordance with the above paragraph, the capacity of a relief valve should be computed on the basis of 1.___
   A. size of boiler
   B. maximum rated capacity of generating equipment
   C. average output of the generating equipment
   D. minimum capacity of generating equipment

2. In accordance with the above paragraph, with a set pressure of 30 p.s.i. gage, the overpressure should not be more than _____ p.s.i. 2.___
   A. 3 B. 6 C. 33 D. 36

3. In accordance with the above paragraph, a relief valve should start relieving at a pressure equal to the 3.___
   A. set pressure
   B. over pressure
   C. over pressure minus set pressure
   D. set pressure plus over pressure

4. In accordance with the above paragraph, the steam relieving capacity can be computed by 4.___
   A. *multiplying* the maximum BTU output by 5
   B. *dividing* the pounds of steam per hour by 1,000
   C. *dividing* the maximum BTU output by the square feet of heating surface
   D. *dividing* the maximum BTU output by 1,000

Questions 5-8.

DIRECTIONS: Questions 5 through 8 are to be answered SOLELY on the basis of the following paragraph.

Air conditioning units requiring a minimum rate of flow of water in excess of one-half ($\frac{1}{2}$) gallon per minute shall be metered. Air conditioning equipment with a refrigeration unit which has a definite rate of capacity in tons or fractions thereof, the charge will be at the rate of $30 per annum per ton capacity from the date installed to the date when the supply is metered. Such units, when equipped with an approved water-conserving device, shall be charged at the rate of $4.50 per annum per ton capacity from the date installed to the date when the supply is metered.

5. A man who was in the market for air conditioning equipment was considering three different units. Unit 1 required a flow of 28 gallons of water per hour; Unit 2 required 30 gallons of water per hour; Unit 3 required 32 gallons of water per hour. The man asked the salesman which units would require the installation of a water meter. According to the above passage, the salesman SHOULD answer: 5.___
   A. All three units require meters
   B. Units 2 and 3 require meters
   C. Unit 3 only requires a meter
   D. None of the units require a meter

6. Suppose that air conditioning equipment with a refrigeration unit of 10 tons was put in operation on October 1; and in the following year on July 1, a meter was installed. According to the above passage, the charge for this period would be _____ the annual rate. 6.____
   A. twice
   B. equal to
   C. three-fourths
   D. one-fourth

7. The charge for air conditioning equipment which has no refrigeration unit 7.____
   A. is $30 per year
   B. is $25.50 per year
   C. is $4.50 per year
   D. cannot be determined from the above passage

8. The charge for air conditioning equipment with a seven-ton refrigeration unit equipped with an approved water-conserving device 8.____
   A. is $4.50 per year
   B. is $25.50 per year
   C. is $31.50 per year
   D. cannot be determined from the above passage

Questions 9-14.

DIRECTIONS: Questions 9 through 14 are to be answered SOLELY on the basis of the following paragraph.

The city makes unremitting efforts to keep the water free from pollution. An inspectional force under a sanitary expert is engaged in patrolling the watersheds to see that the department's sanitary regulations are observed. Samples taken daily from various points in the water supply system are examined and analyzed at the three laboratories maintained by the department. All water before delivery to the distribution mains is treated with chlorine to destroy bacteria. In addition, some water is aerated to free it from gases and, in some cases, from microscopic organisms. Generally, microscopic organisms which develop in the reservoirs and at times impart an unpleasant taste and odor to the water, though in no sense harmful to health, are destroyed by treatment with copper sulfate and by chlorine dosage. None of the supplies is filtered, but the quality of the water supplied by the city is excellent for all purposes, and it is clear and wholesome.

9. According to the above paragraph, microscopic organisms are removed from the water supplied to the city by means of 9.____
   A. chlorine alone
   B. chlorine, aeration, and filtration
   C. chlorine, aeration, filtration, and sampling
   D. copper sulfate, chlorine, and aeration

10. Microscopic organisms in the water supply GENERALLY are 10.____
   A. a health menace
   B. impossible to detect
   C. not harmful to health
   D. not destroyed in the water

11. The MAIN function of the inspectional force, as described in the above paragraph, is to 11.___
    A. take samples of water for analysis
    B. enforce sanitary regulations
    C. add chlorine to the water supply
    D. inspect water-use meters

12. According to the above paragraph, chlorine is added to water before entering the 12.___
    A. watersheds
    B. reservoirs
    C. distribution mains
    D. run-off areas

13. Of the following suggested headings or titles for the above paragraph, the one that BEST tells what the paragraph is about is 13.___
    A. QUALITY OF WATER
    B. CHLORINATION OF WATER
    C. TESTING OF WATER
    D. BACTERIA IN WATER

14. The MOST likely reason for taking samples of water for examination and analysis from various points in the water supply system is: 14.___
    A. The testing points are convenient to the department's laboratories
    B. Water from one part of the system may be made undrinkable by a local condition
    C. The samples can be distributed equally among the three laboratories
    D. The hardness or softness of water varies from place to place

Questions 15-17.

DIRECTIONS: Questions 15 through 17 are to be answered SOLELY on the basis of the following paragraph.

A building measuring 200' x 100' at the street is set back 20' on all sides at the 15th floor, and an additional 10' on all sides at the 30th floor. The building is 35 stories high.

15. The floor area of the 16th floor is MOST NEARLY ____ sq. ft. 15.___
    A. 20,000 B. 14,400 C. 9,600 D. 7,500

16. The floor area of the 35th floor is MOST NEARLY ____ sq. ft. 16.___
    A. 20,000 B. 13,900 C. 7,500 D. 5,600

17. The floor area of the 16th floor, compared to the floor area of the 2nd floor, is MOST NEARLY ____ as much. 17.___
    A. three-fourths (3/4)
    B. two-thirds (2/3)
    C. one-half ($\frac{1}{2}$)
    D. four-tenths (4/10)

Question 18.

DIRECTIONS: Question 18 is to be answered SOLELY on the basis of the following paragraph.

Experience has shown that, in general, a result of the installation of meters on services not previously metered is to reduce the amount of water consumed, but is not necessarily to reduce the peak load on plumbing systems. The permissible head loss through meters at their rated maximum flow is 20 p.s.i. The installation of a meter may therefore appreciably lower the pressures available in fixtures on a plumbing system.

18. According to the above paragraph, a water meter may 18.____
   A. limit the flow in the plumbing system of 20 p.s.i.
   B. reduce the peak load on the plumbing system
   C. increase the overall amount of water consumed
   D. reduce the pressure in the plumbing system

Question 19.

DIRECTIONS: Question 19 is to be answered SOLELY on the basis of the following paragraph.

Spring comes without trumpets to a city. The asphalt is a wilderness that does not quicken overnight; winds blow gritty with cinders instead of merry with the smells of earth and fertilizer. Women wear their gardens on their hats. But spring is a season in the city, and it has its own harbingers, constant as daffodils. Shop windows change their colors, people walk more slowly on the streets, what one can see of the sky has a bluer tone. Pulitzer prizes awake and sing and matinee tickets go-a-begging. But gayer than any of these are the carousels, which are already in sheltered places, beginning to turn with the sound of springtime itself. They are the earliest and the truest and the oldest of all the urban signs.

19. In the passage above, the word *harbingers* means 19.____
   A. storms B. truths C. virtues D. forerunners

Questions 20-22.

DIRECTIONS: Questions 20 through 22 are to be answered SOLELY on the basis of the following paragraph.

Gas heaters include manually operated, automatic, and instantaneous heaters. Some heaters are equipped with a thermostat which controls the fuel supply so that when the water falls below a predetermined temperature, the fuel is automatically turned on. In some types, the hot-water storage tank is well-insulated to economize the use of fuel. Instantaneous heaters are arranged so that the opening of a faucet on the hot-water pipe will increase the flow of fuel, which is ignited by a continuously burning pilot light to heat the water to from 120° to 130°F. The possibility that the pilot light will die out offers a source of danger in the use of automatic appliances which depend on a pilot light. Gas and oil heaters are dangerous, and they should be designed to prevent the accumulation, in a confined space within the heater, of a large volume of an explosive mixture.

20. According to the above passage, the opening of a hot-water faucet on a hot-water pipe connected to an instantaneous hot-water heater will _____ the pilot light. 20.___
   A. *increase* the temperature of
   B. *increase* the flow of fuel to
   C. *decrease* the flow of fuel to
   D. *have a marked effect* on

21. According to the above passage, the fuel is automatically turned on in a heater equipped with a thermostat whenever 21.___
   A. the water temperature drops below 120°F
   B. the pilot light is lit
   C. the water temperature drops below some predetermined temperature
   D. a hot water supply is opened

22. According to the above passage, some hot-water storage tanks are well-insulated to 22.___
   A. accelerate the burning of the fuel
   B. maintain the water temperature between 120° and 130°F
   C. prevent the pilot light from being extinguished
   D. minimize the expenditure of fuel

Question 23.

DIRECTIONS: Question 23 is to be answered SOLELY on the basis of the following paragraph.

Breakage of the piston under high-speed operation has been the commonest fault of disc piston meters. Various techniques are adopted to prevent this, such as *throttling* the meter, cutting away the edge of the piston, or reinforcing it, but these are simply makeshifts.

23. As used in the above paragraph, the word *throttling* means MOST NEARLY 23.___
   A. enlarging
   B. choking
   C. harnessing
   D. dismantling

Questions 24-25.

DIRECTIONS: Questions 24 and 25 are to be answered SOLELY on the basis of the following paragraph.

One of the most common and objectionable difficulties occurring in a drainage system is trap seal loss. This failure can be attributed directly to inadequate ventilation of the trap and the subsequent negative and positive pressures which occur. A trap seal may be lost either by siphonage and/or back pressure. Loss of the trap seal by siphonage is the result of a negative pressure in the drainage system. The seal content of the trap is forced by siphonage into the waste piping of the drainage system through exertion of atmospheric pressure on the fixture side of the trap seal.

24. According to the above paragraph, a positive pressure is a direct result of 24.___
   A. siphonage   B. unbalanced trap seal
   C. poor ventilation   D. atmospheric pressure

25. According to the above paragraph, the water in the trap is forced into the drain pipe by 25.___
   A. atmospheric pressure
   B. back pressure
   C. negative pressure
   D. back pressure on fixture side of seal

KEY (CORRECT ANSWERS)

TEST 1	TEST 2
1. C	1. B
2. B	2. B
3. D	3. D
4. A	4. D
5. D	5. C
6. A	6. C
7. D	7. D
8. D	8. C
9. C	9. D
10. B	10. C
11. B	11. B
12. B	12. C
13. C	13. A
14. D	14. B
15. D	15. C
16. A	16. D
17. C	17. C
18. B	18. D
19. C	19. B
20. C	20. B
21. B	21. C
22. A	22. D
23. B	23. B
24. C	24. C
25. C	25. A

MECHANICAL APTITUDE

TEST OF

MECHANICAL COMPREHENSION

DIRECTIONS

Fill in the requested information on your ANSWER SHEET.

Then look at Sample X on this page. It shows pictures of two rooms and asks, "Which room has more of an echo?" Because it has neither rugs nor curtains, there is more of an echo in room "A"; so blacken the space under "A" on your answer sheet. Now look at Sample Y and answer it yourself. Fill in the space under the correct answer on your answer sheet.

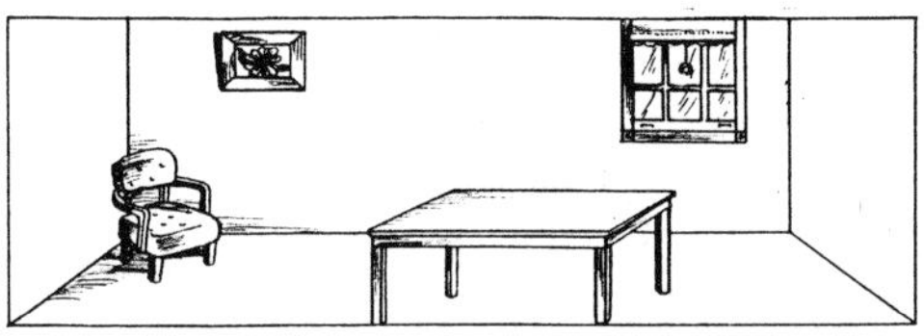

A

B

X

Which room has more of an echo?

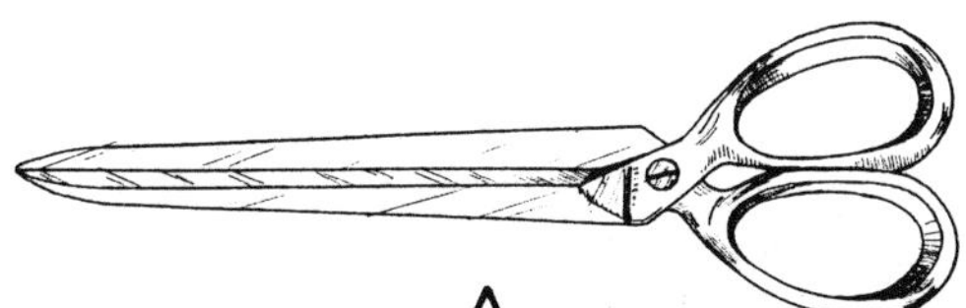

A

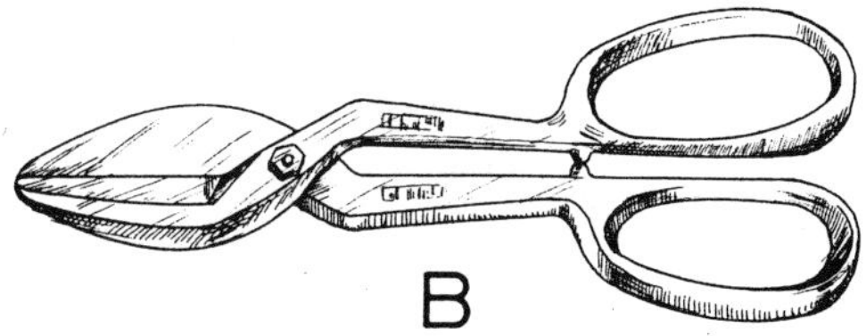

B

Y

Which would be the better shears for cutting metal?

On the following pages there are more pictures and questions.

Each question or incomplete statement is followed by two suggested answers or completions. Select A or B, or C if the two figures have the same value, as the BEST answer that completes the statement or completes the statement. *PRINT THE LETTER OF THE CORRECT ANSWER IN THE SPACE AT THE RIGHT.*

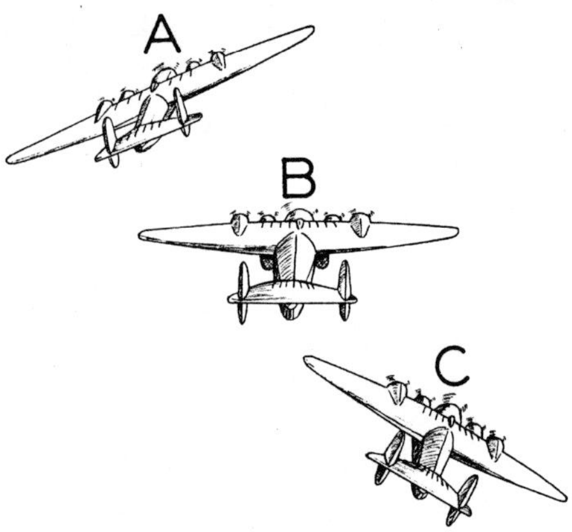

1

Which airplane is turning to the right?

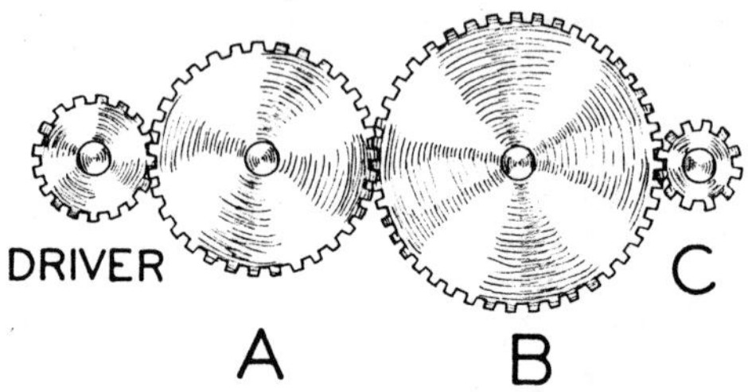

2

Which gear will make the most turns in a minute?

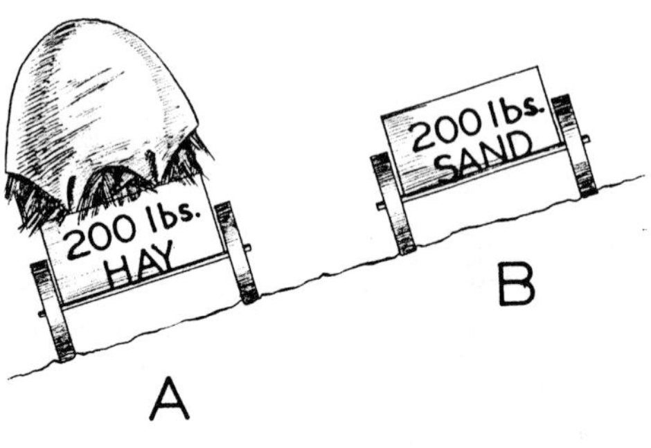

3

Which cart is more likely to tip over on the hillside?

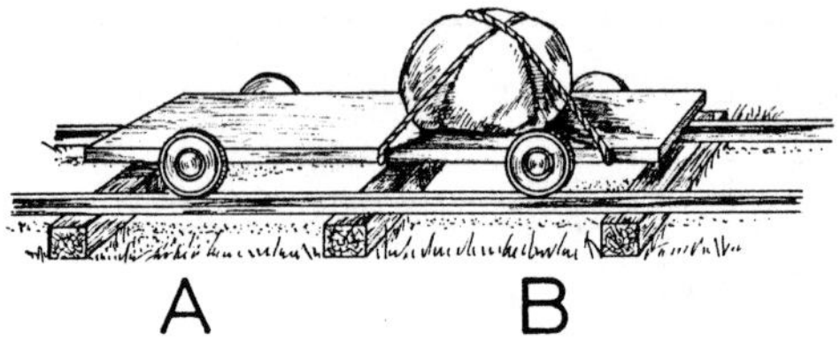

4

Which wheel presses harder against the rail?

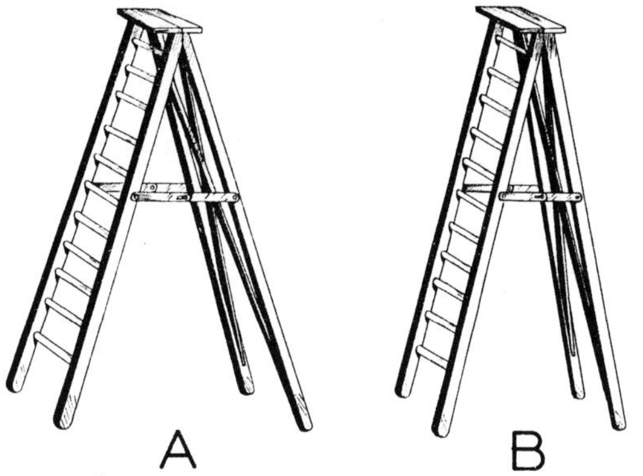

5

Which stepladder is safer to climb on?

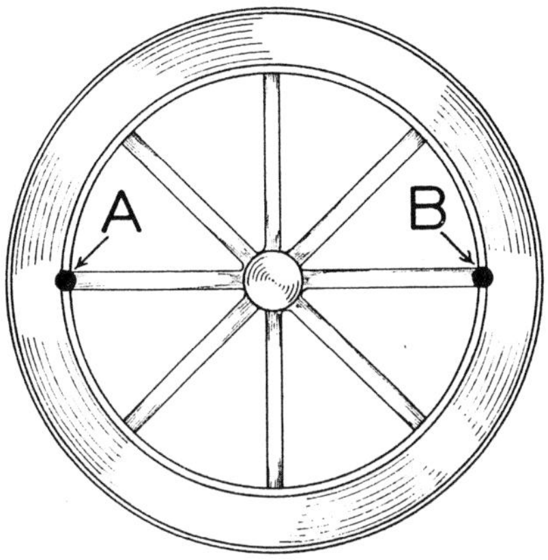

6

Which spot on the wheel travels faster?

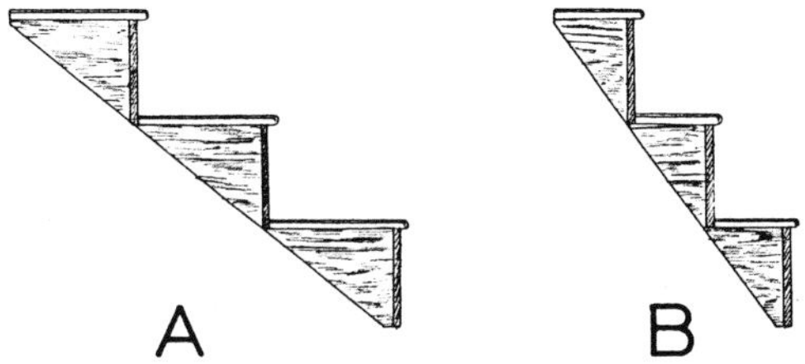

7

Which staircase would take less room?

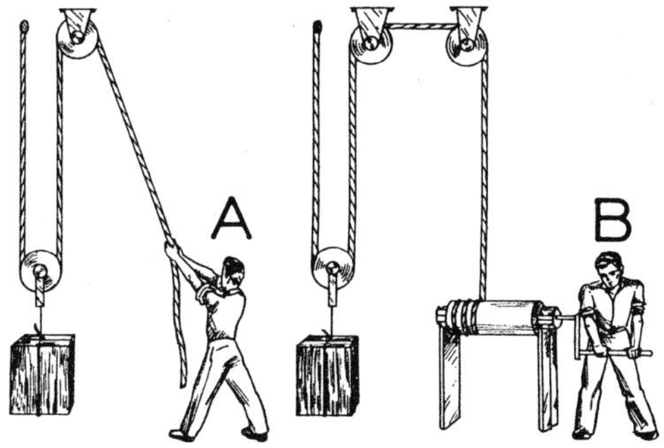

8

Which man can lift more weight?

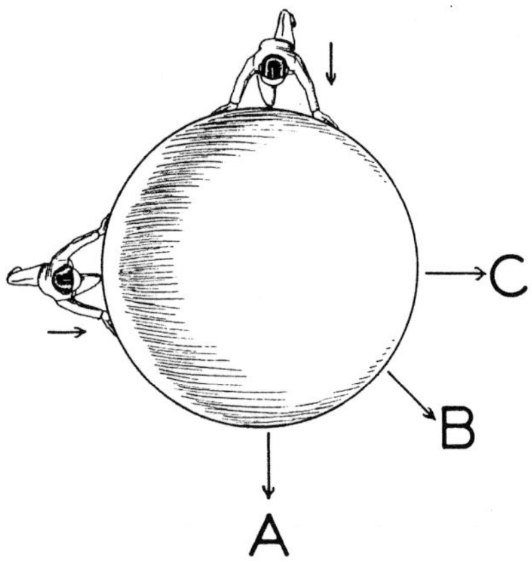

9

If the two men are pushing against the pushball in the directions shown, in which direction is it most likely to go?

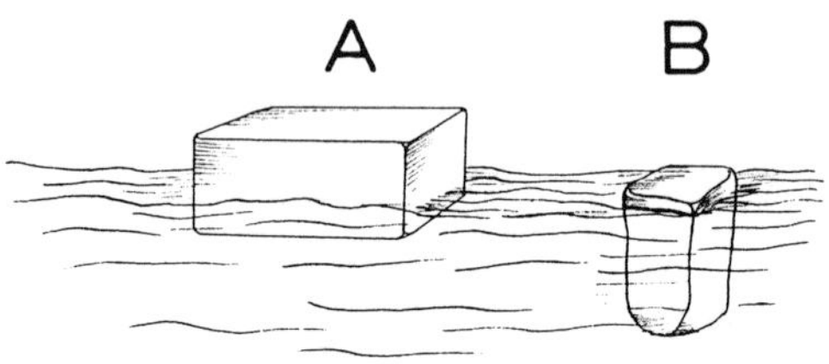

10

Which of these objects is made of the heavier material?

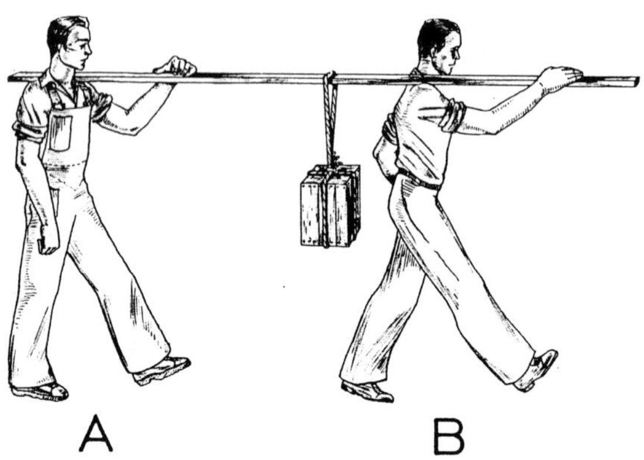

11

Which man carries more weight?

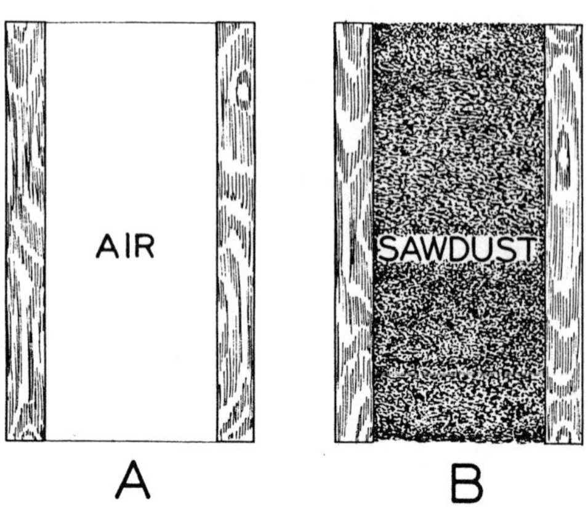

12

Which wall will keep a house warmer in winter?

13

Which horse will be harder to hold?

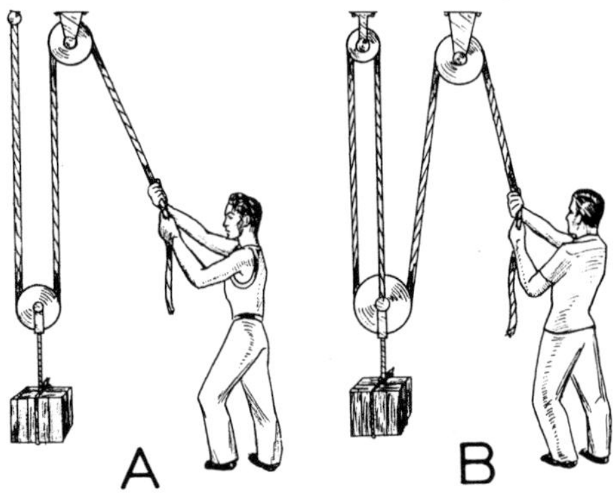

14

Which man has to pull harder?

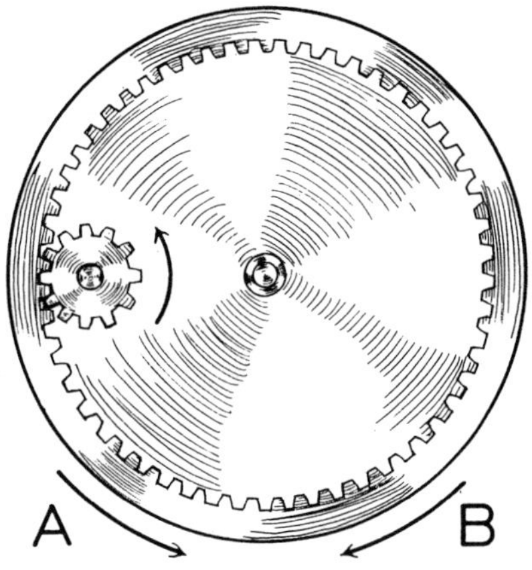

15

If the small wheel goes in the direction shown, in which direction will the large wheel go?

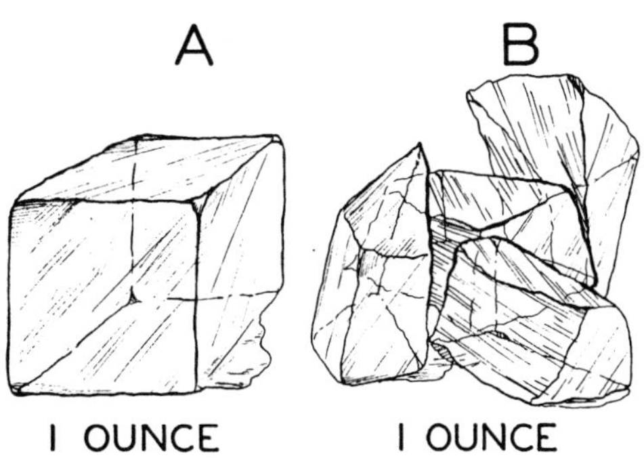

16

Which ounce of ice will cool a drink more quickly?

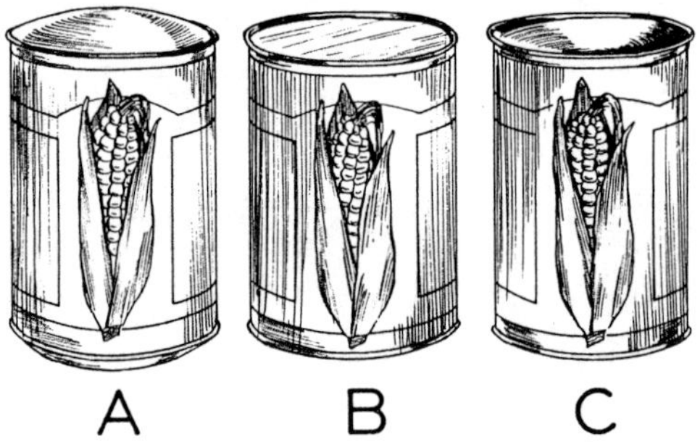

17

If a can is heated it is most likely to look like:

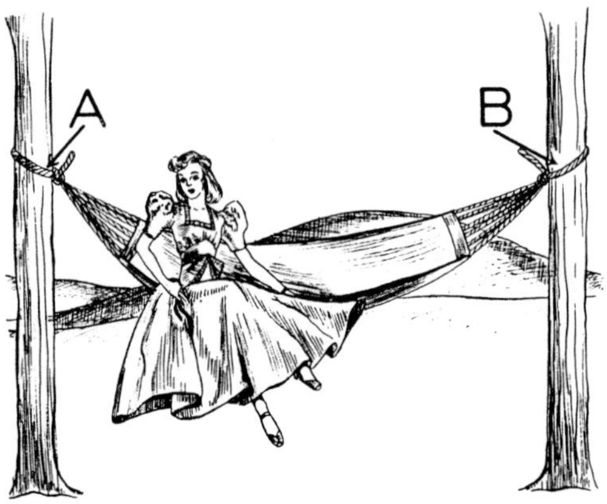

18

Which rope is under more strain?

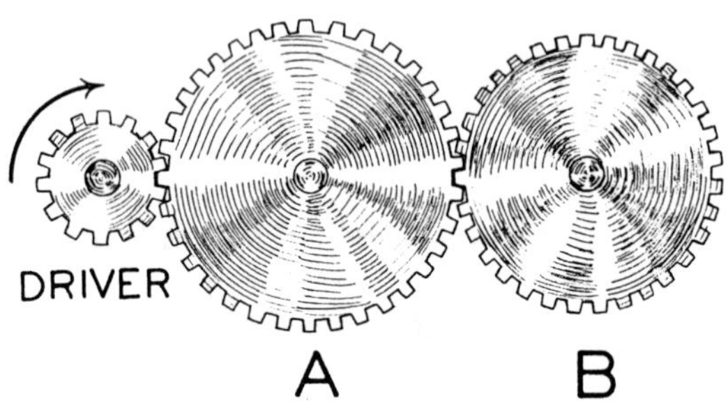

19

Which gear will turn the same way as the driver?

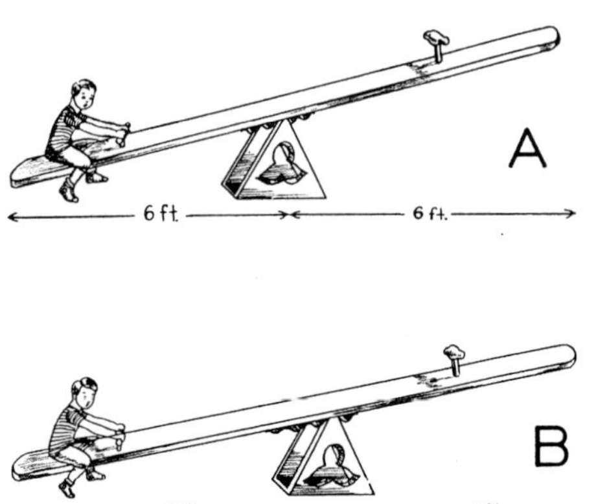

20

If the two boys weigh the same, which of them can balance a heavier boy on the other end of his see-saw?

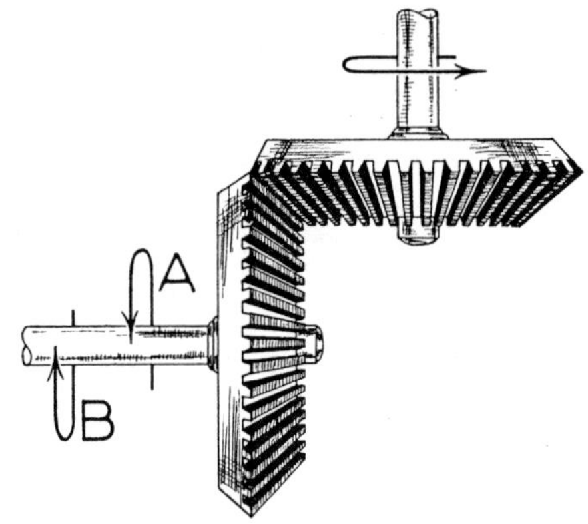

21

If the upper wheel moves in the direction shown, in which direction does the other one move?

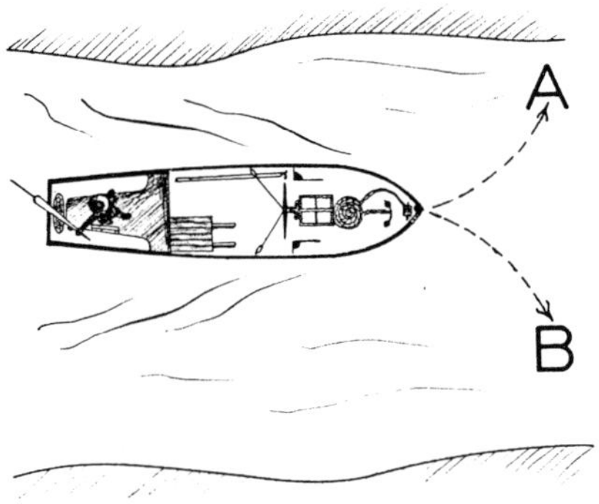

22

Which way will the boat go?

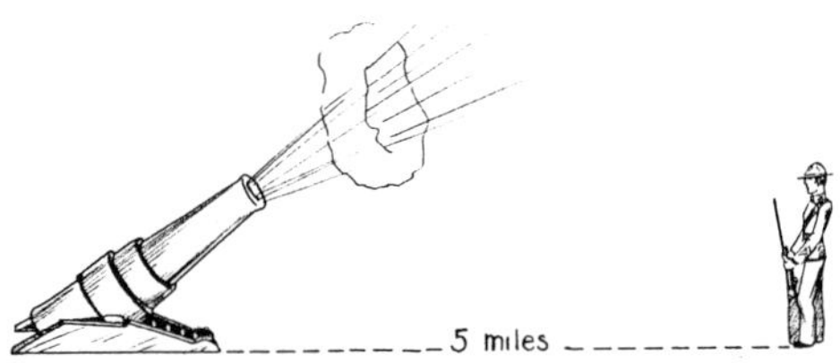

23

The man will hear the sound of the cannon:

A—before he sees the flash,
B—after he sees the flash,
C—at the same time as he sees the flash.

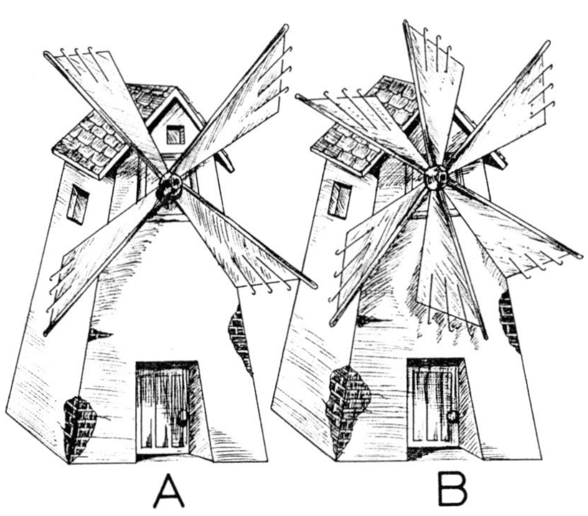

24

Which windmill will do more work?

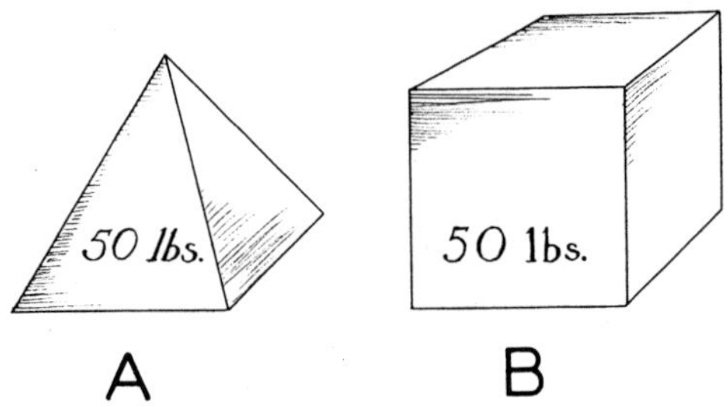

25

Which of these solid blocks will be the harder to tip over?

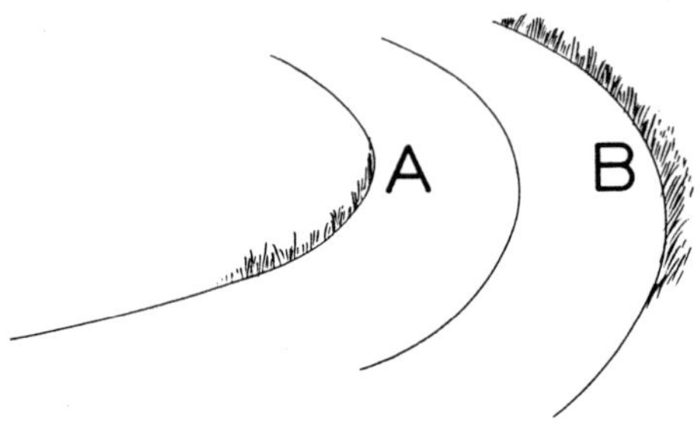

26

Which side of the road should be built higher?

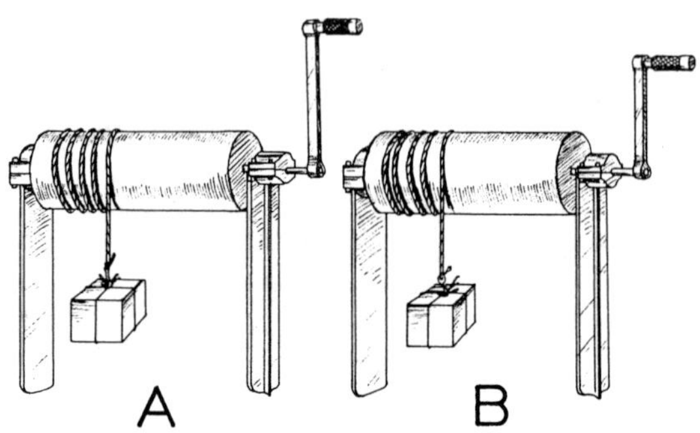

27

With which windlass can a man raise the heavier weight?

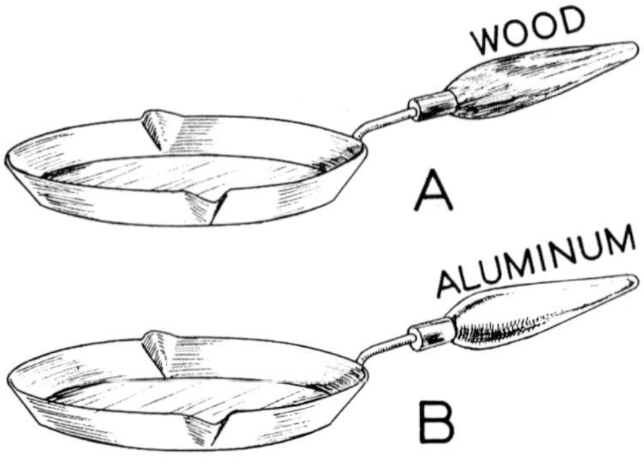

28

Which frying pan will be easier to handle?

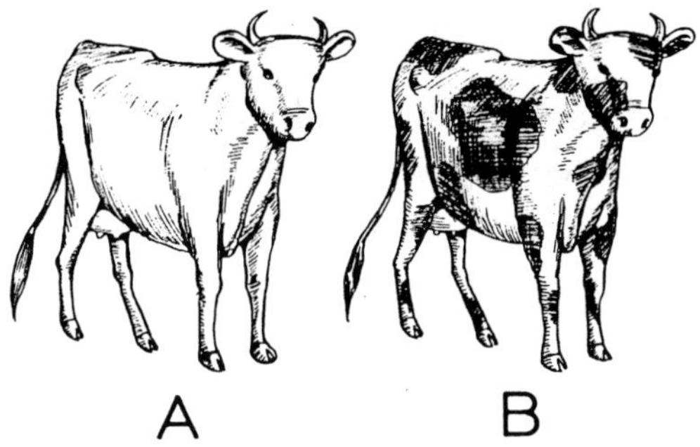

29

Which cow would be harder to see from an airplane?

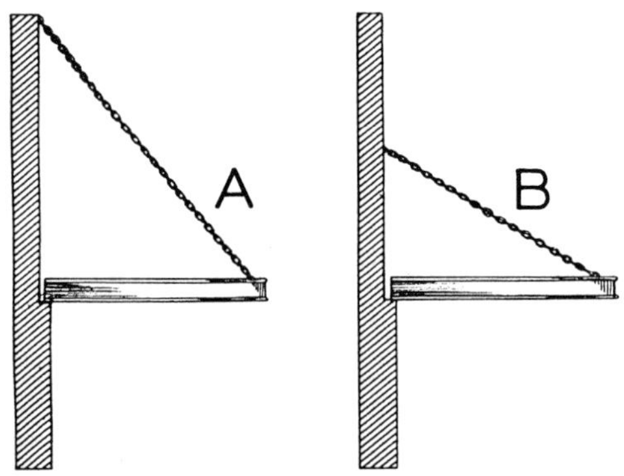

30

Which chain has more strain put upon it?

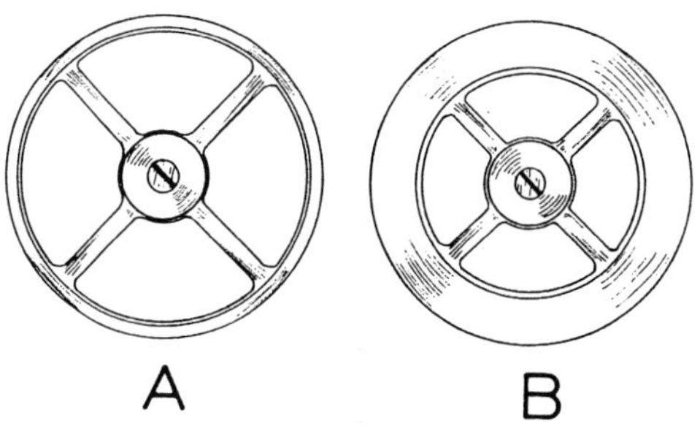

31

Which wheel will keep going longer after the power has been shut off?

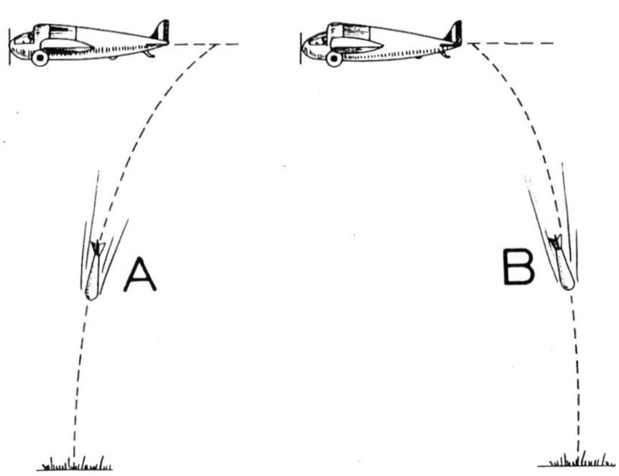

32

Which picture shows the way a bomb falls from a moving airplane if there is no wind?

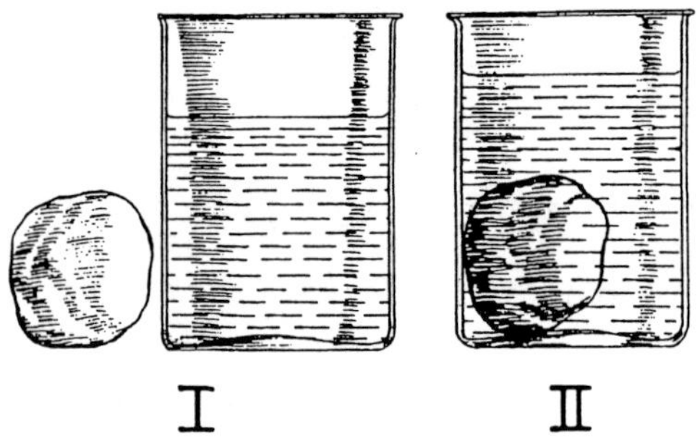

33

If the rock and tank of water together in Picture I weigh 100 pounds, they will weigh __________ in Picture II.

A. more
B. less
C. the same

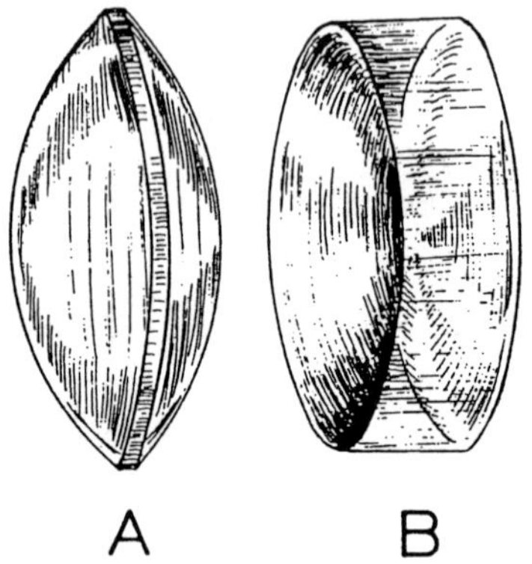

34

If light travels more slowly through glass than through air, which shape lens will make objects look larger?

35

If a man were lifting a stone with this crowbar, at which point would the bar be most likely to break?

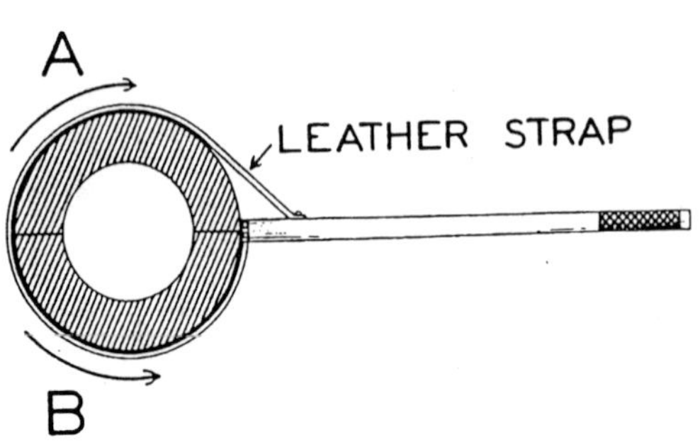

36

This wrench can be used to turn the pipe in direction:

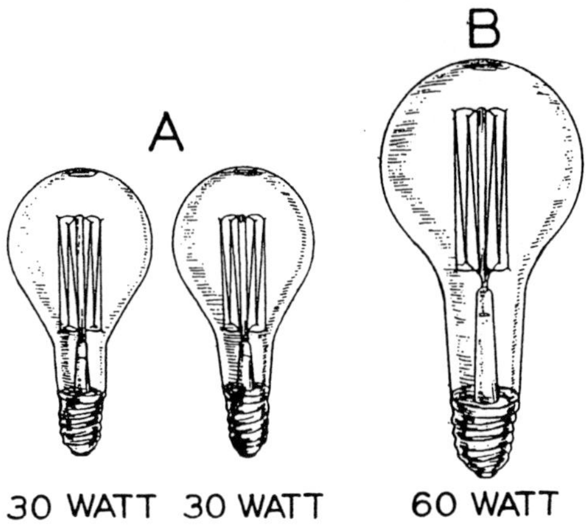

37

Which will use more current; the two bulbs at A, or the bulb at B? If the same, mark C.

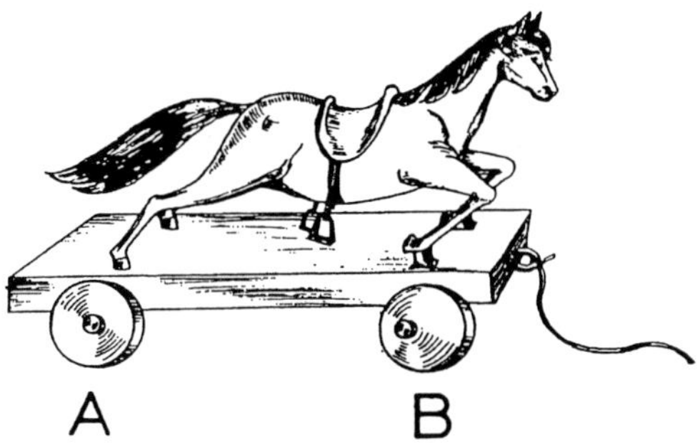

38

Which end of the toy horse will buck more when it is pulled along the floor?

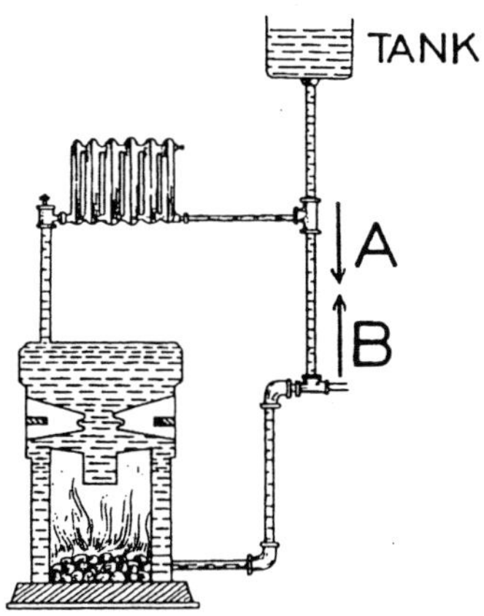

39

In which direction does the water in the right hand pipe go?

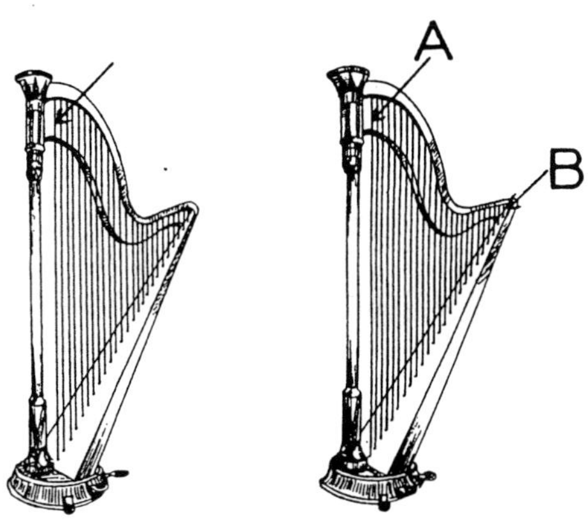

40

If the string shown by the arrow is plucked on the first harp, which string on the second harp will be more likely to sound?

41

Which of these clocks will tick faster?

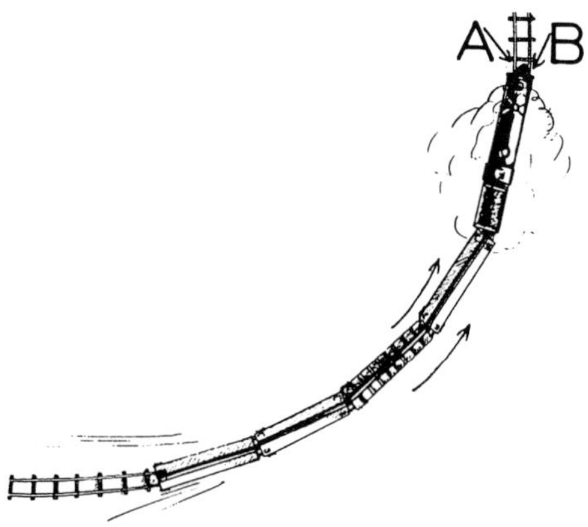

42

If the track is exactly level, on which rail does more pressure come?

43

If there are no clouds, on which night will you be able to see more stars?

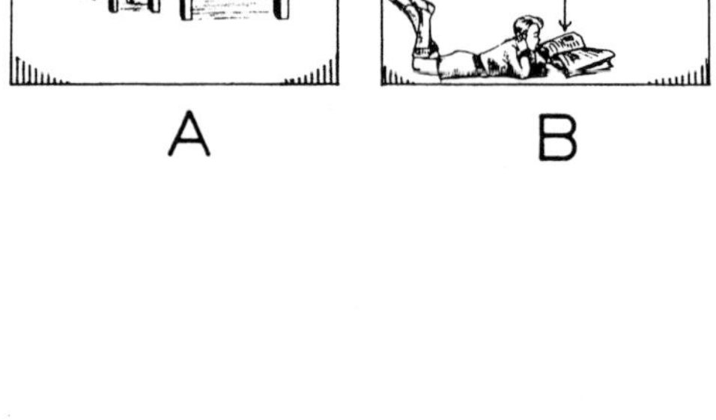

44

Which boy gets more light on the pages of his book?

45

Which rock will get hotter in the sun?

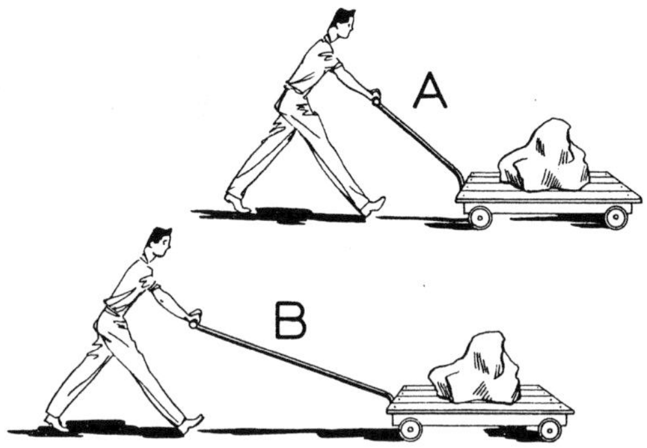

46

Which way can the man push the heavier load?

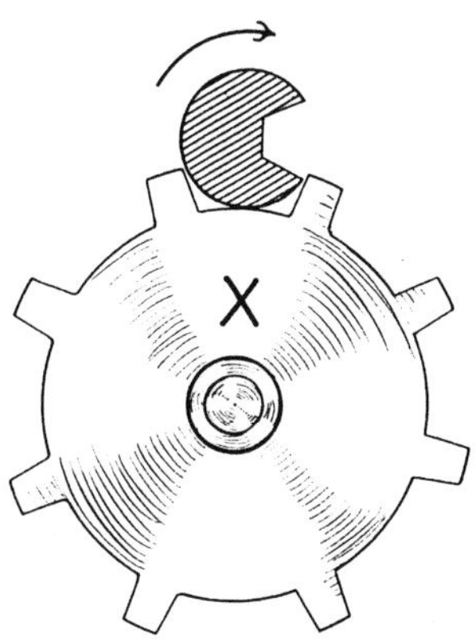

47

The top of the wheel "X" will go:

A—steadily to the right,

B—steadily to the left,

C—by jerks to the left.

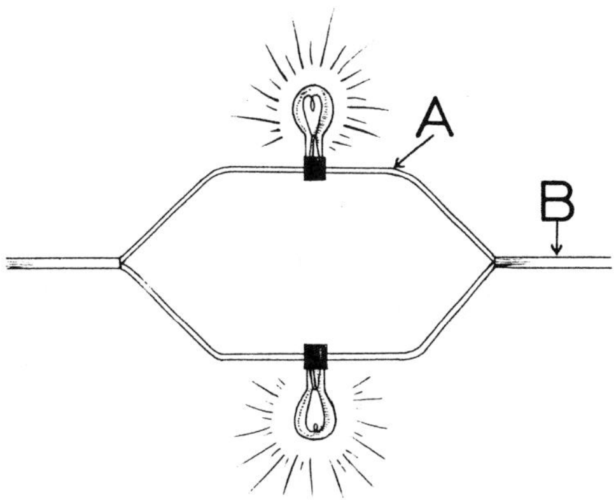

48

Which wire carries more current?

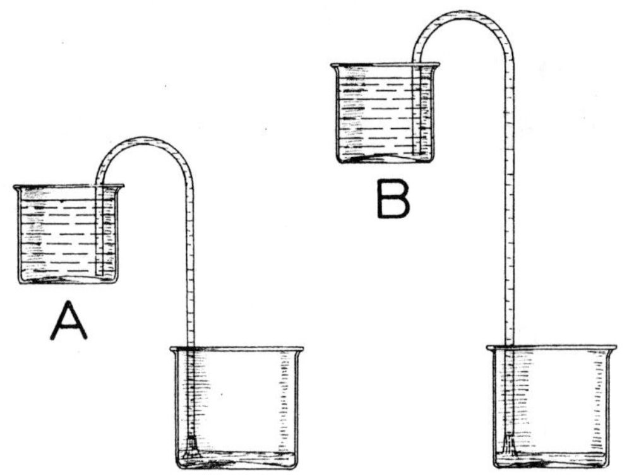

49

Which tank will empty faster?

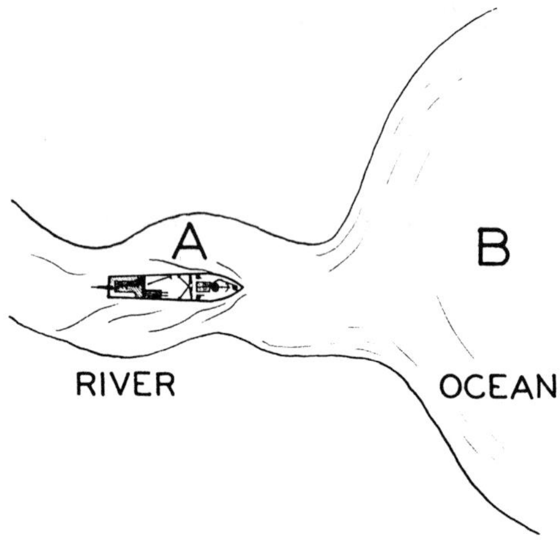

50

At which point will the boat be lower in the water?

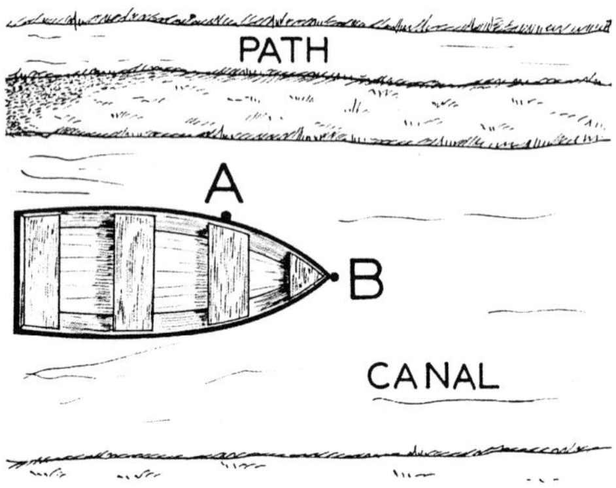

51

To pull this boat along the canal, at which point is it better to attach the rope?

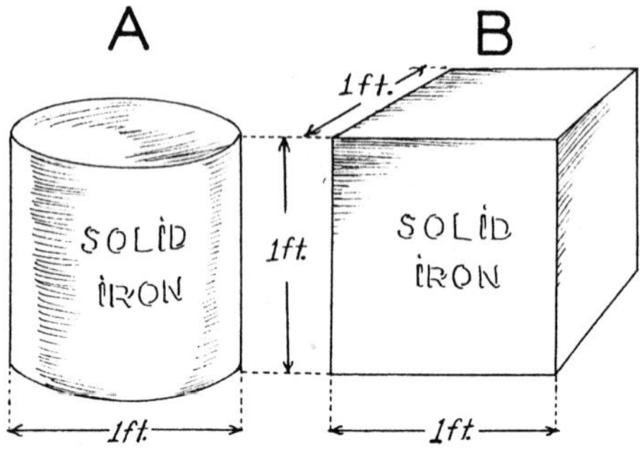

52

Which weighs more?

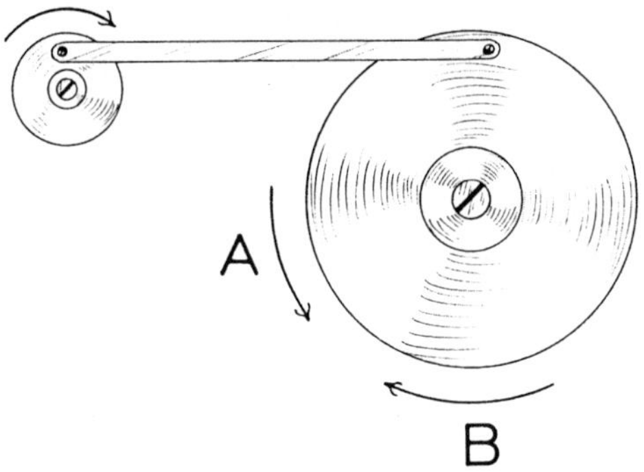

53

When the little wheel turns around, the big wheel will:

A—turn in direction A,

B—turn in direction B,

C—move back and forth.

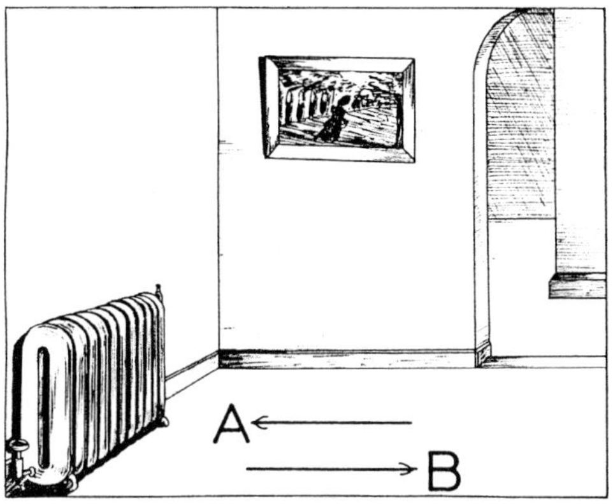

54

Which arrow shows the way the air will move along the floor when the radiator is turned on?

55

Which weighs more?

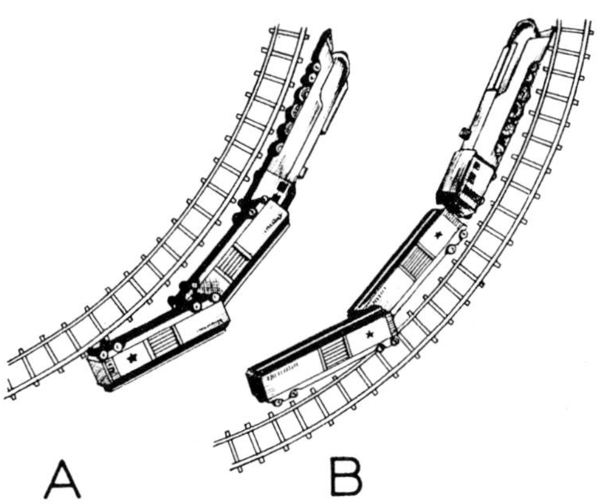

56

Which of these is the more likely picture of a train wreck?

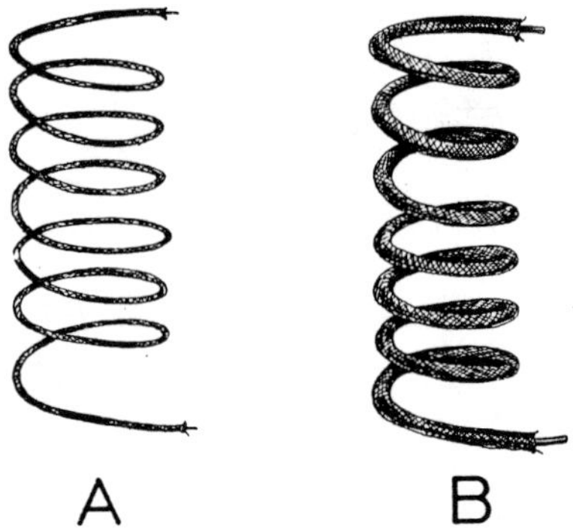

57

Which of these wires offers more resistance to the passage of an electric current?

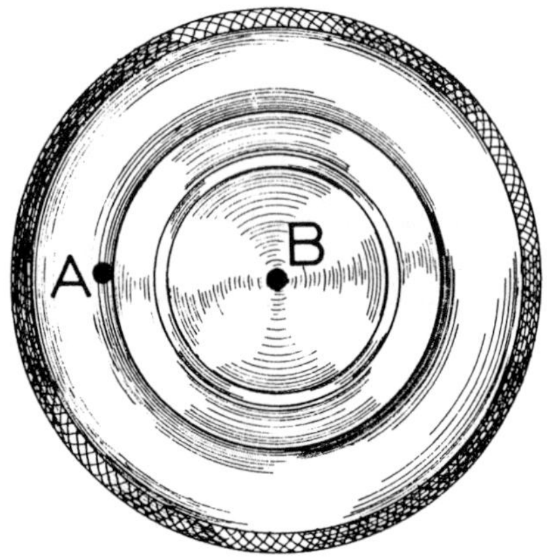

58

Which spot on the wheel travels faster?

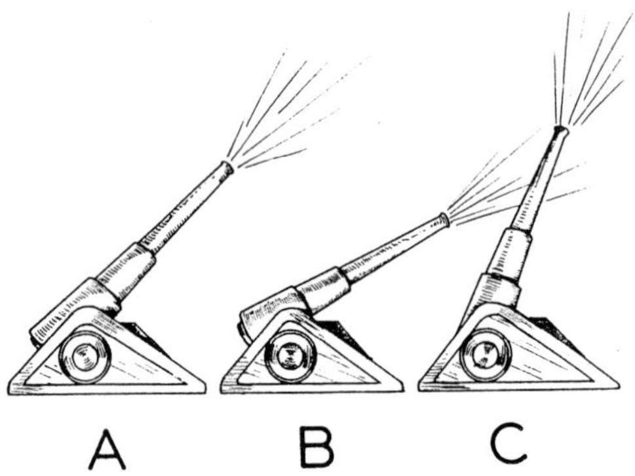

59

Which cannon will shoot farthest?

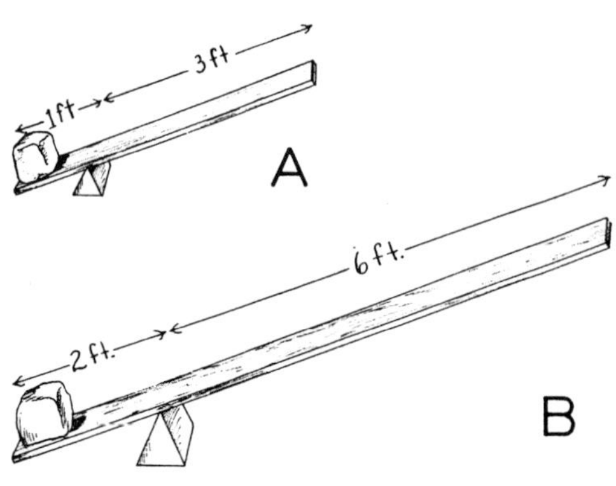

60

With which arrangement can a man lift the heavier weight?

KEY (CORRECT ANSWERS)

1.	C	41.	A
2.	C	42.	B
3.	A	43.	B
4.	B	44.	A
5.	A	45.	A
6.	C	46.	A
7.	B	47.	C
8.	B	48.	B
9.	B	49.	B
10.	B	50.	A
11.	B	51.	B
12.	C	52.	B
13.	A	53.	C
14.	B	54.	A
15.	B	55.	B
16.	B	56.	A
17.	A	57.	A
18.	A	58.	B
19.	B	59.	B
20.	A	60.	B
21.	A		
22.	A		
23.	B		
24.	B		
25.	A		
26.	B		
27.	A		
28.	A		
29.	B		
30.	B		
31.	B		
32.	A		
33.	C		
34.	A		
35.	A		
36.	A		
37.	C		
38.	B		
39.	A		
40.	A		

SPATIAL RELATIONS

EXAMINATION SECTION
TEST 1

DIRECTIONS: In each of questions 1 to 16 the front and top views of an object are given. Of the views labelled 1, 2, 3 and 4, select the one that CORRECTLY represents the right side view of each object for third angle projection.

1. 1.___

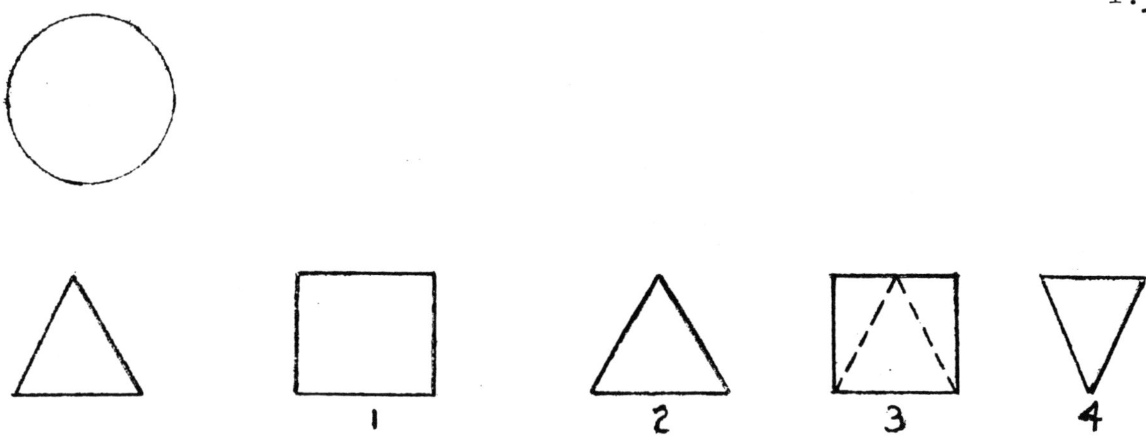

A. 1 B. 2 C. 3 D. 4

2. 2.___

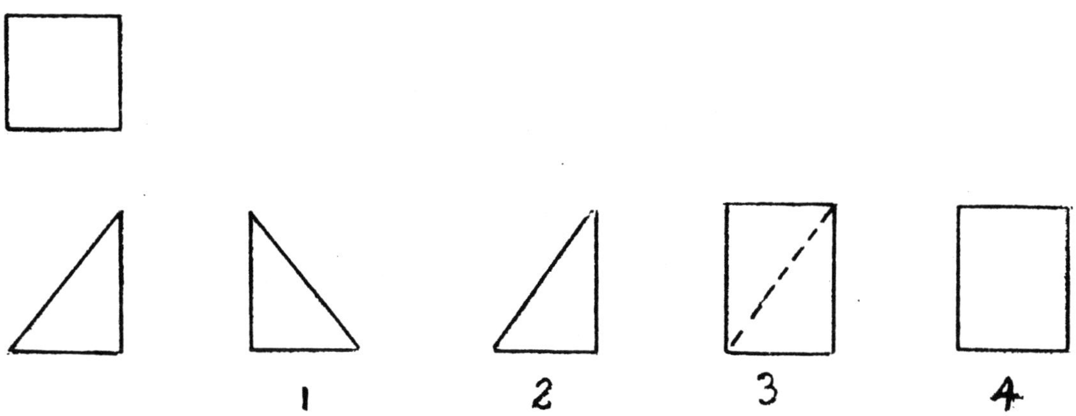

A. 1 B. 2 C. 3 D. 4

3. 3.____

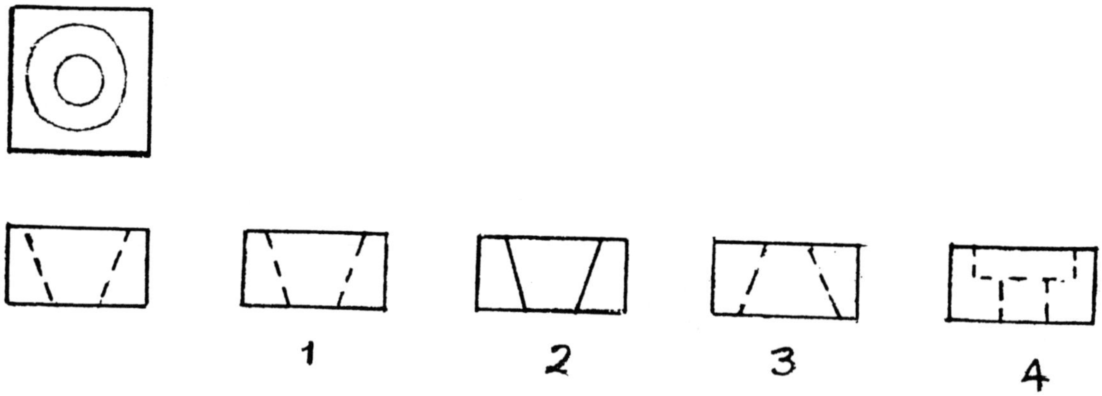

A. 1 B. 2 C. 3 D. 4

4. 4.____

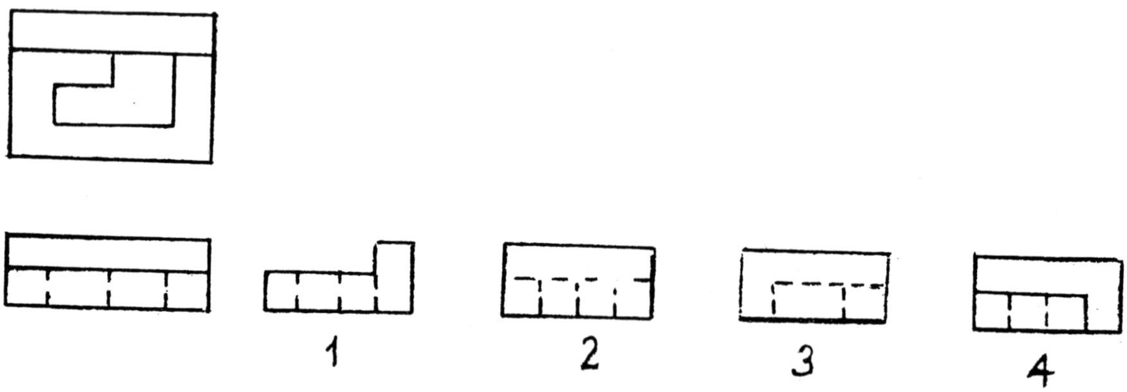

A. 1 B. 2 C. 3 D. 4

5. 5.___

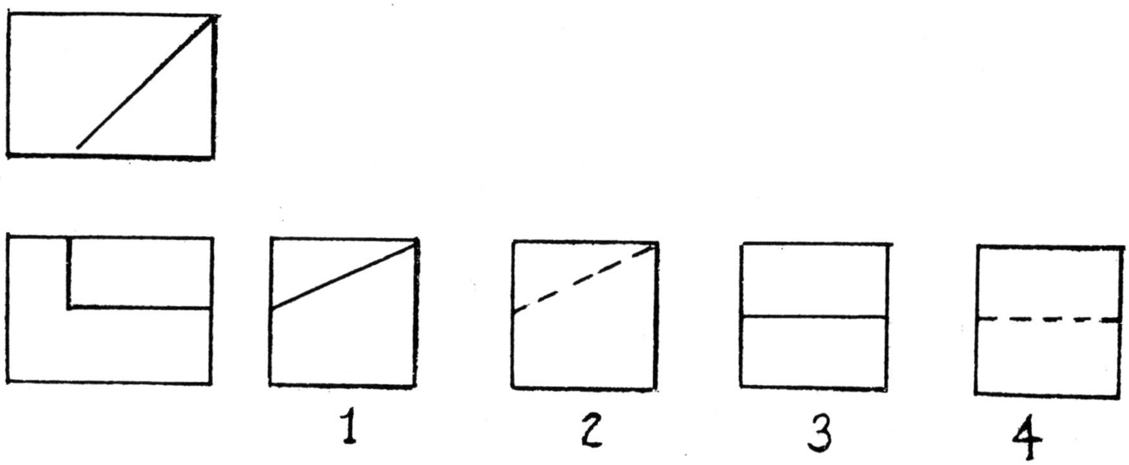

A. 1 B. 2 C. 3 D. 4

6. 6.___

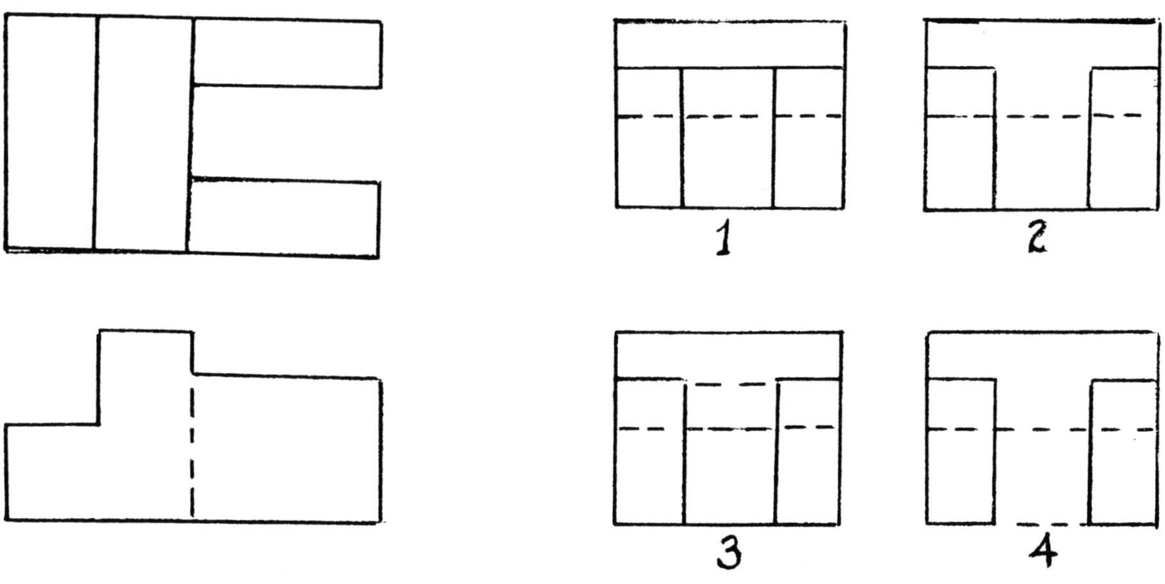

A. 1 B. 2 C. 3 D. 4

7. 7.___

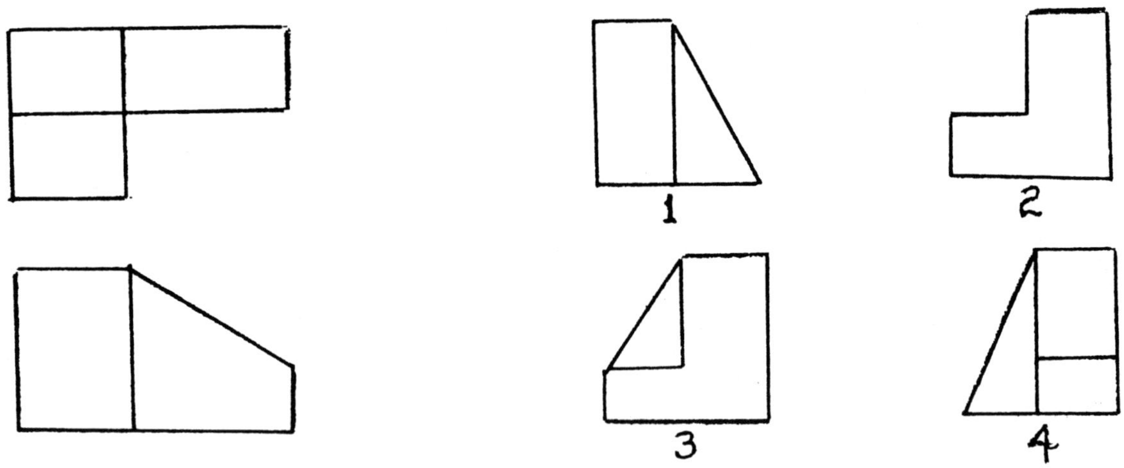

A. 1 B. 2 C. 3 D. 4

8. 8.___

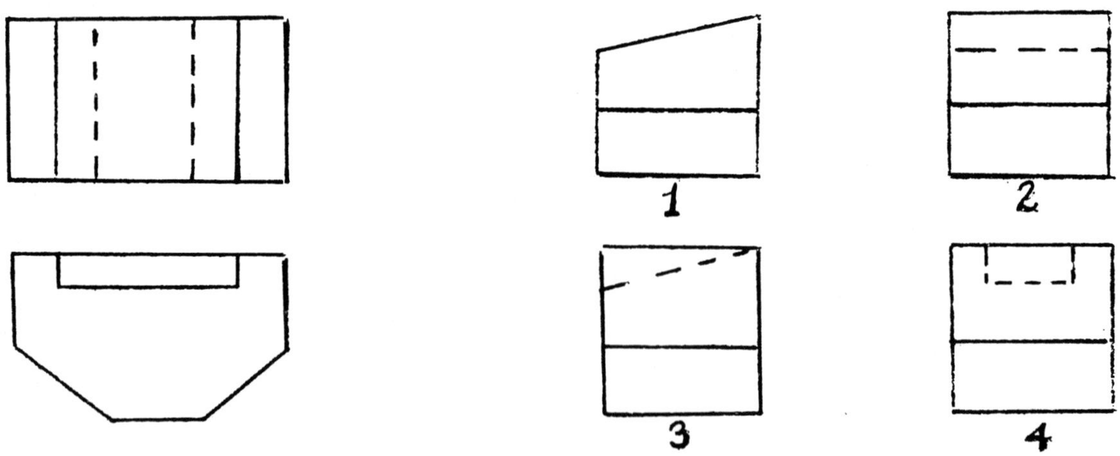

A. 1 B. 2 C. 3 D. 4

9. 9.___

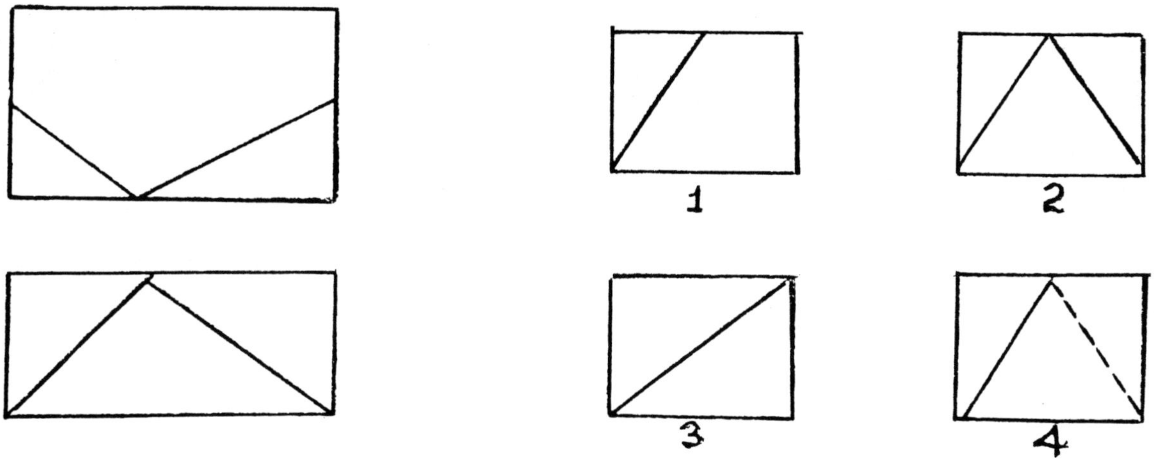

A. 1 B. 2 C. 3 D. 4

10. 10.___

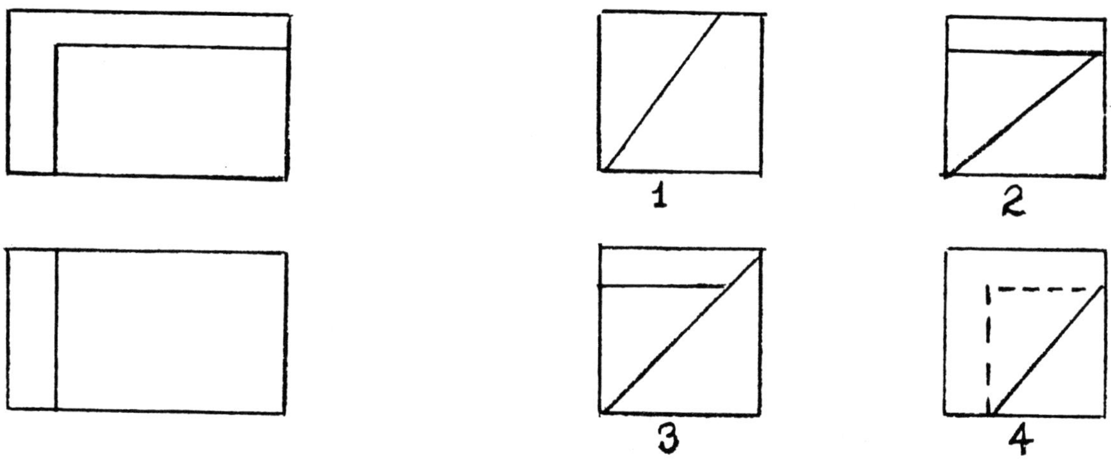

A. 1 B. 2 C. 3 D. 4

11. 11.___

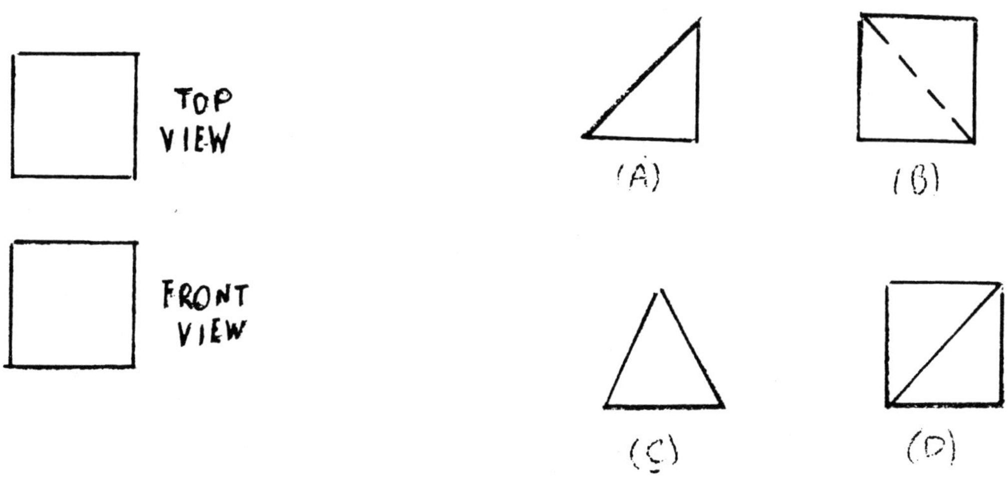

A. 1 B. 2 C. 3 D. 4

Questions 12 - 16

DIRECTIONS: In each of questions 12 to 16 inclusive, two views of an object are given. Of the views labelled 1, 2, 3, and 4, select the one that CORRECTLY represents the right side view of each object.

12. 12.___

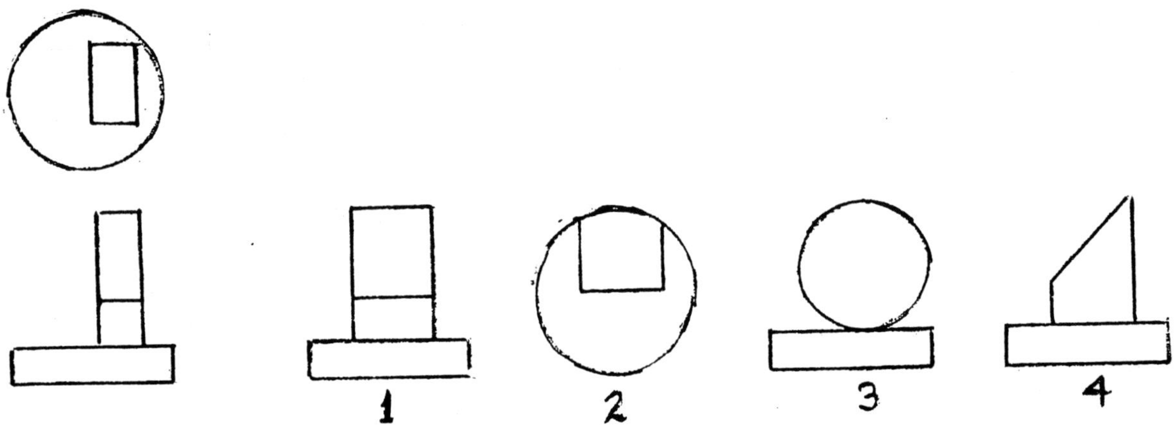

A. 1 B. 2 C. 3 D. 4

13. 13.___

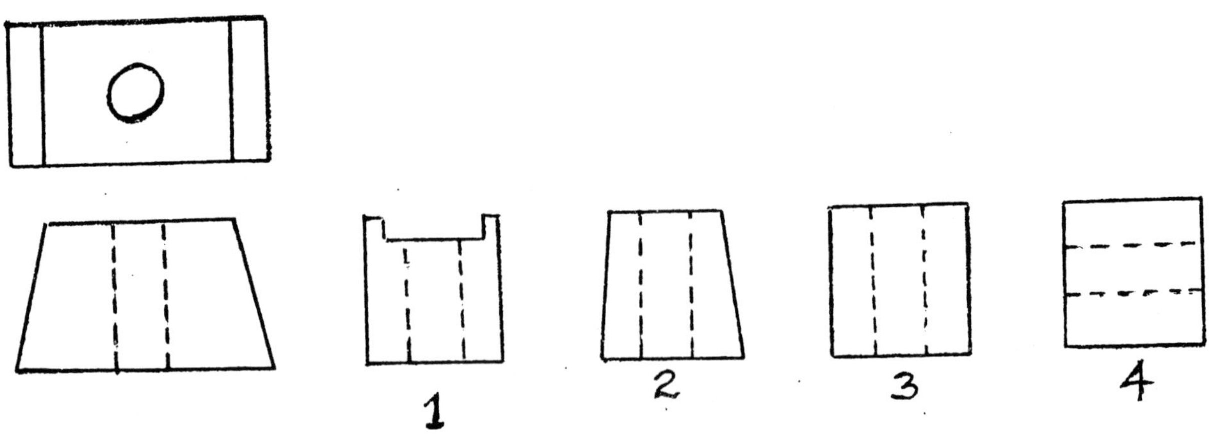

A. 1 B. 2 C. 3 D. 4

14. 14.___

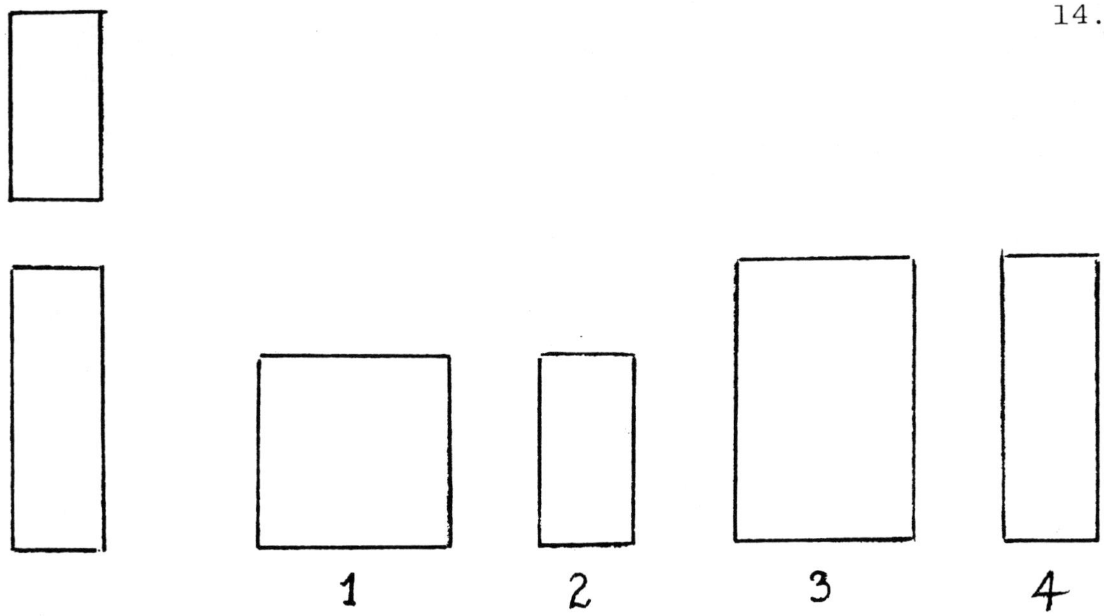

A. 1 B. 2 C. 3 D. 4

15. 15.___

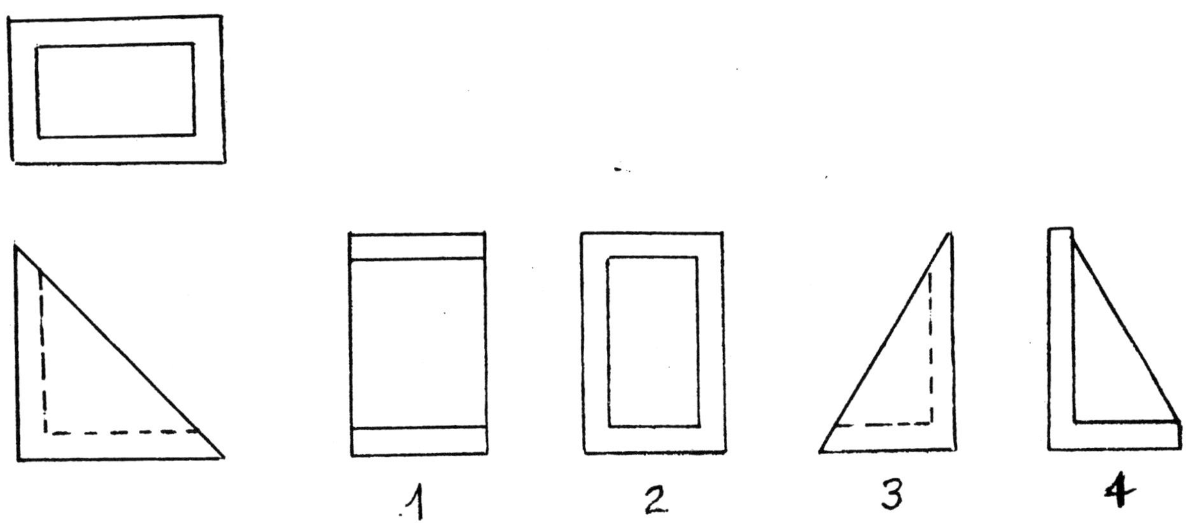

A. 1 B. 2 C. 3 D. 4

16. 16.___

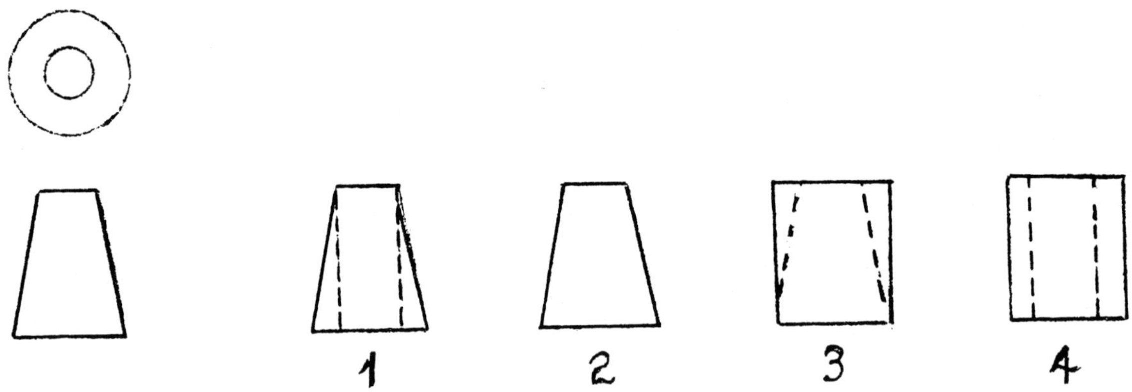

A. 1 B. 2 C. 3 D. 4

17. 17.___

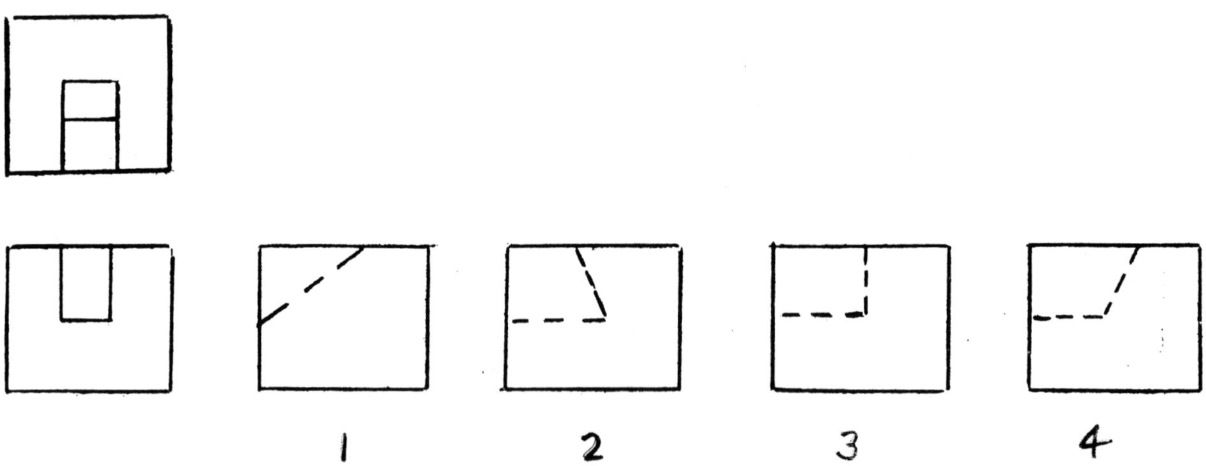

A. 1 B. 2 C. 3 D. 4

18. 18.___

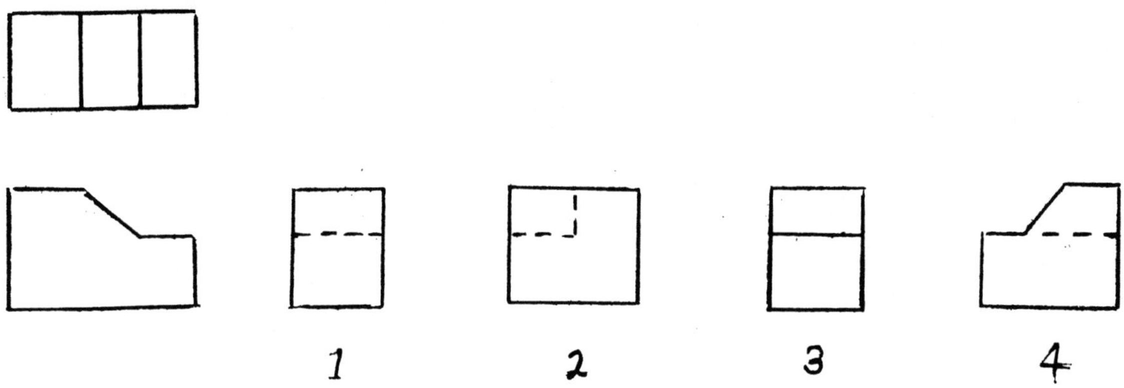

A. 1 B. 2 C. 3 D. 4

19. 19.___

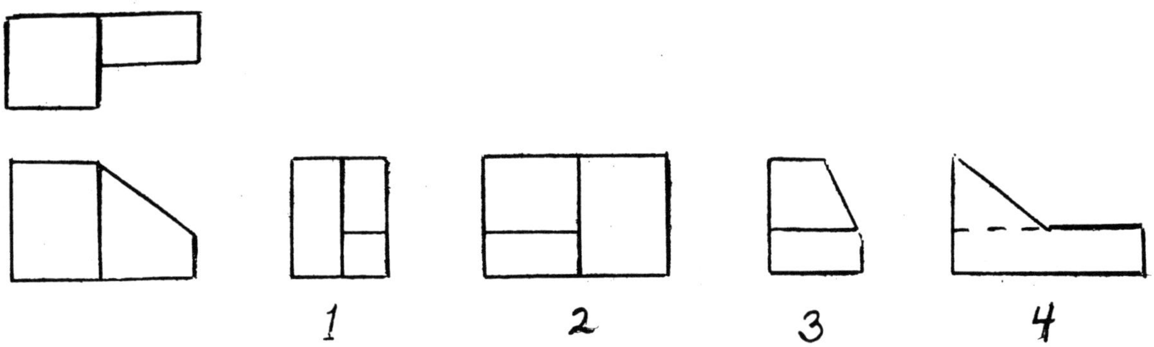

A. 1 B. 2 C. 3 D. 4

20. 20.___

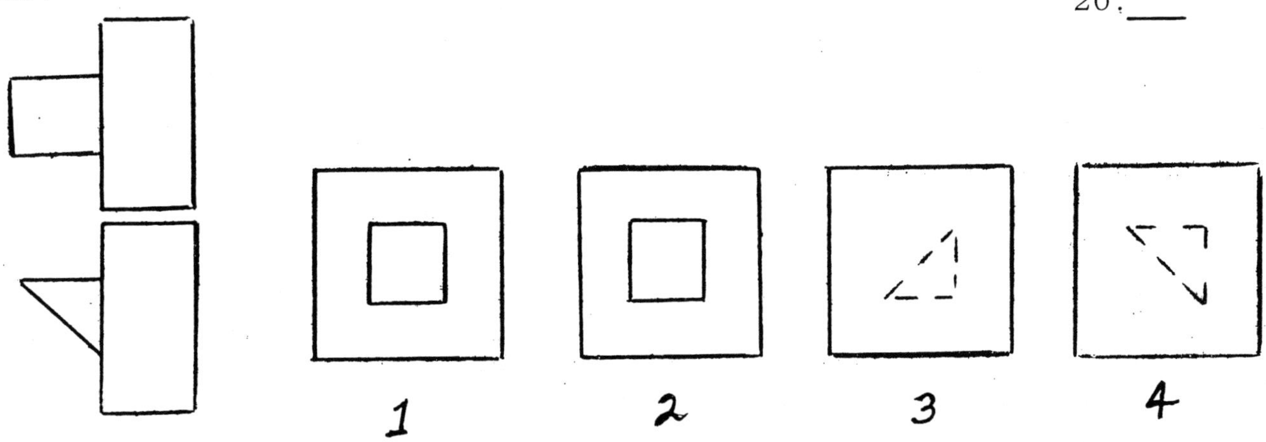

A. 1 B. 2 C. 3 D. 4

21. 21.___

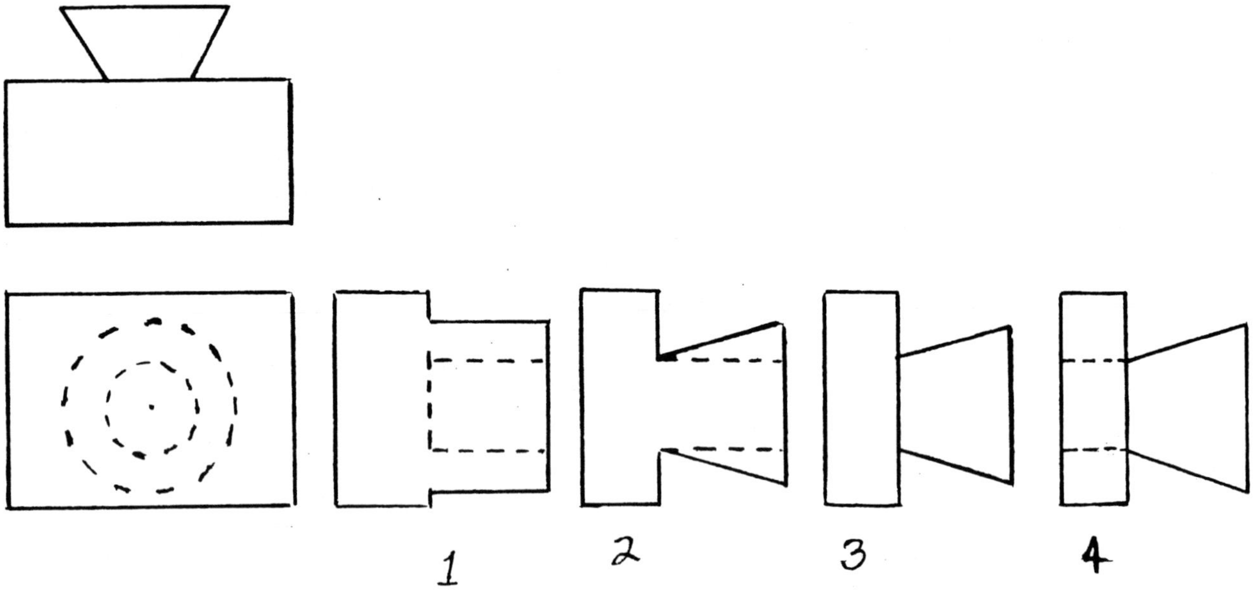

A. 1 B. 2 C. 3 D. 4

22. 22.___

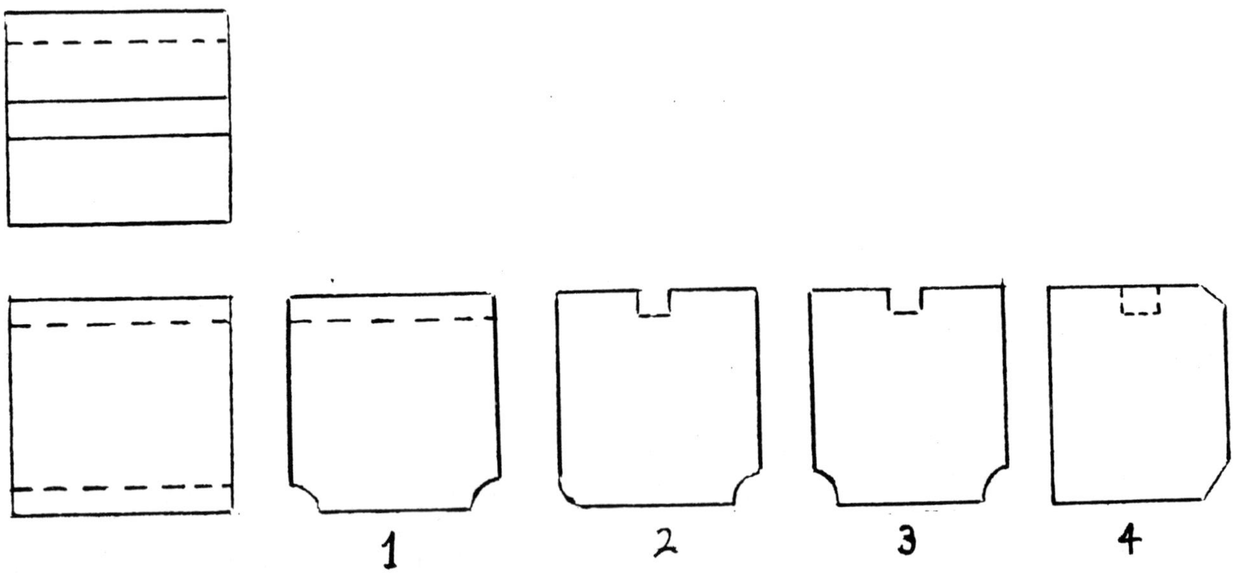

A. 1 B. 2 C. 3 D. 4

23. 23.___

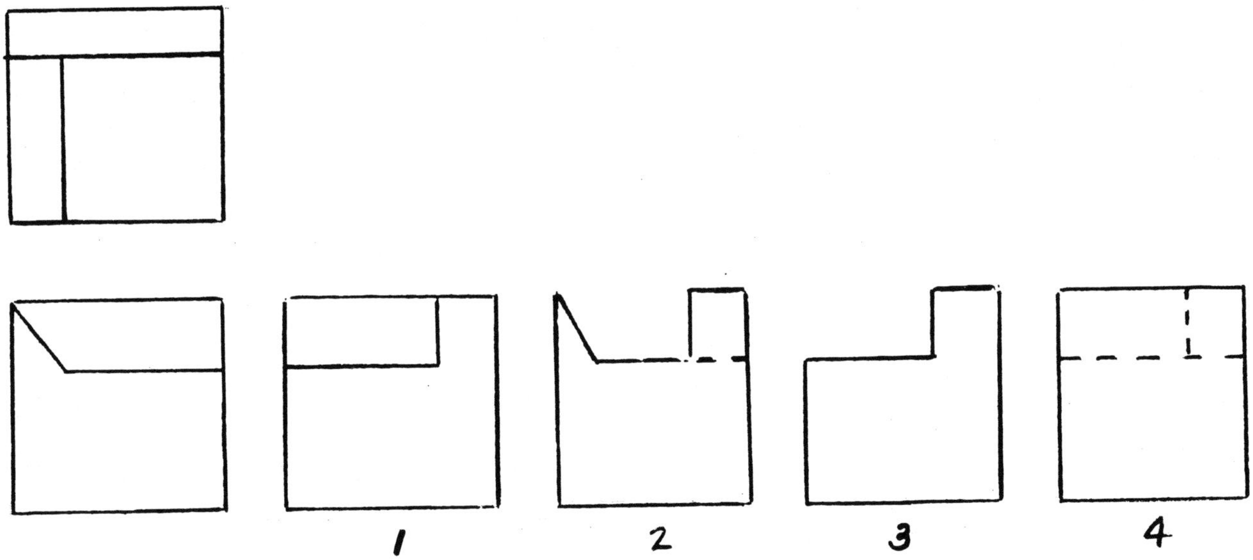

A. 1 B. 2 C. 3 D. 4

24. 24.___

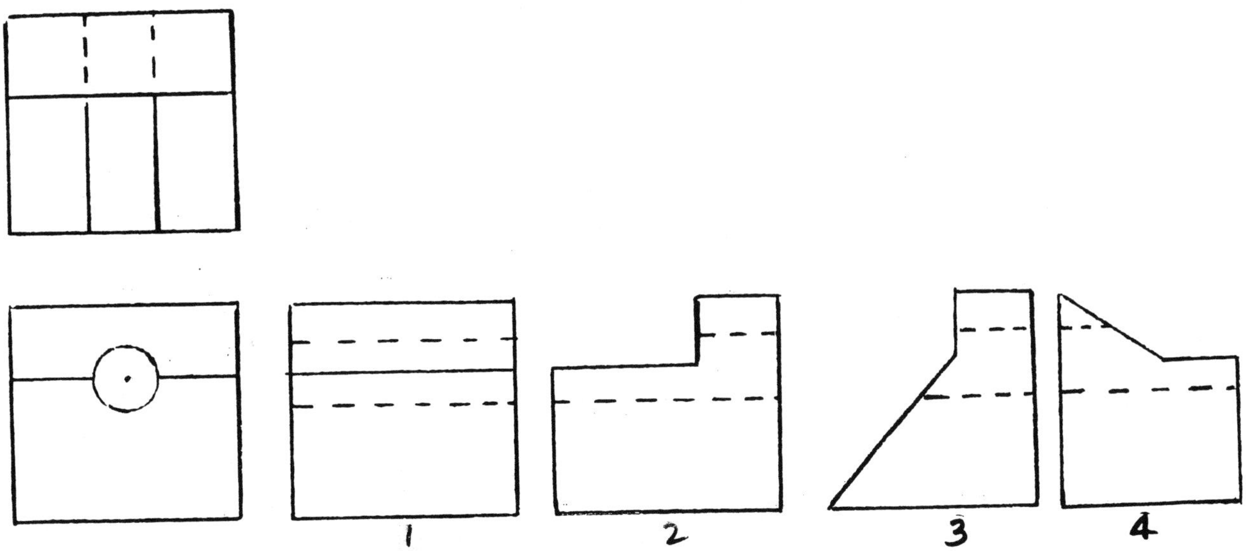

A. 1 B. 2 C. 3 D. 4

25. 25.___

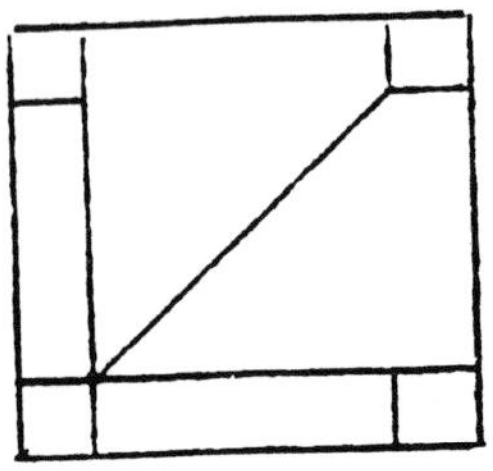

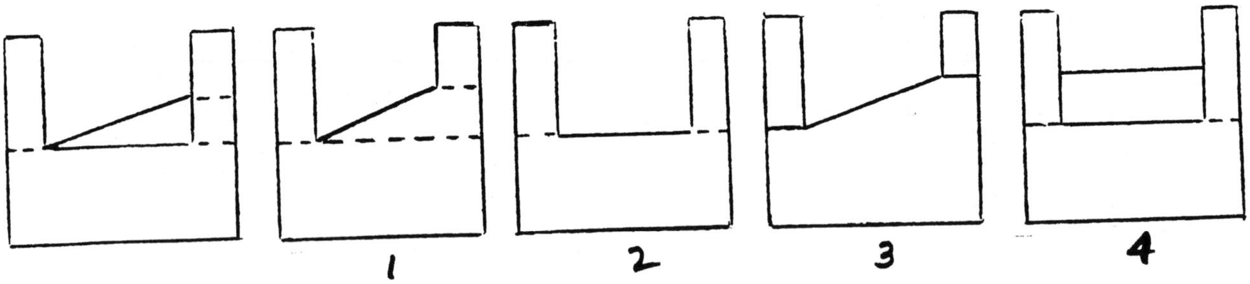

A. 1 B. 2 C. 3 D. 4

DIRECTIONS: In questions 26 through 30 which follow, the plan and front elevation of an object are shown on the left, and on the right are shown four figures one of which, and only one, represents the right side elevation. Mark in the space at the right the letter which represents the right side elevation. In the sample below, which figure correctly represents the right side elevation?

A. 1 B. 2 C. 3 D. 4

The correct answer is A

<u>Sample Question</u>

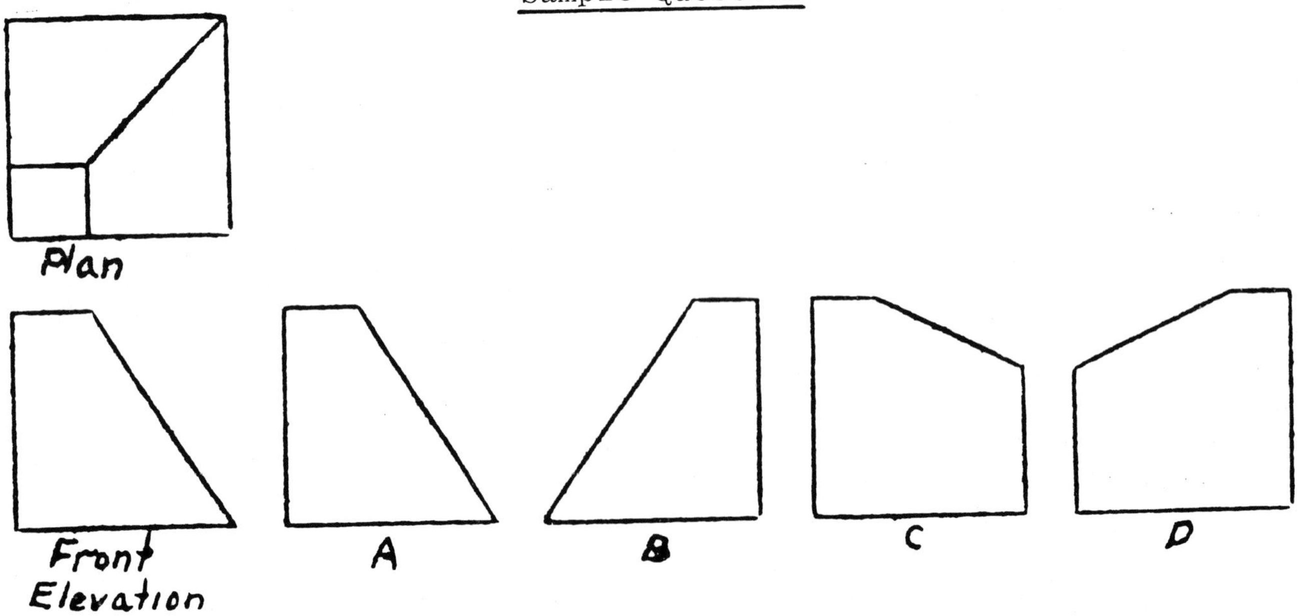

26. 26.____

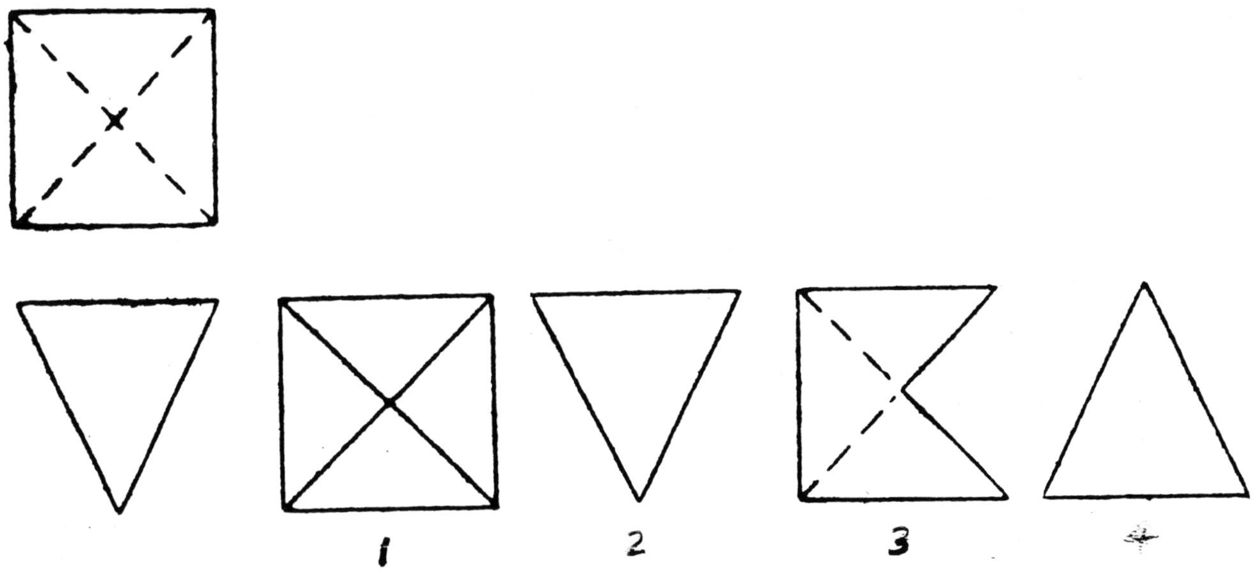

A. 1 B. 2 C. 3 D. 4

27. 27.____

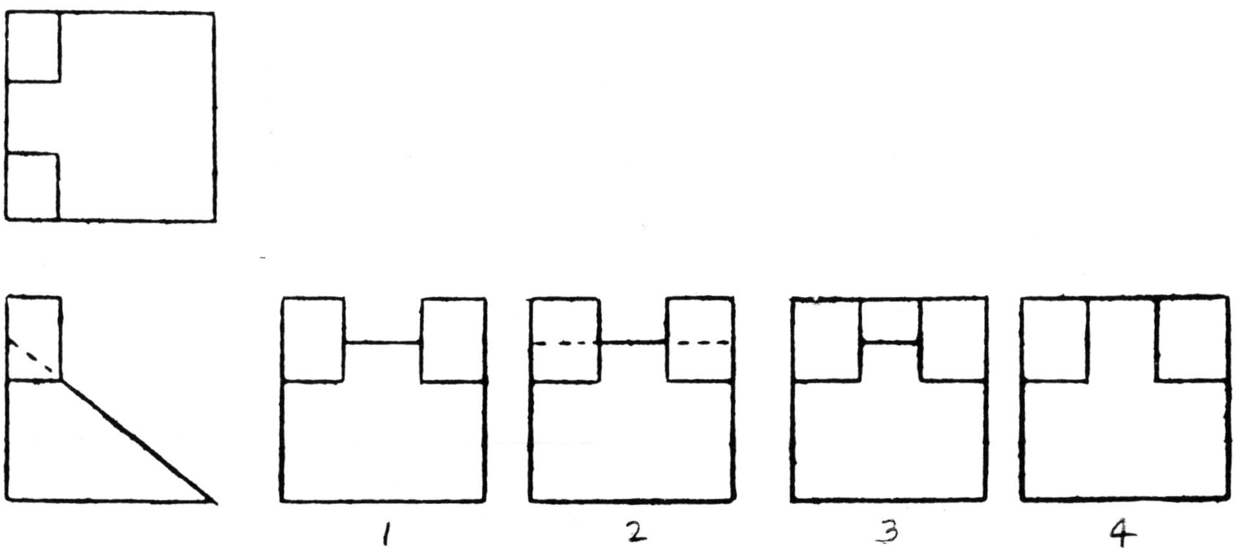

A. 1 B. 2 C. 3 D. 4

28. 28. ____

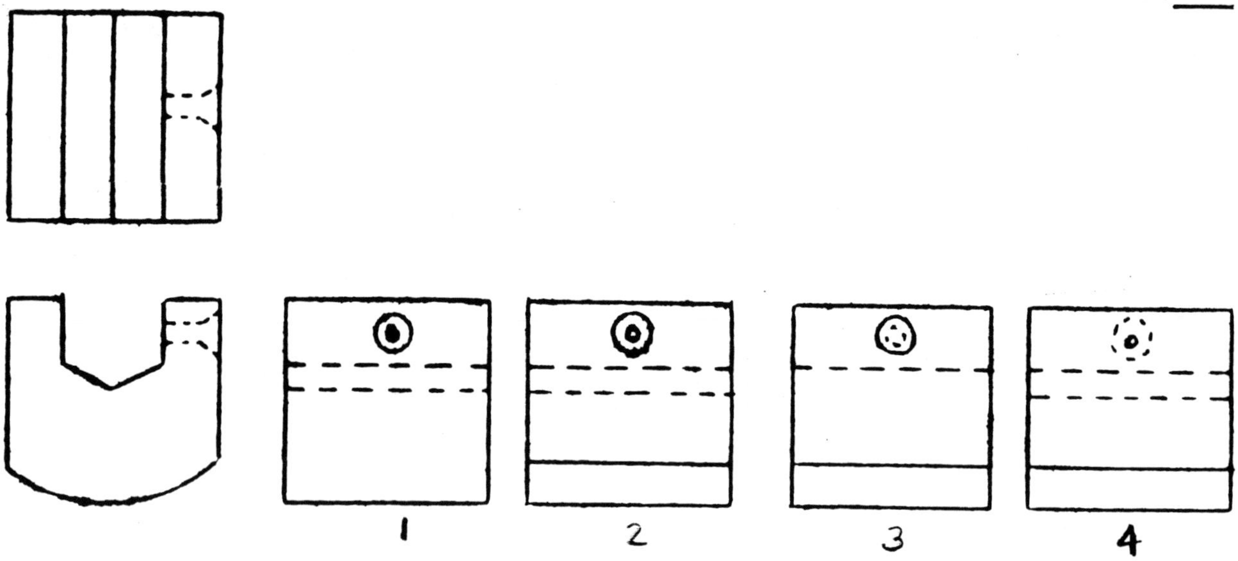

A. 1 B. 2 C. 3 D. 4

29. 29. ____

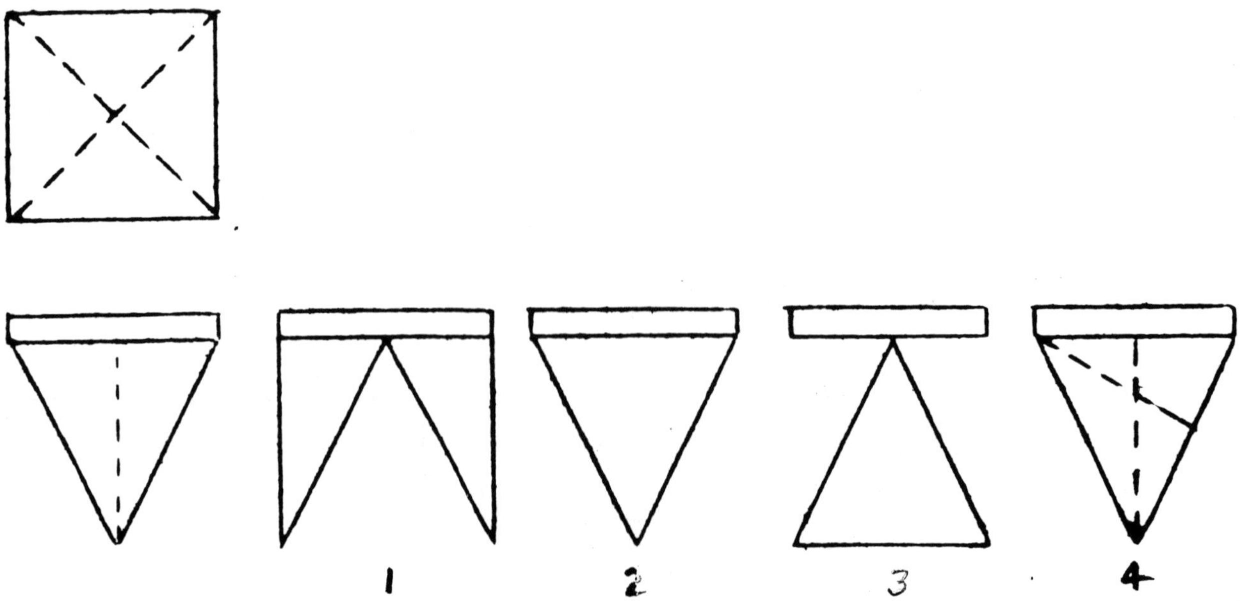

A. 1 B. 2 C. 3 D. 4

30. 30.___

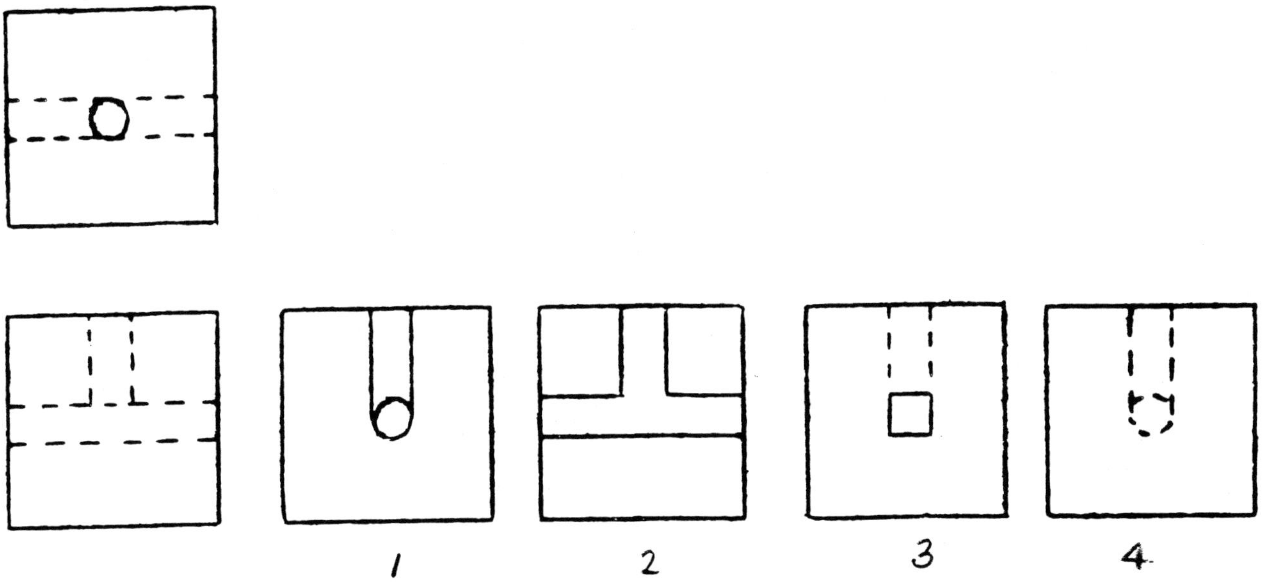

A. 1 B. 2 C. 3 D. 4

KEY (CORRECT ANSWERS)

1. B
2. D
3. A
4. A
5. C

6. B
7. D
8. C
9. A
10. A

11. A
12. D
13. C
14. C
15. B

16. B
17. D
18. C
19. A
20. B

21. C
22. B
23. A
24. B
25. A

26. B
27. A
28. B
29. A
30. C

TEST 2

DIRECTIONS: Questions 1 through 10 deal with relationships between sets of figures. For each question, select that choice (A, or B, or C, or D) which has the SAME relationship to Figure 3 that Figure 2 has to Figure 1.

SAMPLE: Study Figures 1 and 2 in the SAMPLE. Notice that Figure 1 has been turned clockwise 1/4 of a turn to get Figure 2. Taking Figure 3 and turning it clockwise 1/4 of a turn, we get choice A, the correct answer.

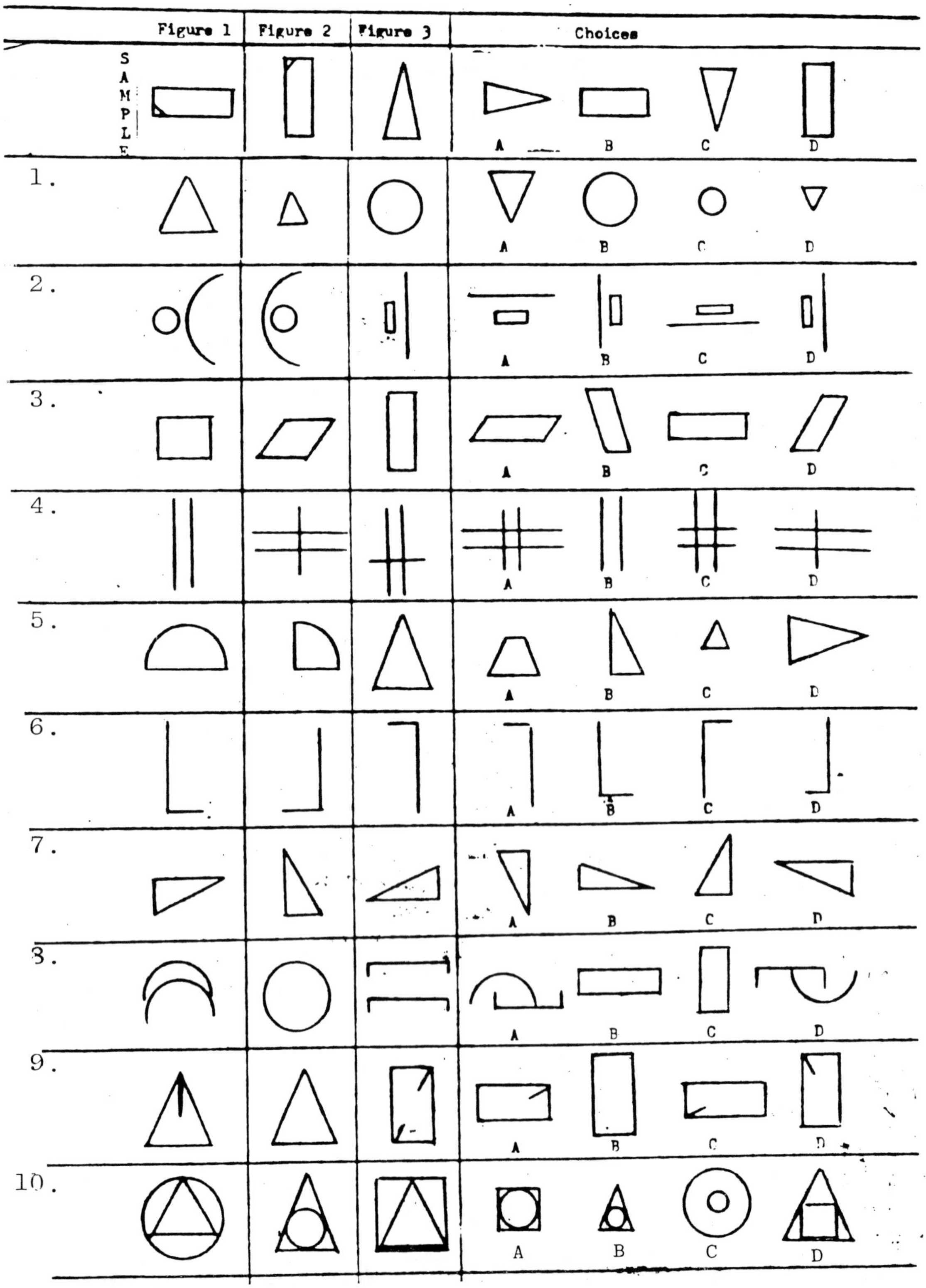

1. ____
2. ____
3. ____
4. ____
5. ____
6. ____
7. ____
8. ____
9. ____
10. ____

DIRECTIONS: Questions 11 through 16 show the top view of an object in the first column, the front view of the same object in the second column and four drawings in the third column, one of which correctly represents the RIGHT side view of the object.
Select the CORRECT right side view.
As a guide, the first one is an illustrative example, the correct answer of which is C.

TOP VIEW | FRONT VIEW | RIGHT SIDE VIEW

SAMPLE

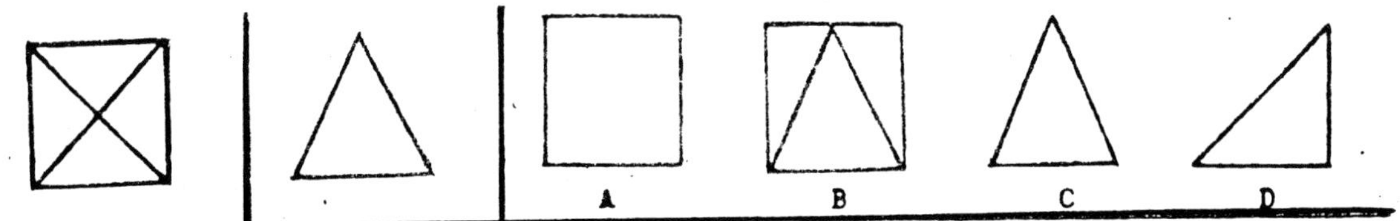

11\. 11. ____

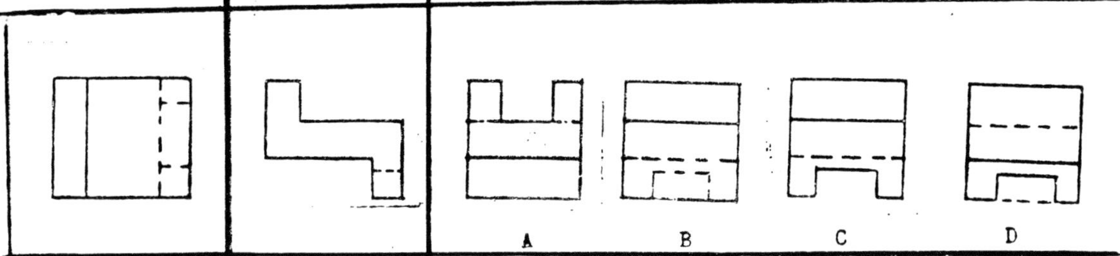

12\. 12. ____

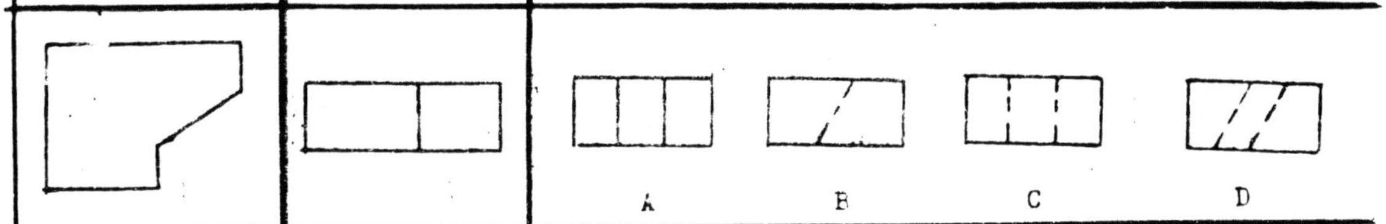

13\. 13. ____

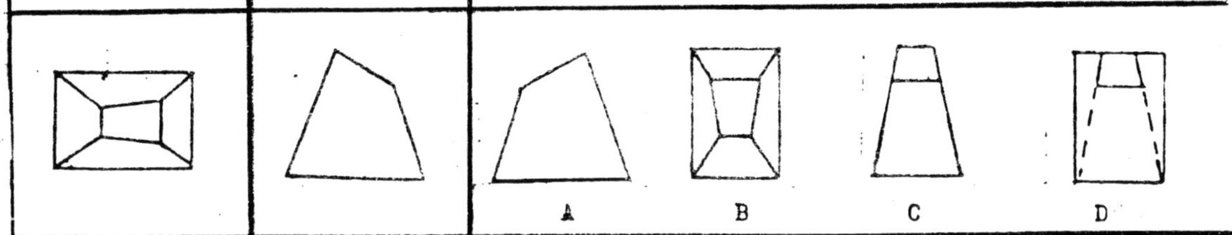

14\. 14. ____

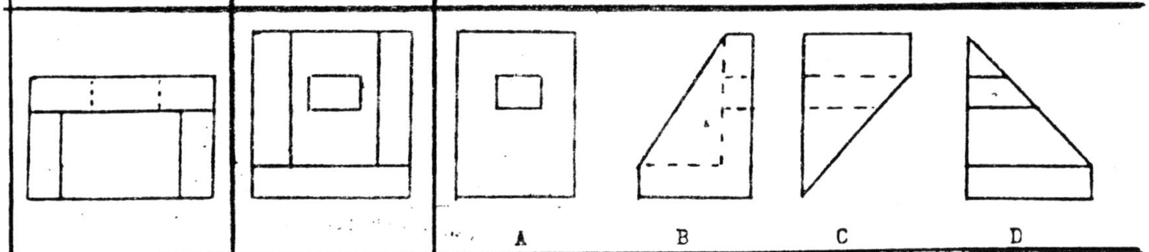

15\. 15. ____

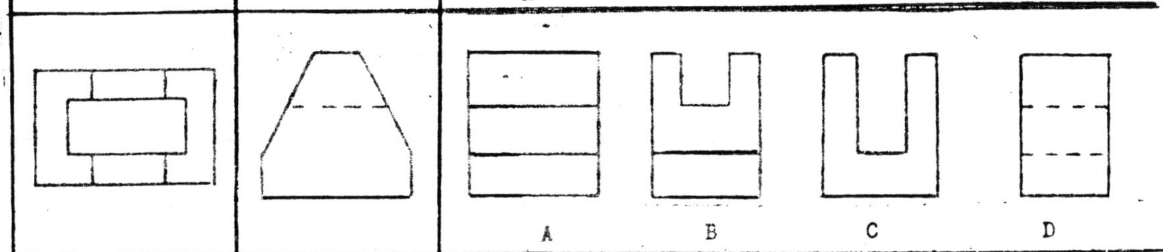

16\. 16. ____

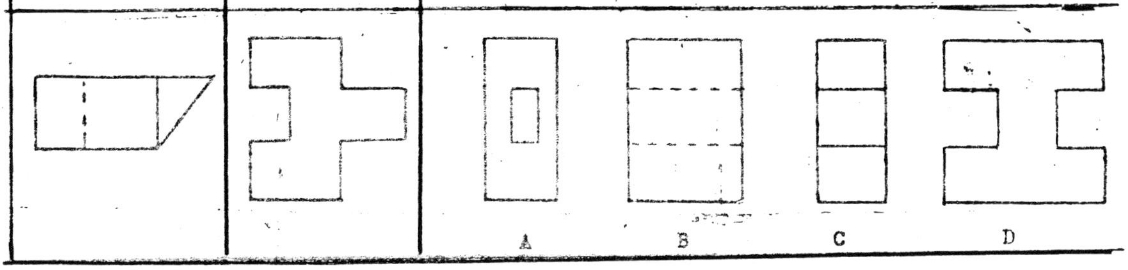

DIRECTIONS: Each question or incomplete statement is followed by several suggested answers or completions. Select the one that *BEST* answers the question or completes the statement. *PRINT THE LETTER OF THE CORRECT ANSWER IN THE SPACE AT THE RIGHT.*

Questions 17 - 20

DIRECTIONS: In each of the following groups of drawings, the top view and front elevation of an object are shown on the left. At the right are four drawings, one of which represents the end elevation of the object as seen from the right. Select the drawing which represents the correct end elevation and mark in the answer space the letter of this drawing

The first group is shown as an example only.

The correct answer in this group is C.

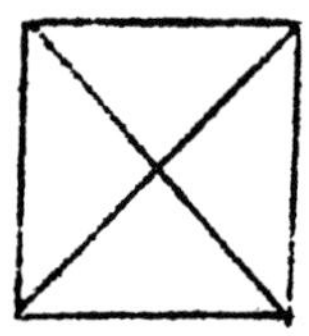

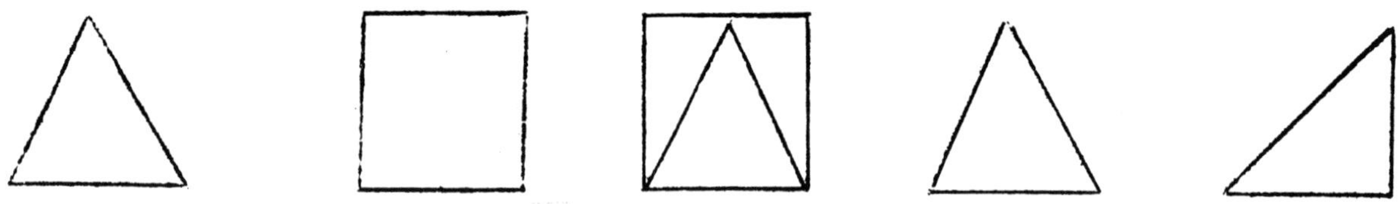

A. B. C. D.

17. 17.____

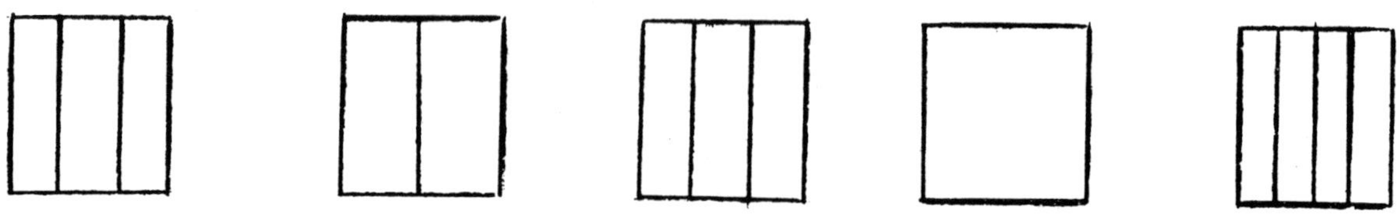

A. B. C. D.

18. 18.____

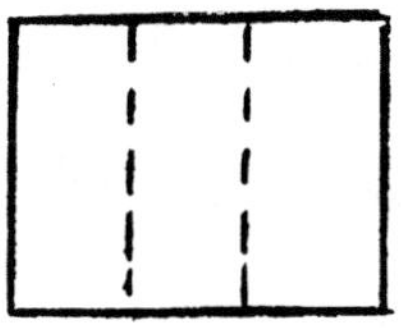

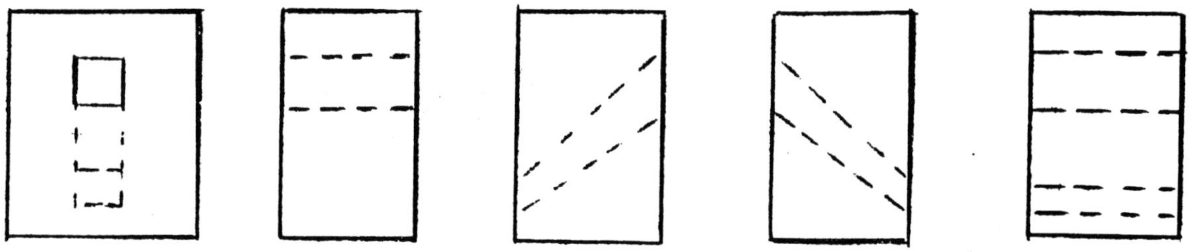

A. B. C. D.

19. 19.____

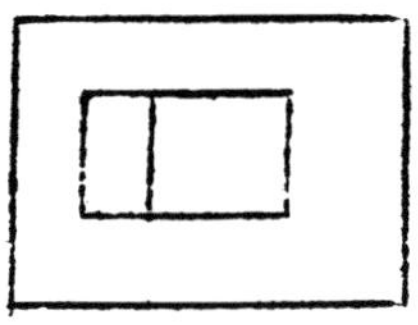

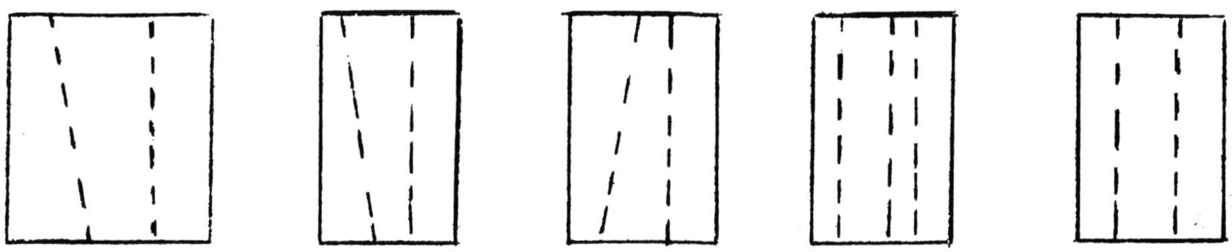

A. B. C. D.

20. 20.____

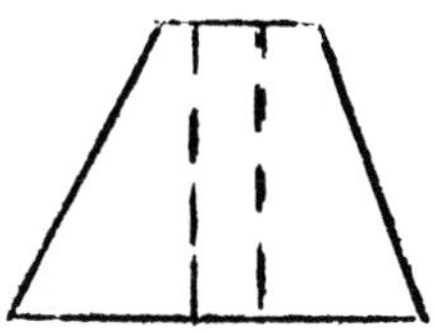

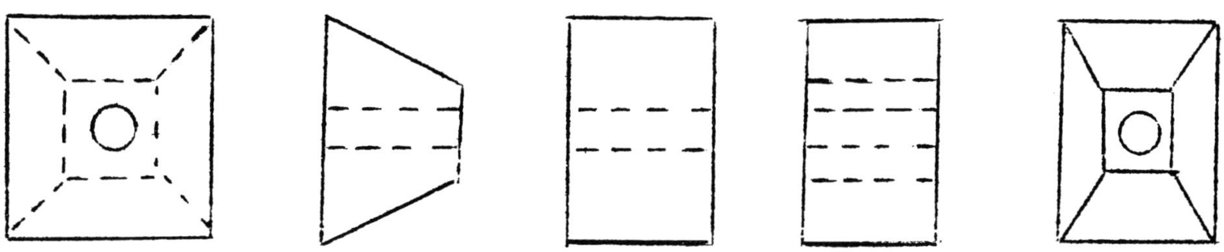

A. B. C. D.

KEY (CORRECT ANSWERS)

1.	C	11.	C
2.	B	12.	A
3.	D	13.	C
4.	A	14.	B
5.	B	15.	B
6.	C	16.	C
7.	A	17.	A
8.	B	18.	C
9.	B	19.	D
10.	D	20.	A

ABSTRACT REASONING
EXAMINATION SECTION

COMMENTARY

Since intelligence exists in many forms or phases and the theory of differential aptitudes is now firmly established in testing, other manifestations and measurements of intelligence than verbal or purely arithmetical must be identified and measured.

Classification inventory, or figure classification, involves the aptitude of form perception, i.e., the ability to perceive pertinent detail in objects or in pictorial or graphic material. It involves making visual comparisons and discriminations and discerning slight differences in shapes and shading figures and widths and lengths of lines.

Leading examples of presentation are the figure analogy and the figure classification. The Section that follows presents progressive and varied samplings of this type of question.

SAMPLE QUESTIONS

DIRECTIONS: In each of these sample questions, look at the symbols in the first two boxes. Something about the three symbols in the first box makes them alike; something about the two symbols in the other box with the question mark makes them alike. Look for some characteristic that is common to all symbols in the same box, yet makes them different from the symbols in the other box. Among the five answer choices, find the symbol that can BEST be substituted for the question mark, because it is *like* the symbols in the second box, and, *for the same reason,* different from those in the first box.

1.

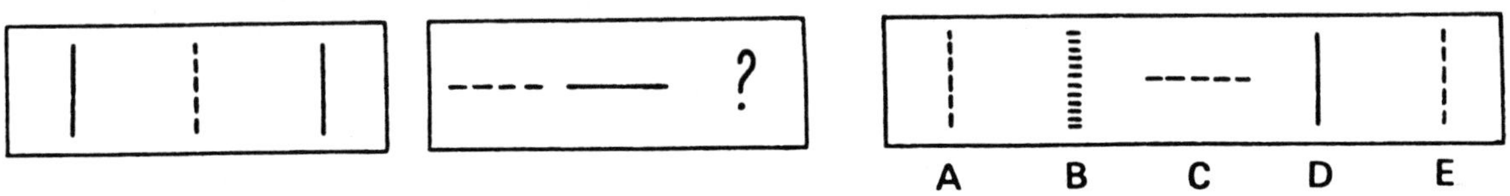

In sample question 1, all the symbols in the first box are vertical lines. The second box has two lines, one broken and one solid. Their *likeness* to each other consists in their being horizontal; and their being horizontal makes them *different* from the vertical lines in the other box. The answer must be the only one of the five lettered choices that is a horizontal line, either broken or solid. Therefore, the CORRECT answer is C.

2.

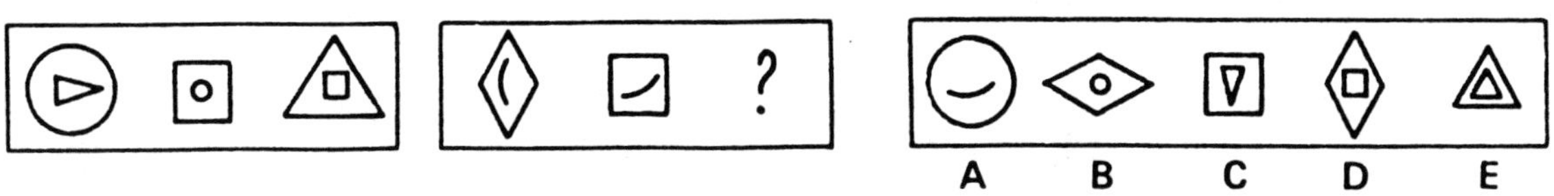

The CORRECT answer is A.

EXAMINATION SECTION

DIRECTIONS FOR THIS SECTION: In each of these questions, look at the symbols in the first two boxes. Something about the three symbols in the first box makes them alike; something about the two symbols in the other box with the question mark makes them alike. Look for some characteristic that is common to all symbols in the same box, yet makes them different from the symbols in the other box. Among the five answer choices, find the symbol that can BEST be substituted for the question mark, because it is *like* the symbols in the second box, and, *for the same reason,* different from those in the first box.

TEST 1

1. 

A B C D E

2. 

A B C D E

3. 

A B C D E

4. 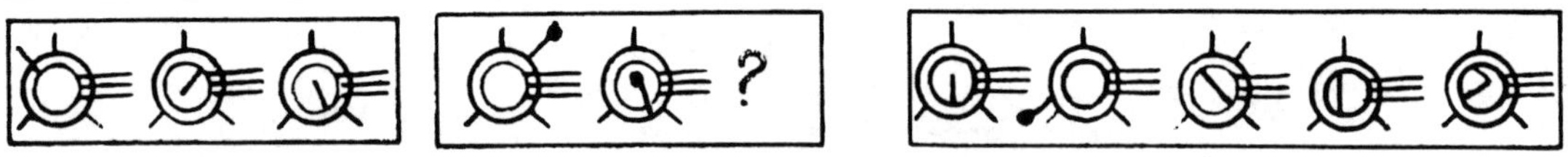

A B C D E

5. 

A B C D E

6. 

A B C D E

7. 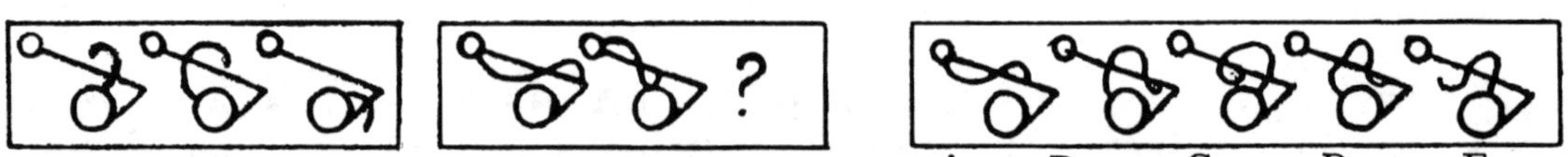

A B C D E

8. 

A B C D E

9. 

A B C D E

10. 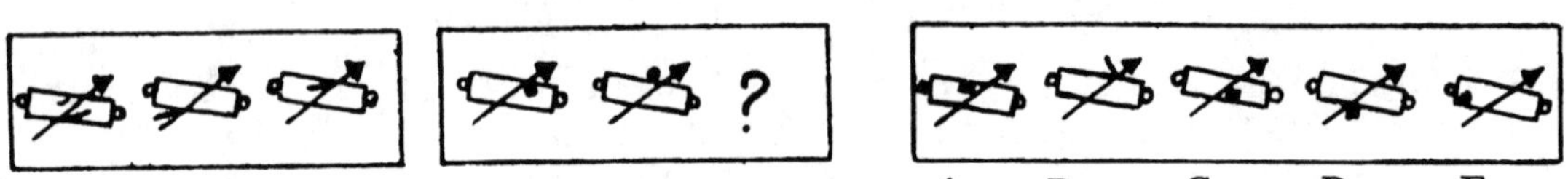

A B C D E

TEST 2

1. 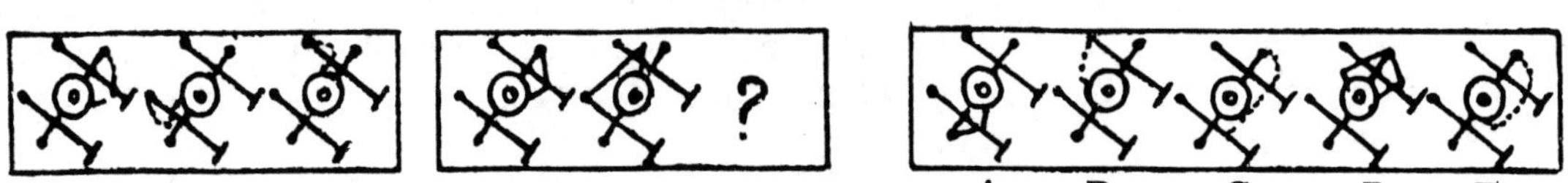

A B C D E

2. 

A B C D E

3. 

A B C D E

4. 

A B C D E

5. 

A B C D E

6. 

A B C D E

7. 

A B C D E

8. 

A B C D E

9. 

A B C D E

10. 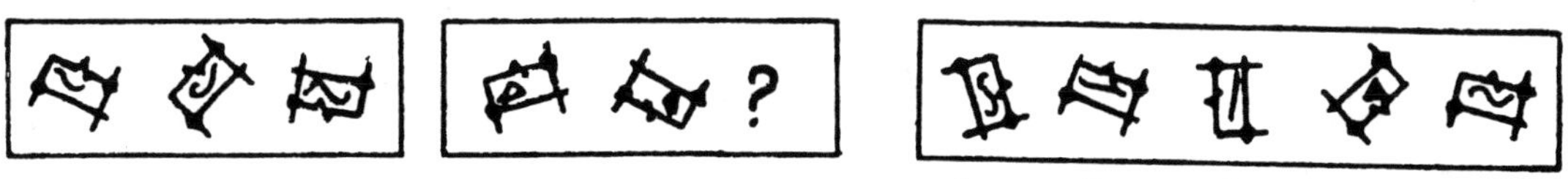

A B C D E

TEST 3

1. 

A B C D E

2. 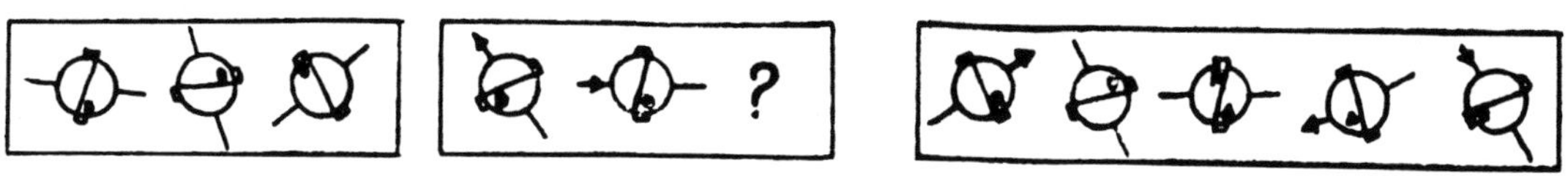

A B C D E

3.

4.

5.

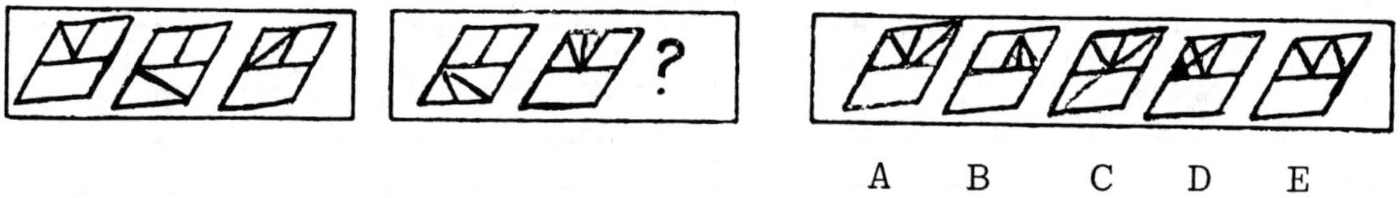

6.

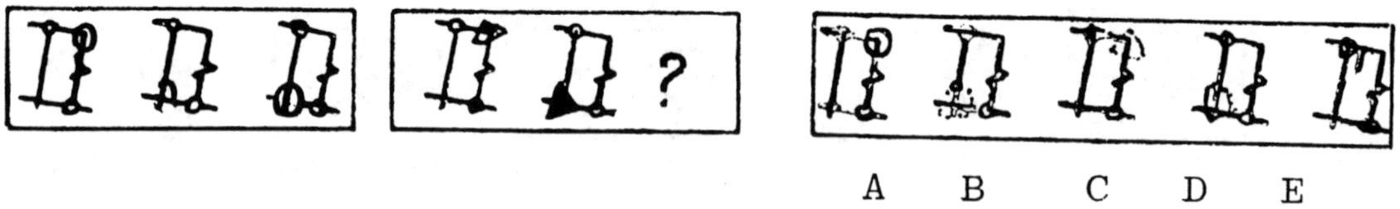

7.

8.

9.

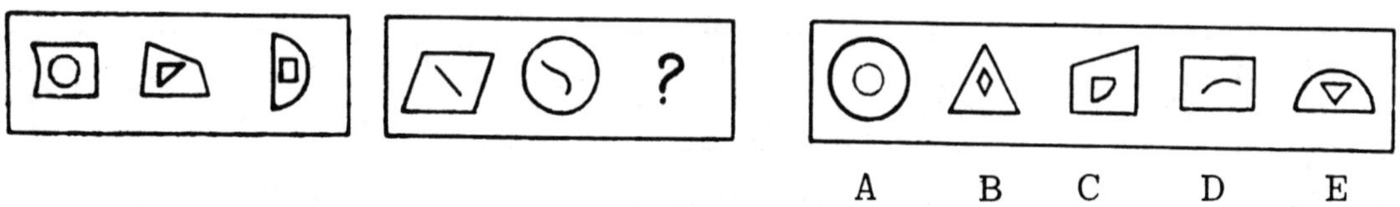

10.

KEY (CORRECT ANSWERS)

TEST 1		TEST 2		TEST 3	
1.	B	1.	A	1.	B
2.	C	2.	A	2.	E
3.	C	3.	A	3.	C
4.	B	4.	D	4.	A
5.	D	5.	E	5.	B
6.	B	6.	D	6.	C
7.	A	7.	D	7.	C
8.	C	8.	C	8.	B
9.	B	9.	E	9.	D
10.	D	10.	D	10.	B

BOOLEAN ALGEBRA

CONTENTS

BOOLEAN ALGEBRA

The father of Boolean algebra was George Boole, who was an English logician and mathematician. In the spring of 1847, he wrote a pamphlet on symbolic logic. Later he wrote a much larger text on which are founded the mathematical theories of logic. He did not regard logic as a branch of mathematics, but he did point out that a close analogy between symbols of algebra and those symbols which he devised to represent logical forms does exist.

Boolean algebra lay almost dormant until 1937 when Boole's algebra was used to write symbolic analyses of relay and switching circuits. Boolean algebra has now become an important subject to be learned in order to understand electronic computer circuits.

CLASSES AND ELEMENTS

We have previously determined that in our universe we can logically visualize two divisions; all things of interest in any discussion are in one division, and all other things not of interest are in the other division. These two divisions comprise a set or class called the UNIVERSAL CLASS. All objects contained in the universal class are called ELEMENTS. We also identify a set or class containing no elements; this class is called the NULL CLASS.

If we group some elements of the universal class together to form the combinations which are possible in a particular discussion, we call each of these combinations a class. In Boolean logic, these combinations called classes should not be confused with the null class or universal class. Actually, these classes are subclasses of the universal class. It should also be noted that the elements and classes in Boolean algebra are the sets and subsets previously discussed.

Each class is dependent upon its elements and the possible states (stable, nonstable, or both) that the elements can take.

Boolean algebra is that algebra which is based on Boolean logic and concerned with all elements having only two possible stable states and no unstable states.

To determine the number of classes or combinations of elements in Boolean algebra, we solve for the numerical value of 2^n where n equals the number of elements. If we have two elements (each element has two possible states) then we have 2^n or 2^2 possible classes. If we let the elements be A and B, then A may be true or false and B may be true or false. The classes which could be formed are as follows:

A true and B false
A true and B true
A false and B true
A false and B false

where we use the connective word "and." We could also form classes by use of the connective word "or" which would result in a different form of classes.

VENN DIAGRAMS

Since the Venn diagram is a topographical picture of logic, composed of the universal class divided into classes depending on the n number of elements, we show this logic as follows.

We may consider the universal class as containing submarines and atomic powered sound sources. Let A equal submarines and B equal atomic powered sound sources. Therefore, we have four classes which are:

(1) Submarines and not atomic
(2) Submarines and atomic
(3) Atomic and not submarines
(4) Not submarines and not atomic

A diagram of these classes is

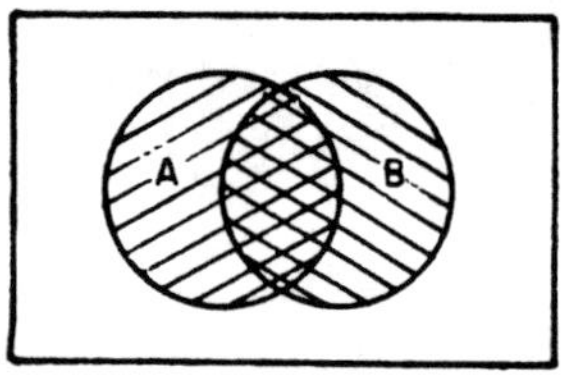

We may show these classes separately by

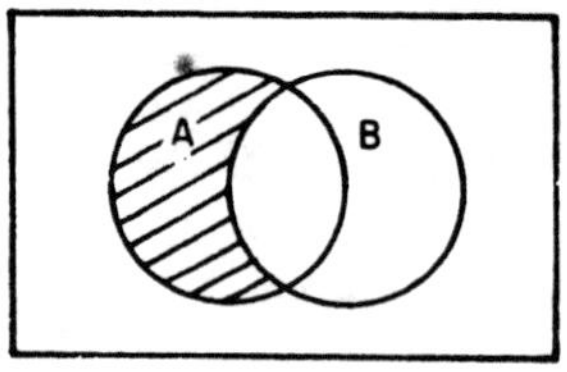

A and not B

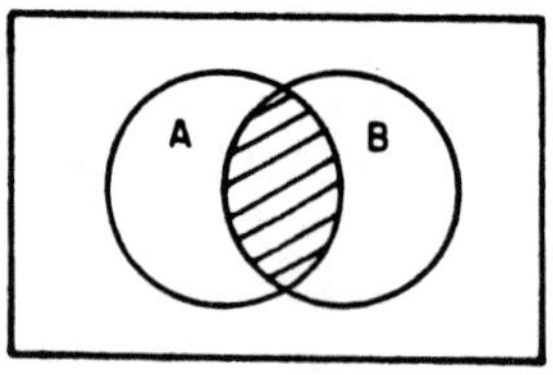

A and B

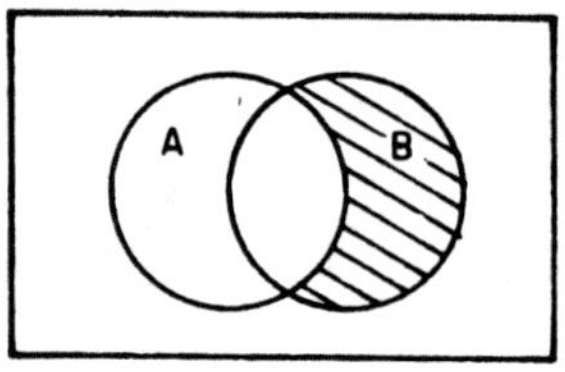

B and not A

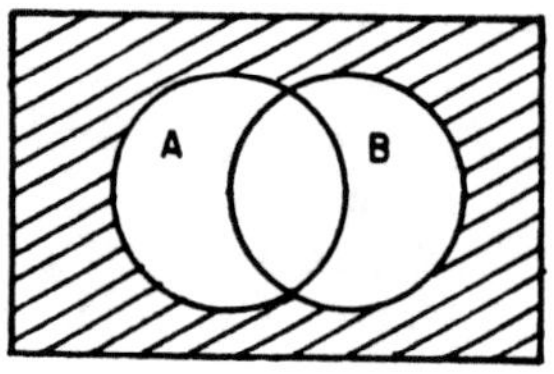

not A and not B

These four classes are called minterms because they represent the four minimum classes. The opposite of the minterms are called maxterms and are shown by

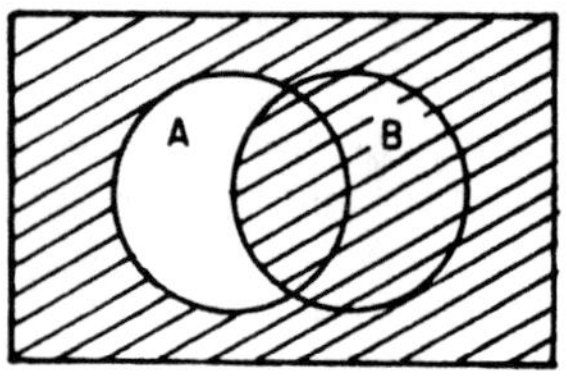

B or not A

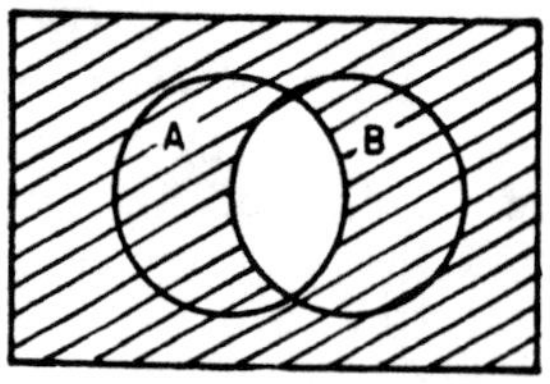

not A or not B

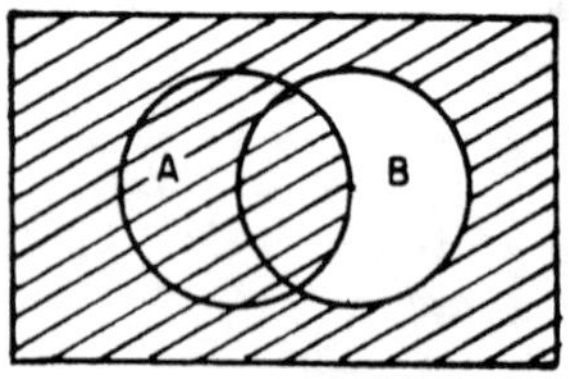

A or not B

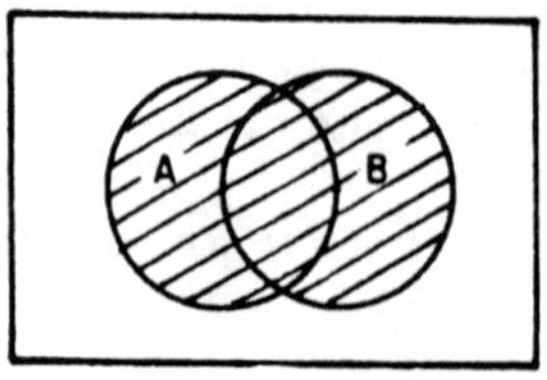

A or B

We will discuss minterms and maxterms in more detail later in the chapter.

BASIC EXPRESSIONS

It has been seen that the Venn diagram may be used to represent a picture of logic. The logic previously used was written in longhand, and used the words "and," "or," and "not." We used these words as a basis for combining elements to form classes in Boolean algebra logic descriptions. The symbols from sets and subsets are $\cap$ for "and," $\cup$ for "or," and $-$ for "not." The relationships of symbols are given by the following:

Sets and subsets	Words	Boolean algebra
$\cap$	and	$\cdot$
$\cup$	or	$+$
$-$	not	$-$

The following are examples of these relationships:

(1) $A \cdot B$ reads A and B
(2) $A + B$ reads A or B
(3) $\overline{A}$ reads not A

Relationships to the previously indicated classes, about submarines and atomic powered sound, are

(1) A and not B = $A \cdot \overline{B}$
(2) A and B = $A \cdot B$
(3) B and not A = $B \cdot \overline{A}$
(4) not A and not B = $\overline{A} \cdot \overline{B}$

also

(1) B or not A = $B + \overline{A}$
(2) not A or not B = $\overline{A} + \overline{B}$
(3) A or not B = $A + \overline{B}$
(4) A or B = $A + B$

Notice that

$A \cdot \overline{B}$
$A \cdot B$
$B \cdot \overline{A}$
$\overline{A} \cdot \overline{B}$

are called minterms. As related to algebra, there is a minimum number of terms in each; that is, one. Notice also that

$B + \overline{A}$
$\overline{A} + \overline{B}$
$A + \overline{B}$
$A + B$

are called maxterms. As related to algebra, there is a maximum number of terms in each. That is, two.

A further relationship may be made to sets and subsets as follows:

$A \cdot \overline{B} = A \cap \overline{B}$
$A \cdot B = A \cap B$
$B \cdot \overline{A} = B \cap \overline{A}$
$\overline{A} \cdot \overline{B} = \overline{A} \cap \overline{B}$

If we take any of these minterms, such as $A \cdot B$, and find its complement we have, according to DeMorgan's theorem

$$\overline{(A \cdot B)} = \overline{(A \cap B)} = \overline{A} \cup \overline{B} = \overline{A} + \overline{B}$$

which is a maxterm; therefore the complement of a minterm is a maxterm.

APPLICATIONS TO SWITCHING CIRCUITS

Since Boolean algebra is based upon elements having two possible stable states, it becomes very useful in representing switching circuits. The reason for this is that a switching circuit can be in only one of two possible states. That is, it is either open or it is closed. We may represent these two states as 0 and 1, respectively. Since the binary number system consists of only the symbols 0 and 1, we employ these symbols in Boolean algebra and call this "binary Boolean algebra."

THE "AND" OPERATION

Let us consider the Venn diagram in figure 1 (A). Its classes are labeled using the basic expressions of Boolean algebra. Note that there are two elements, or variables, A and B. The shaded area represents the class of elements that are $A \cdot B$ in Boolean notation and is expressed as:

$$f(A,B) = A \cdot B$$

The other three classes are also indicated in figure 1 (A). This expression is called an AND operation because it represents one of the four minterms previously discussed. Recall that AND indicates class intersection and both A and B must be considered simultaneously.

We can conclude then that a minterm of n variables is a logical product of these n variables with each variable present in either its noncomplemented or its complemented form, and is considered an AND operation.

For any Boolean function there is a corresponding truth table which shows, in tabular form, the true conditions of the function for each way in which conditions can be assigned its variables. In Boolean algebra, 0 and 1 are the symbols assigned to the variables of any function. Figure 1 (B) shows the AND operation function of two variables and its corresponding truth table.

This function can be seen to be true if one thinks of the logic involved: AB is equal to A and B which is the function f(A,B). Thus, if either A or B takes the condition of 0, or both take this condition, then the function f(A,B) equal AB is equal to 0. But if both A and B take the condition of 1 then the AND operation function has the condition of 1.

Figure 1 (C) shows a switching circuit for the function f(A,B) equal AB in that there will be an output only if both A and B are closed. An output in this case equals 1. If either switch is open, 0 condition, then there will be no output or 0.

In any digital computer equipment, there will be many circuits like the one shown in figure 1 (C). In order to analyze circuit operation, it is necessary to refer frequently to these circuits without looking at their switch arrangements. This is done by logic diagram mechanization as shown in figure 1 (D). This indicates that there are two inputs, A and B, into an AND operation circuit producing the function in Boolean algebra form of AB. These diagrams simplify equipment circuit diagrams by indicating operations without drawing all the circuit details.

It should be understood that while the previous discussion concerning the AND operation dealt with only two variables that any number of variables will fit the discussion. For example, in figure 2 three variables are shown along with their Venn diagram, truth table, switching circuit, and logic diagram mechanization.

THE "OR" OPERATION

We will now consider the Venn diagram in figure 3 (A). Note that there are two elements, or variables, A and B. The shaded area represents the class of elements that are $A + B$

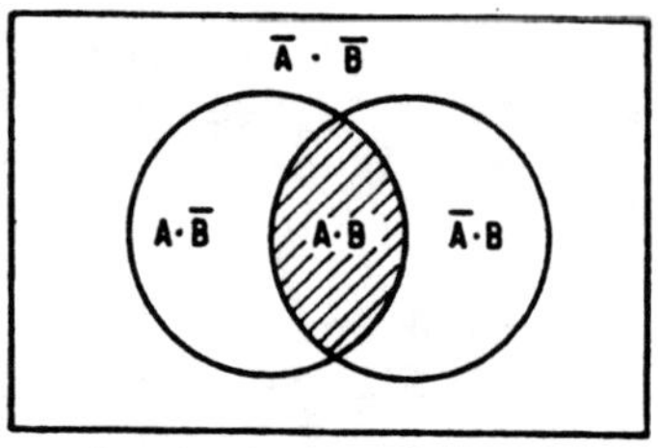

(A) VENN DIAGRAM

A	B	f(A,B) = AB
0	0	0
0	1	0
1	0	0
1	1	1

(B) TRUTH TABLE

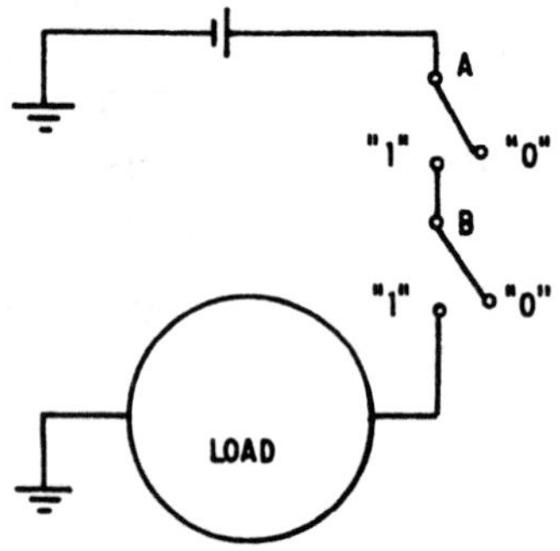

(C) AND SWITCHING CIRCUIT

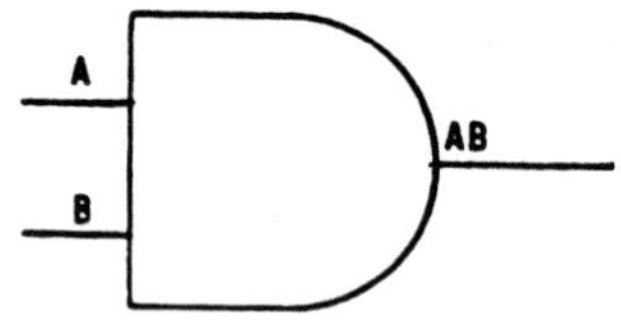

(D) LOGIC DIAGRAM
MECHANIZATION OF f(A,B) = AB

Figure 1.—The AND operation.

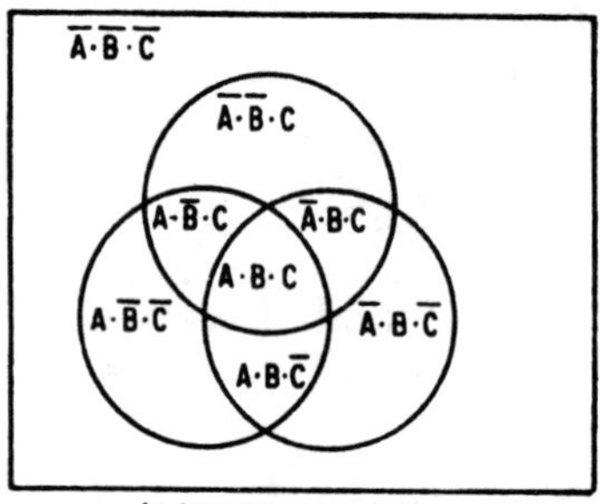

(A) VENN DIAGRAM

A	B	C	f(A, B, C) = ABC
0	0	0	0
0	0	1	0
0	1	0	0
0	1	1	0
1	0	0	0
1	0	1	0
1	1	0	0
1	1	1	1

(B) TRUTH TABLE

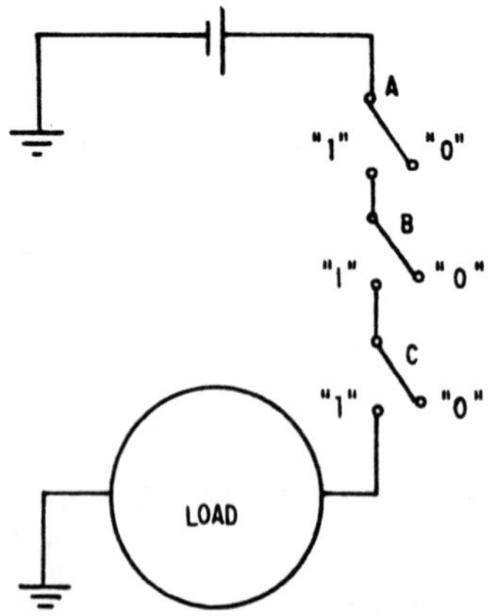

(C) AND SWITCHING CIRCUIT

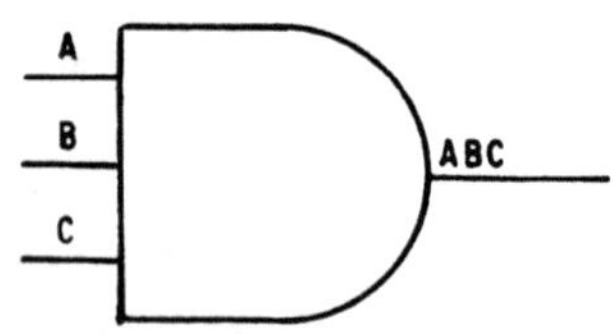

(D) LOGIC DIAGRAM
MECHANIZATION OF f(A,B,C) = ABC

Figure 2.—The AND operation (three variables).

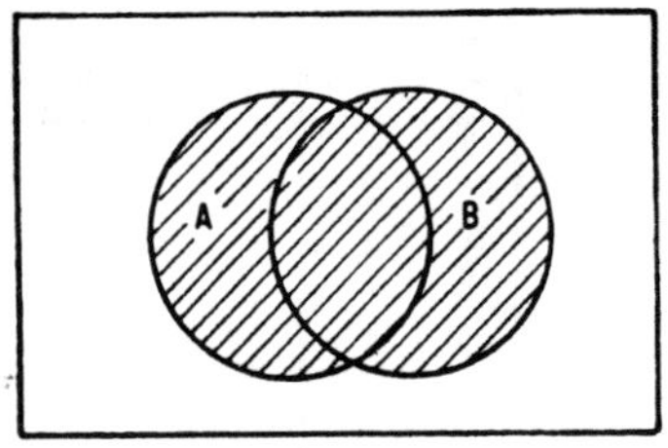

(A) VENN DIAGRAM

A	B	f(A,B) = A + B
0	0	0
0	1	1
1	0	1
1	1	1

(B) TRUTH TABLE

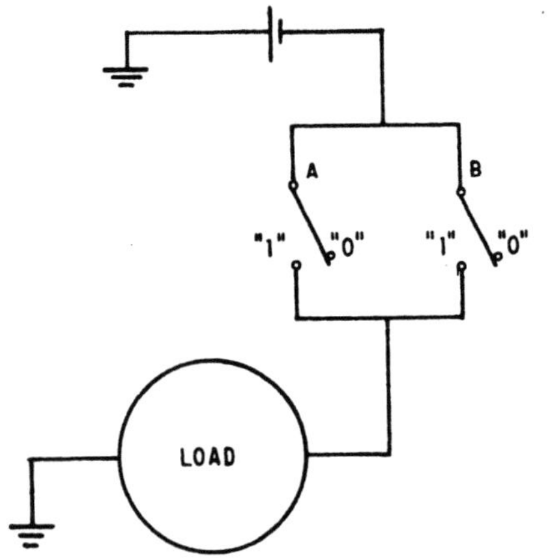

(C) OR SWITCHING CIRCUIT

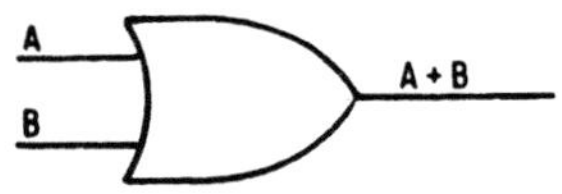

(D) LOGIC DIAGRAM MECHANIZATION OF f(A,B) = A + B

Figure 3.—The OR operation.

in Boolean notation and is expressed in Boolean algebra as:

$$f(A,B) = A + B$$

This expression is called an OR operation for it represents one of the four maxterms previously discussed. Recall that OR indicates class union and either A or B or both must be considered.

We can conclude then that a maxterm of n variables is a logical sum of these n variables where each variable is present in either its noncomplemented or its complemented form.

In figure 3 (B) the truth table of an OR operation is shown. This truth table can be seen to be true if one thinks of A + B being equal to A or B which is the function f(A,B). Thus if A or B takes the value 1, then f(A,B) must equal 1. If not, then the function equals zero.

Figure 3 (C) shows a switching circuit for the OR operation which is two or more switches in parallel. It is apparent that the circuit will transmit if either A or B is in a closed position; that is, equal to 1. If, and only if, both A and B are open, equal to 0, the circuit will not transmit.

The logic diagram for the OR operation is given in figure 3 (D). This means that there are two inputs, A and B, into an OR operation circuit producing the function in Boolean form of A + B. Note the difference in the diagram from that of figure 2 (D).

As in the discussion of the AND operation the OR operation may also be used with more than two inputs. Figure 4 shows the OR operation with three inputs.

THE "NOT" OPERATION

The shaded area in figure 5 (A) represents the complement of A which in Boolean algebra is $\bar{A}$ and read as "NOT A." The expression f(A) equals $\bar{A}$ is called a NOT operation. The

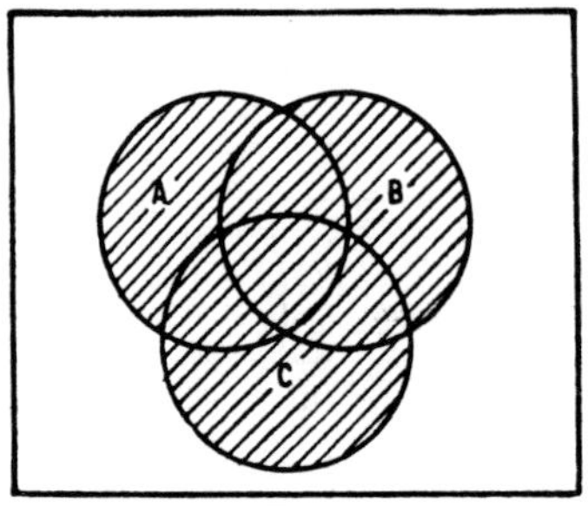

(A) VENN DIAGRAM

A	B	C	f(A,B,C) = A+B+C
0	0	0	0
0	0	1	1
0	1	0	1
0	1	1	1
1	0	0	1
1	0	1	1
1	1	0	1
1	1	1	1

(B) TRUTH TABLE

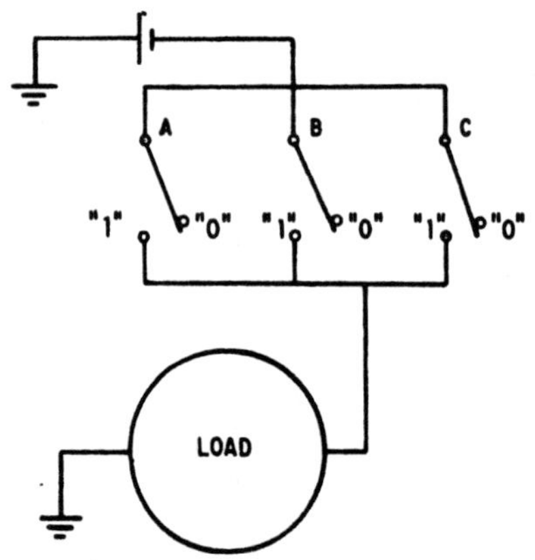

(C) OR SWITCHING CIRCUIT

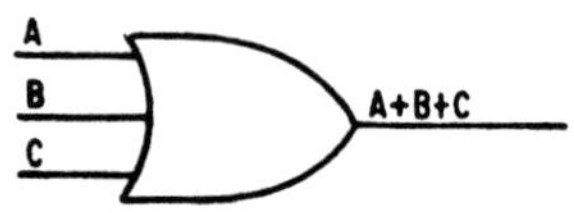

(D) LOGIC DIAGRAM
MECHANIZATION OF f(A,B,C) = A+B+C

Figure 4.—The OR operation (three variables).

truth table for the NOT operation is explained by the NOT switching circuit. The requirement of a NOT circuit is that a signal injected at the input produce the complement of this signal at the output. Thus, in figure 5 (C) it can be seen that when switch A is closed, that is, equal to 1, the relay opens the circuit to the load. When switch A is open, that is, equal to 0, the relay completes a closed circuit to the load. The logic diagram for the NOT operation is given in figure 5 (D). This means that A is the input to a NOT operation circuit and gives an output of $\overline{A}$. The NOT operation may be applied to any operation circuit such as AND or OR. This is discussed in the following section.

THE "NOR" OPERATION

The shaded area in figure 6 (A) represents the quantity, A OR B, negated. If reference is made to the preceding chapter it will be found that this figure is identical to the minterm expression $\overline{A}\,\overline{B}$; that is, A OR B negated is $\overline{A \text{ OR } B}$ and by application of DeMorgan's theorem is equal to $\overline{A}\,\overline{B}$.

The truth table for the NOR operation is shown in figure 6 (B). The table shows that if either A or B is equal to 1, then f(A,B) is equal to 0. Furthermore, if A and B equal 0, then f(A,B) equals 1.

The NOR operation is a combination of the OR operation and the NOT operation. The NOR switching circuit in figure 6 (C) is the OR circuit placed in series with the NOT circuit. If either switch A, switch B, or both are in the closed position, equal to 1, then there is no transmission to the load. If both switches A and B are open, equal to 0, then current is transmitted to the load.

The logic diagram mechanization of f(A,B) equal $\overline{A+B}$ (NOR operation) is shown in figure 6 (D). It uses both the OR logic diagrams and the NOT logic diagrams. The NOR logic diagram mechanization shows there are two

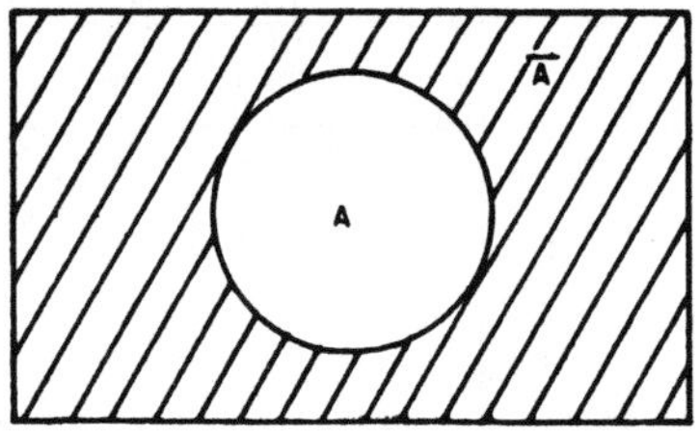

(A) Venn Diagram

A	$f(A) = \overline{A}$
1	0
0	1

(B) Truth Table

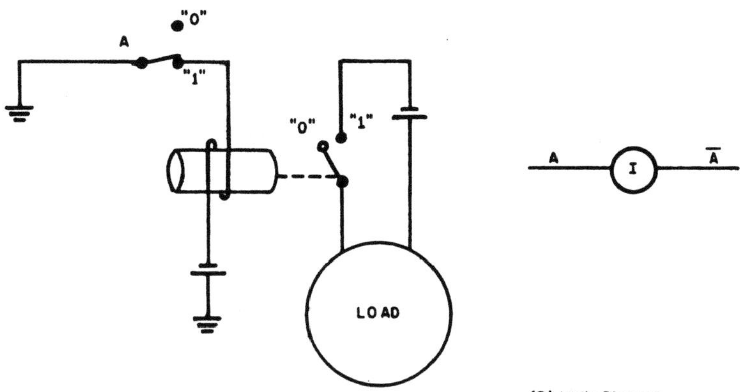

(C) NOT Switching Circuit

(D) Logic Diagram Mechanization of $f(A)=\overline{A}$

Figure 5.—The NOT operation.

inputs, A and B, into an OR circuit producing the function in Boolean form of $A+B$. This function is the input to the NOT (inverter) which gives the output, in Boolean form, of $\overline{A+B}$. Note that the whole quantity of $A+B$ is complemented and not the separate variables.

THE "NAND" OPERATION

The shaded area in figure 7 (A) represents the quantity A AND B negated (NOT), and is a maxterm expression. Notice that $\overline{AB}$ is equal to the maxterm expression $\overline{A}+\overline{B}$.

The truth table is shown for the NAND operation in figure 7 (B). When A and B equal 1, then f(A,B) is equal to 0. In all other cases, the function is equal to 1.

The NAND operation is a combination of the AND operation and the NOT operation. The NAND switching circuit in figure 7 (C) is the AND circuit put in series with the NOT circuit. If either switch A or B is open, equal to 0, then current is transmitted to the load. If both switch A and B are closed, equal to 1, then there is no transmission to the load.

The logic diagram mechanization of f(A,B) equal $\overline{AB}$ (NAND operation) is shown in figure 7 (D). The AND operation logic diagram and the NOT logic diagram mechanization shows that there are two inputs, A and B, into the AND circuit producing the function in Boolean form of AB. This function is the input to the NOT circuit which gives the output, in Boolean form, of $\overline{AB}$. Note that the entire quantity AB is complemented and not the separate variables.

It should be noted that in the previously discussed logic diagrams that each input signal represents the operation of a switch, circuit, or other component part.

Generally, a Boolean expression that has been inverted is said to be NOTTED. While we have previously used the inverter symbol

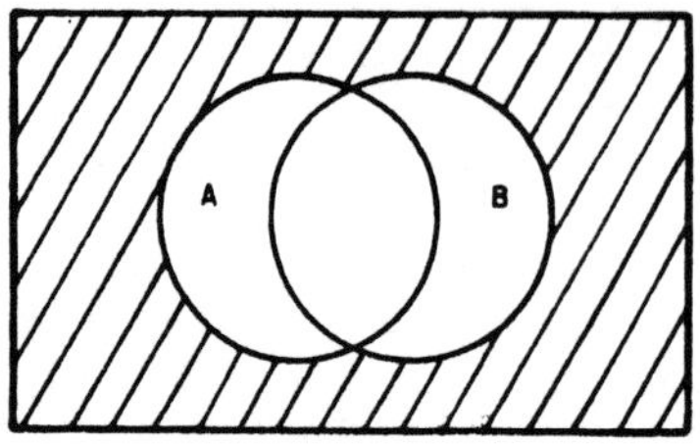

(A) Venn Diagram

A	B	A + B	$f(A,B) = \overline{A+B}$
0	0	0	1
0	1	1	0
1	0	1	0
1	1	1	0

(B) Truth Table

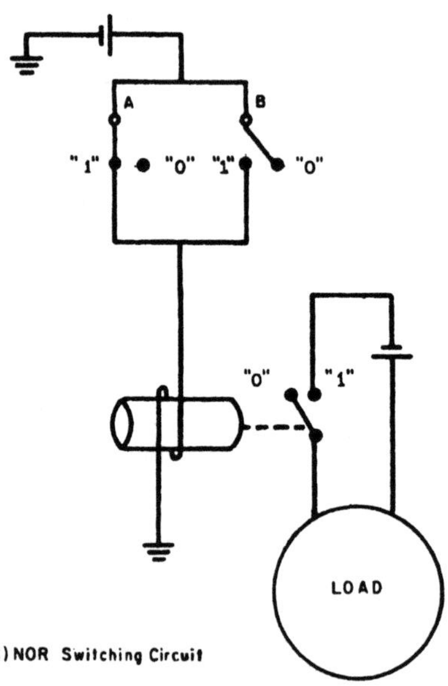

(C) NOR Switching Circuit

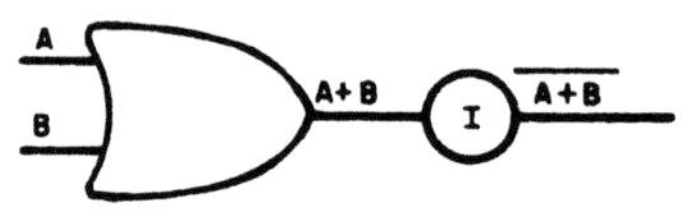

(D) Logic Diagram Mechanization of $f(A,B)=\overline{A+B}$

Figure 6.—The NOR operation.

separate from the AND or OR logic diagram it is common practice to show the NAND or NOR logic diagrams as indicated in figure 8, in accordance with American Standard for Graphic Symbols for Logic Diagrams.

The output of a NAND or a NOR gate is a NOTTED expression. The vinculum is used to indicate that such an expression has been NOTTED. Therefore, the output of a NAND gate having inputs A,B will appear as $\overline{AB}$ and the output of a NOR gate having inputs A,B will appear as $\overline{A+B}$. If any of the inputs to a logic gate are themselves NOTTED a vinculum will appear over the letter representing an input. Examples are shown in figure 8.

OUTPUT USED AS INPUT

The output from one gate may be an input to another gate. If so, that input will contain two or more letters. Figure 9 (A) shows an OR gate feeding into an OR gate. There are four possible combinations of inputs and logic symbols. These are shown in figure 9 (B,C,D,E). Notice that signs of grouping occur in all outputs except the AND input to the OR gate. The AB, in this case, is naturally grouped because the letters are written together and are separated from C by the OR sign. Figure 10 shows several different cases along with the proper output expressions.

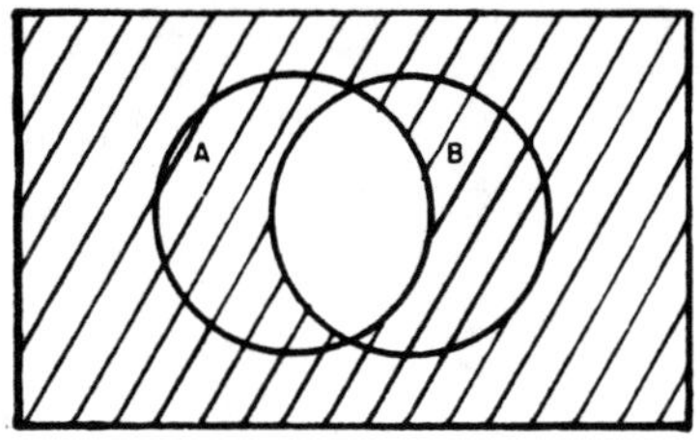

(A) Venn Diagram

A	B	AB	f(A,B)= $\overline{AB}$
0	0	0	1
0	1	0	1
1	0	0	1
1	1	1	0

(B) Truth Table

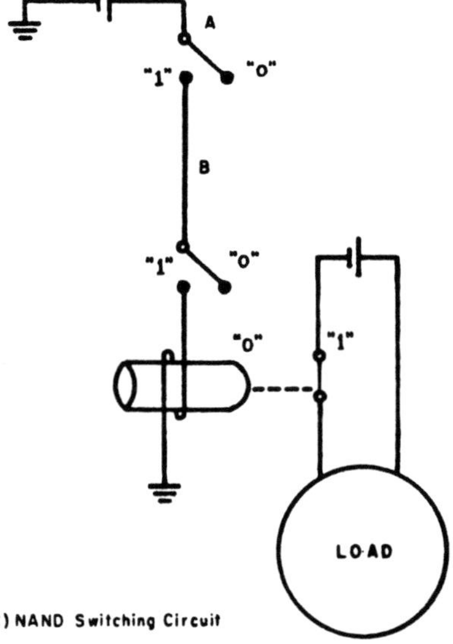

(C) NAND Switching Circuit

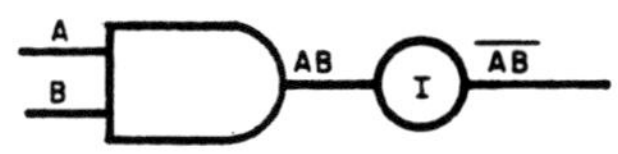

(D) Logic Diagram
Mechanization of f(A,B)= $\overline{AB}$

Figure 7.—The NAND operation.

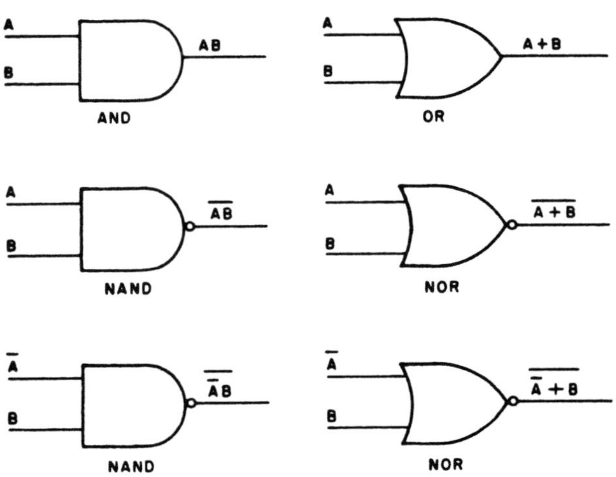

Figure 8.—American Standard Logic Symbols.

Although the vinculum is not used in place of parentheses or brackets, it is also a grouping sign. Consider the NOR symbol of figure 11 (A). The AB and C are the inputs to the OR circuit and form AB+C. The AB+C is then inverted to form $\overline{AB+C}$. The vinculum groups whatever portion or portions of the output expression that has been inverted. Figure 11 (B,C,D) gives examples of this type output.

To determine the output of a logic diagram, find the output of each logic symbol in the diagram. You should begin with the inputs at the left and move right, using the output of each logic symbol as an input to the following symbol, as illustrated in figure 12.

When determining the output of a logic diagram, one should be careful of the two most

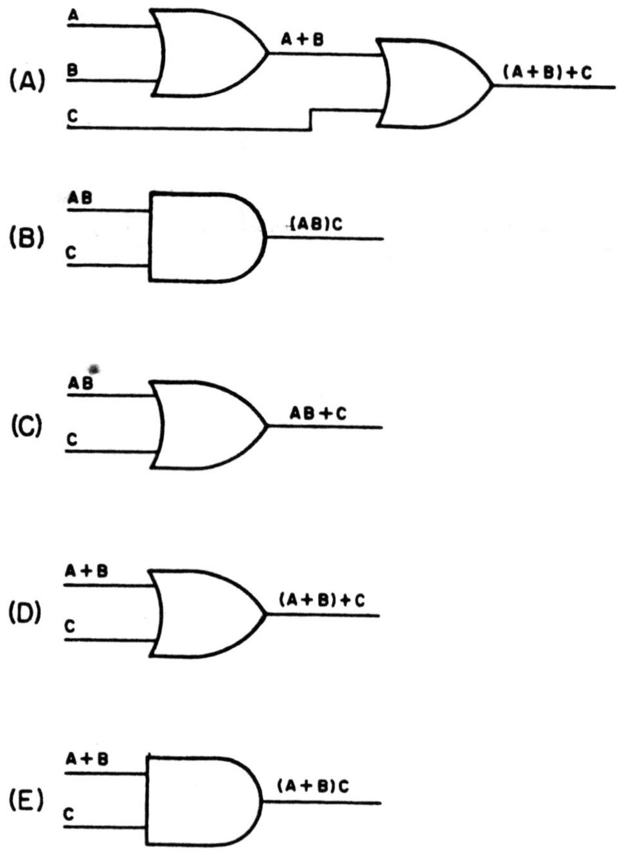

Figure 9.—Output as an input.

common mistakes which are leaving out vincula and leaving out grouping signs.

PROBLEMS: Find the outputs of the following logic diagrams.

1.

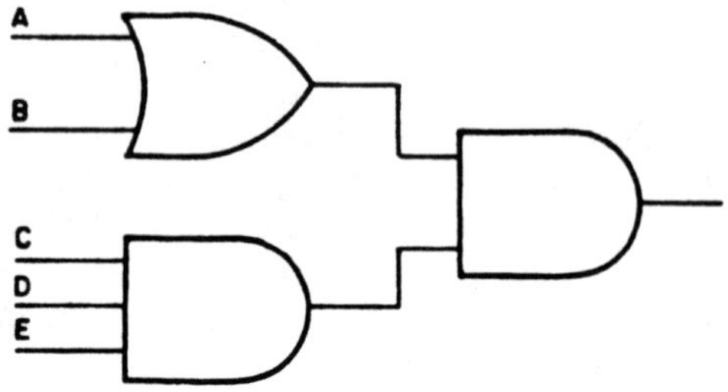

2.

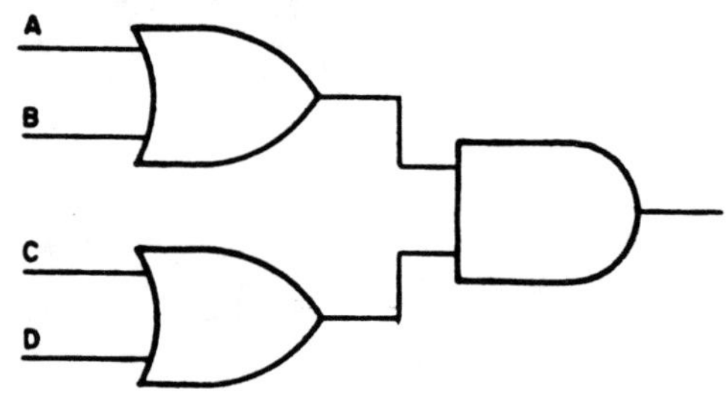

3.

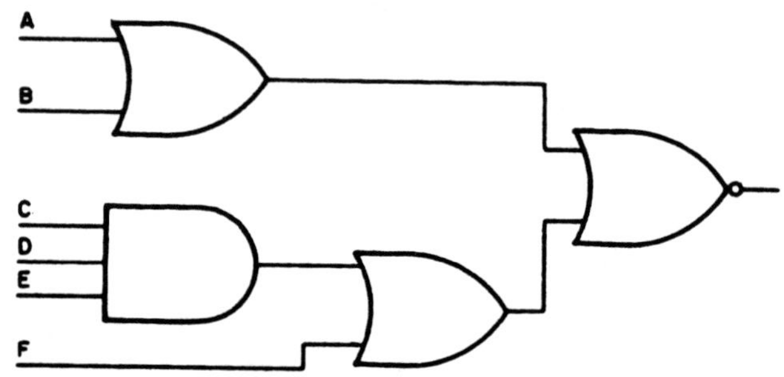

4.

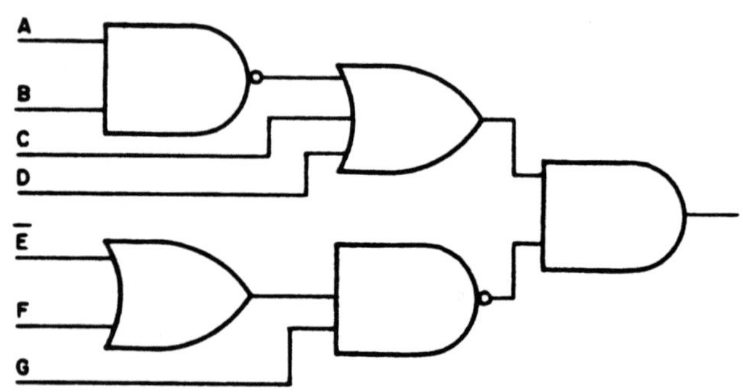

ANSWERS:

1. $(A+B)(CDE)$

2. $(A+B)(C+D)$

3. $\overline{(A+B)+(CDE+F)}$

4. $(\overline{AB}+C+D)[G(\overline{E}+F)]$

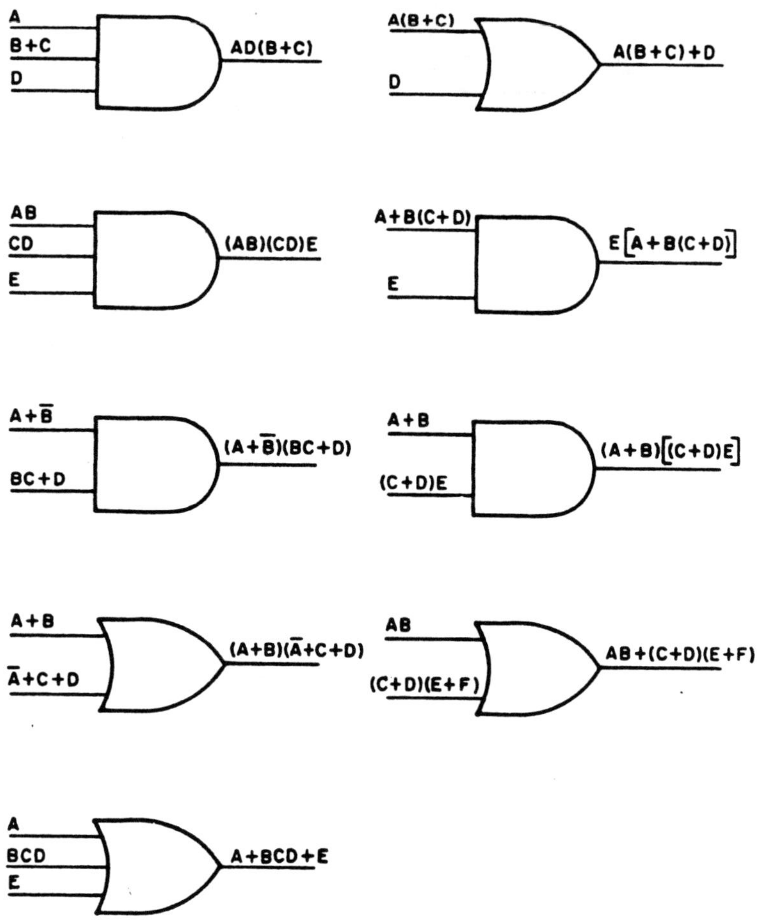

Figure 10.—Examples of grouping.

DEDUCING INPUTS FROM OUTPUTS

In order to draw a logic diagram from an output expression you should start with the output and work toward the input. Separate, in steps, the output expression until you have all single-letter inputs. If letters are grouped, first separate the group from other groups or letters, then separate the letters within groups.

To diagram the input that produces A + BC, you would first separate A from BC by using an OR logic symbol; that is,

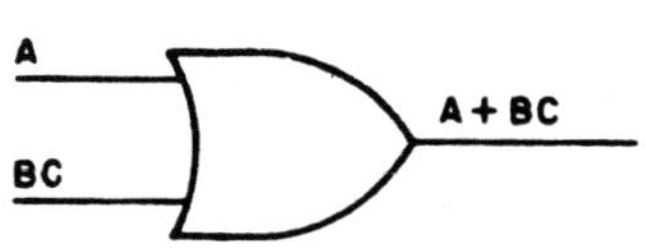

You now draw an AND logic symbol to separate B from C, and extend all lines to a common column on the left. This is shown by the following diagram.

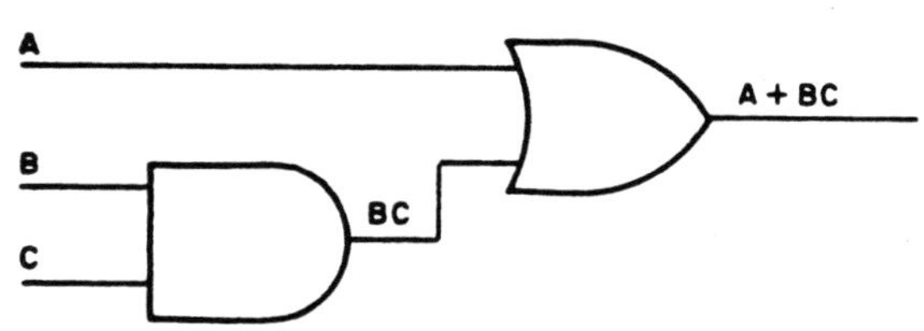

One common mistake in drawing the simplest possible diagram from an output expression is

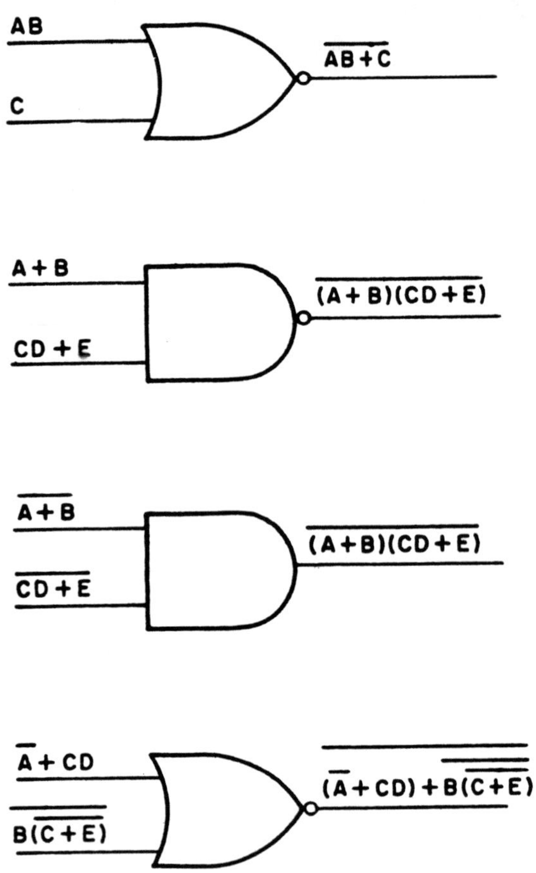

Figure 11.—Vinculum as grouping sign.

when the expression is similar to AB(C + D). The mistake is made by drawing

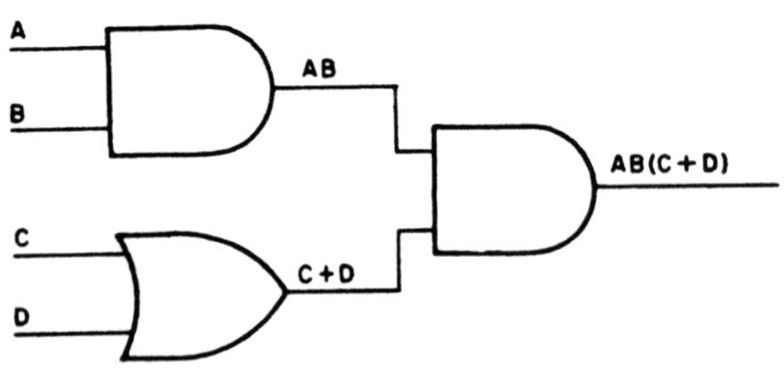

If the foregoing were your results, you would have failed to notice that A, B, and (C + D) were all ANDed together. You should have drawn

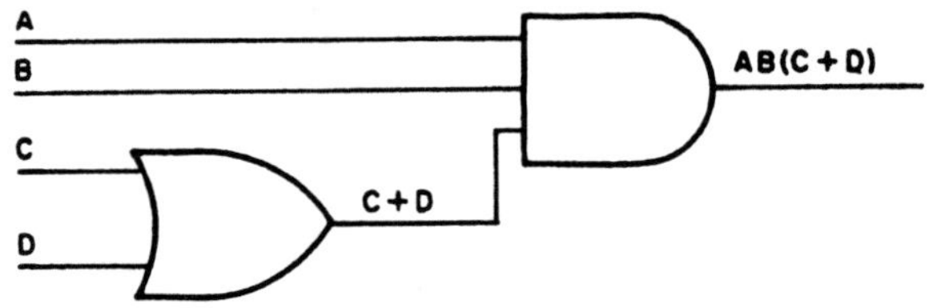

which would have saved the use of one gate. A gate is considered one circuit such as OR, AND, NOR, etc.

To diagram the expression $A(B+\bar{C})(D+\bar{E}F)$ write

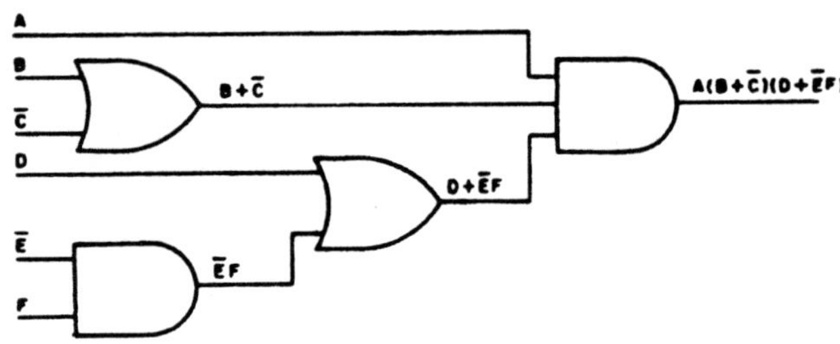

PROBLEMS: Draw the logic diagrams for the following expressions.

1. A + B(C + D)
2. (A + B + C)D + E
3. (A + B + C)DE
4. ABC(D + E)

ANSWERS:

1.

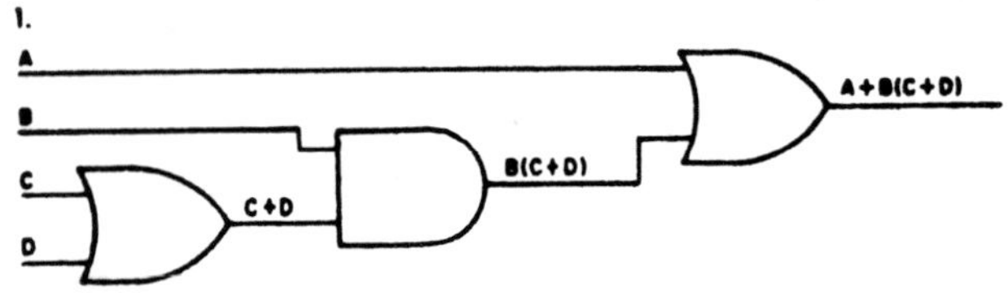

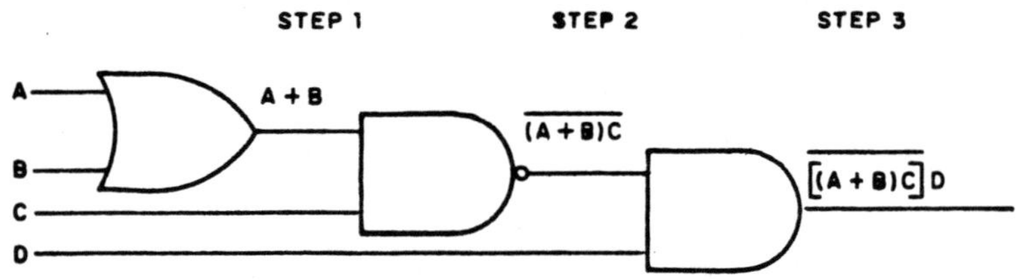

Figure 12.—Steps for determining output.

2.

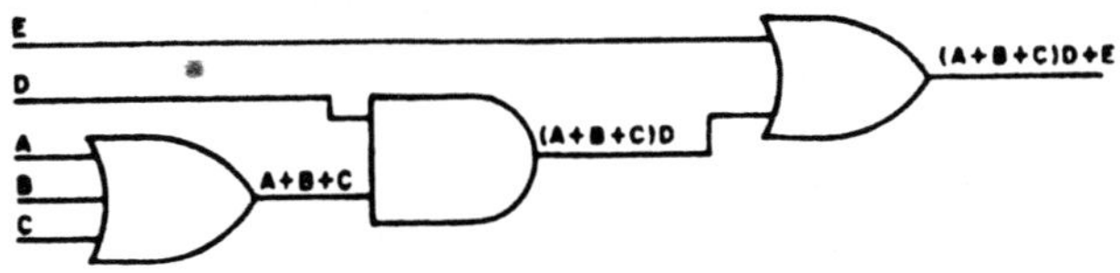

3.

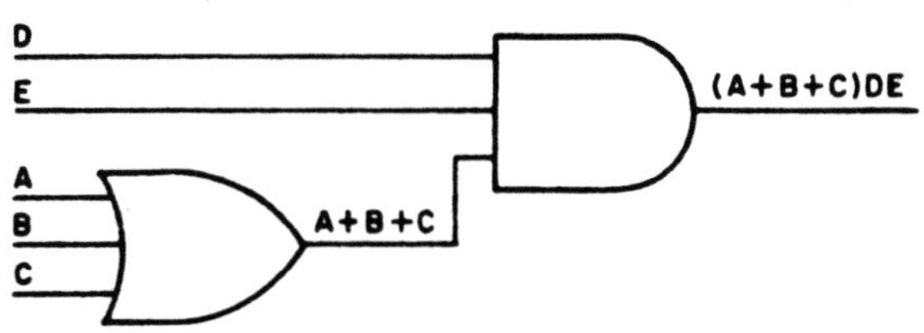

4.

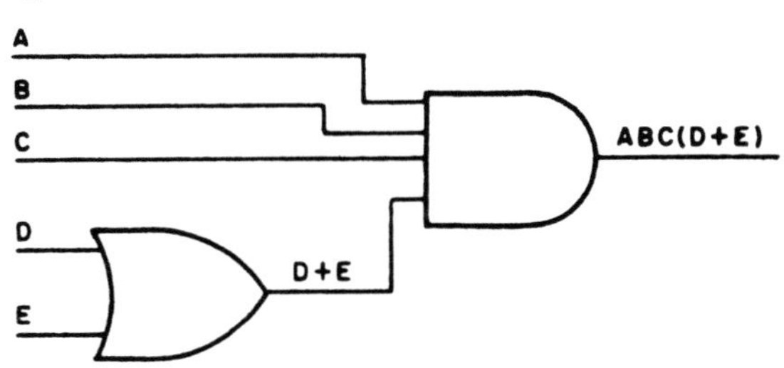

POSTULATES AND THEOREMS

In this section we will discuss the basic laws of Boolean algebra which enables one to simplify many Boolean expressions. By applying the basic laws, the digital systems designer can be sure a circuit is in the simplest possible algebraic form.

The laws of Boolean algebra may also be used for trouble-shooting defective components or for locating errors in computer programs. It should be understood that not all of the laws are similar to the laws of ordinary algebra.

LAW OF IDENTITY

This law is shown as

$$A = A$$

$$\bar{A} = \bar{A}$$

and indicates that any letter, number, or expression is equal to itself. The law of identity is shown in figure 13.

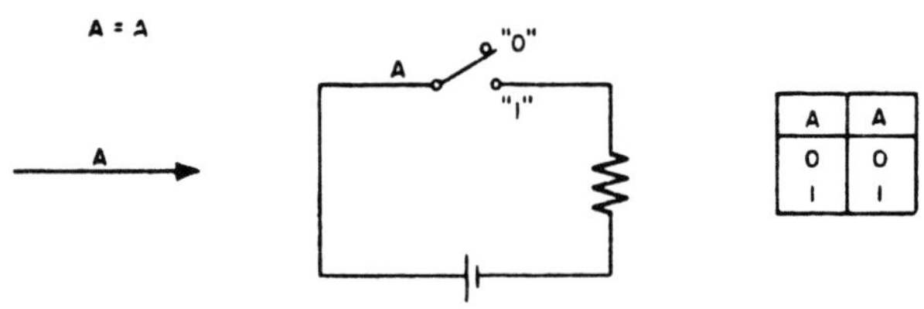

Figure 13.—Law of Identity.

COMMUTATIVE LAW

The commutative law is:

$$AB = BA$$

and

$$A + B = B + A$$

which is shown in figure 14. This indicates that when inputs to a logic symbol are ANDed

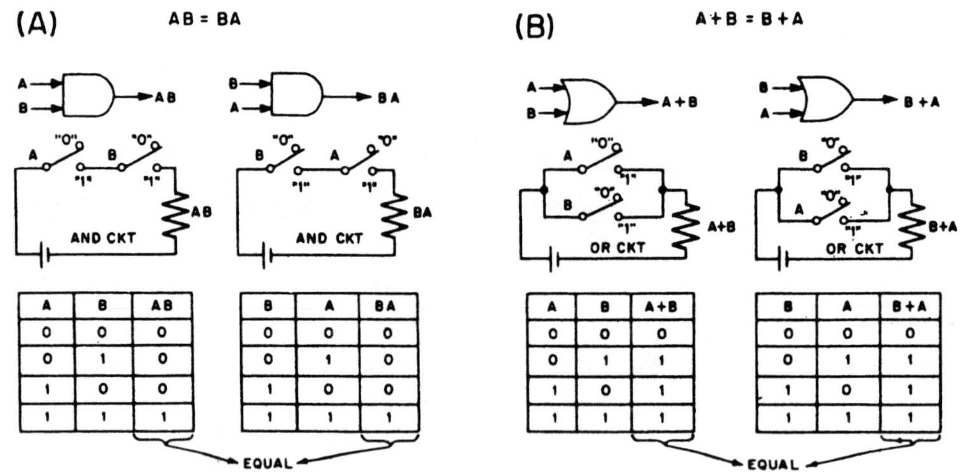

Figure 14.—Commutative Law.

or ORed, the order in which they are written does not affect the binary value of the output; that is,

$$R(S+T) = (S+T)R$$

and

$$A(BC+D+E) = (E+BC+D)A$$

ASSOCIATIVE LAW

The associative law is:

$$A(BC) = (AB)C$$

and

$$A + (B+C) = (A+B) + C$$

which is shown in figure 15. This indicates that when inputs to a logic symbol are ANDed or ORed, the order in which they are grouped does not affect the binary value of the output; that is,

$$ABC + D(EF) = (AB)C + DEF$$

and

$$C + (D+E) + (F+G) = C + D + E + F + G$$

IDEMPOTENT LAW

As seen in figure 16, if A is ANDed with A or if A is ORed with A, the output will equal A; that is,

$$AA = A$$

$$A + A = A$$

and

$$(RS)(RS) = RS$$

LAW OF DOUBLE NEGATION

This law is:

$$\overline{\overline{A}} = A$$

which indicates that when two bars of equal length cover the same letter or expression, both may be removed. This is shown in figure 17. Examples are

$$\overline{\overline{AB}} = AB$$

and

$$\overline{\overline{\overline{AB}}} + \overline{\overline{X}} = \overline{AB} + X$$

(A)

$(A \cdot B) \cdot C = A \cdot (B \cdot C)$

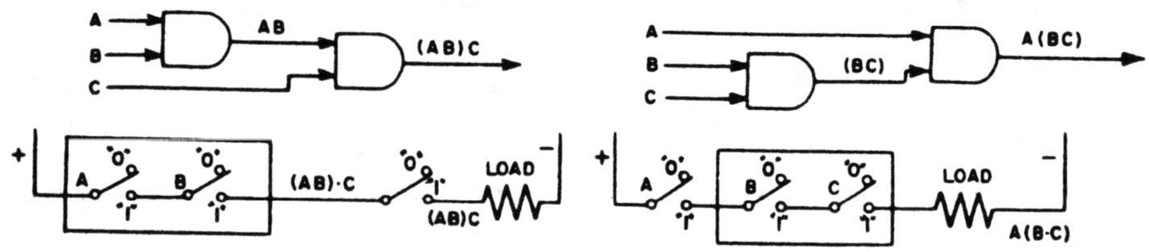

A	B	C	AB	(AB)·C	(BC)	A(BC)
0	0	0	0	0	0	0
0	0	1	0	0	0	0
0	1	0	0	0	0	0
0	1	1	0	0	0	0
1	0	0	0	0	0	0
1	0	1	0	0	0	0
1	1	0	1	0	0	0
1	1	1	1	1	1	1

(AB)·C and A(BC): EQUAL

(B)

$(A+B)+C = A+(B+C)$

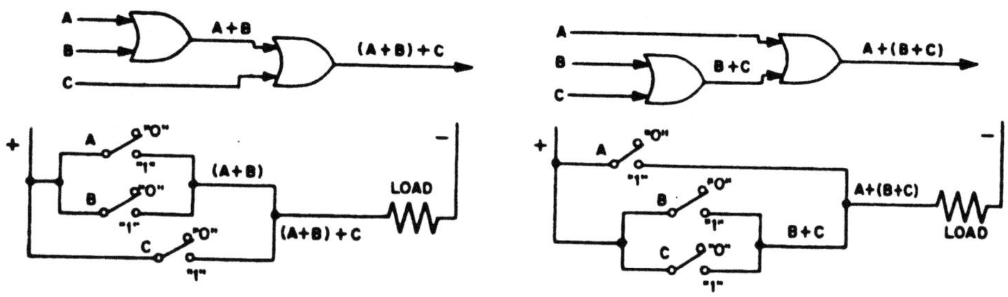

A	B	C	A+B	(A+B)+C	B+C	A+(B+C)
0	0	0	0	0	0	0
0	0	1	0	1	1	1
0	1	0	1	1	1	1
0	1	1	1	1	1	1
1	0	0	1	1	0	1
1	0	1	1	1	1	1
1	1	0	1	1	1	1
1	1	1	1	1	1	1

(A+B)+C and A+(B+C): EQUAL

Figure 15.—Associative Law.

(A)

$AA = A$

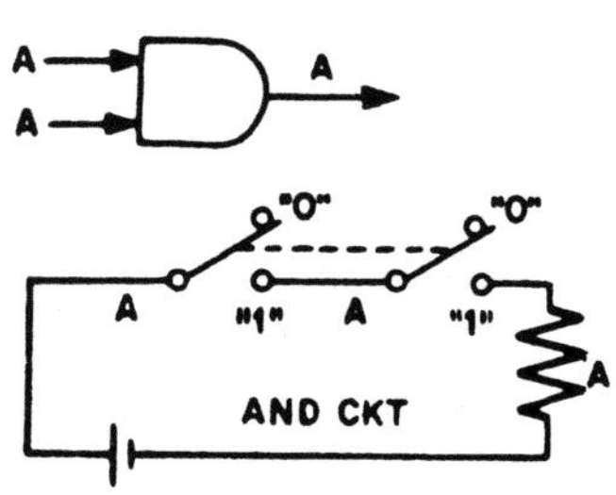

A	A	AA
0	0	0
1	1	1

EQUAL

(B)

$A + A = A$

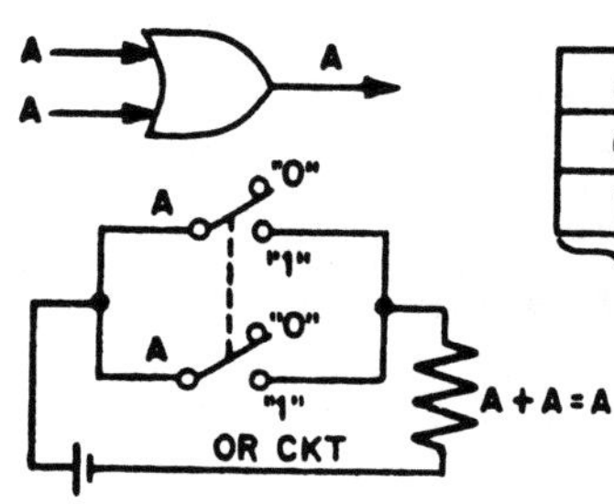

A	A	A+A
0	0	0
1	1	1

EQUAL

Figure 16.—Idempotent Law.

$$\bar{\bar{A}} = A$$

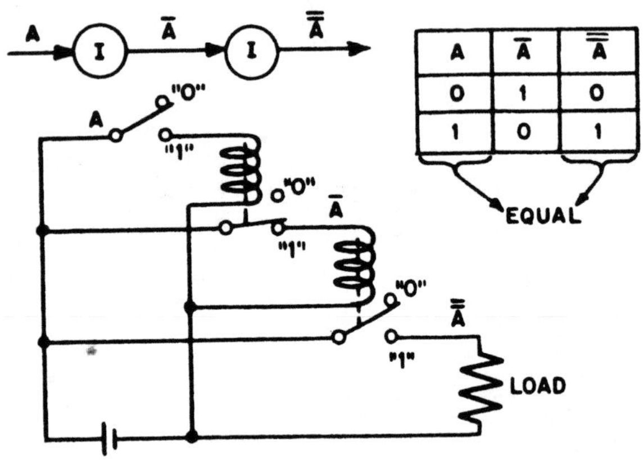

A	Ā	A̿
0	1	0
1	0	1

Figure 17.—Law of Double Negation.

COMPLEMENTARY LAW

This law is stated as:

$$A\bar{A} = 0$$

and

$$A + \bar{A} = 1$$

which indicates that when any letter or expression is ANDed with its complement, the output is 0. Also, when any letter or expression is ORed with its complement, the output is 1. This is shown in figure 18. Examples are:

$$CD\overline{CD} = 0$$

and

$$\overline{ABC} + ABC = 1$$

LAW OF INTERSECTION

As shown in figure 19, if one input to an AND circuit has a value of 1 the output will take the value of the other input. That is, if the two inputs to an AND circuit are 1 and A, then when A is 1 the output will be 1 and when A is 0 the output will be 0. If the inputs are 0 and A, then the output will always be 0.

The law of intersection is given by the following:

$$A \cdot 1 = A$$

and

$$A \cdot 0 = 0$$

Examples are:

$$AB \cdot 1 = AB$$

and

$$CD \cdot 0 = 0$$

LAW OF UNION

As shown in figure 20, if one input to an OR circuit has a binary value of 1, the output will be 1. If the inputs are 0 and A, the output will be the same as the value of A.

The law of union is given by the following:

$$A + 1 = 1$$

and

$$A + 0 = A$$

Examples of this law are as follows:

$$1 + ABC = 1$$

and

$$E + 0(AB) = E$$

LAW OF DUALIZATION
(DeMorgan's Theorem)

To split a vinculum that extends over more than one letter, and to join separate vincula into one vinculum requires the use of the law of dualization. This law is commonly referred to as DeMorgan's theorem. This law is shown in figure 21.

DeMorgan's theorem may be written as follows:

$$\overline{AB} = \bar{A} + \bar{B}$$

and

$$\overline{A + B} = \bar{A}\,\bar{B}$$

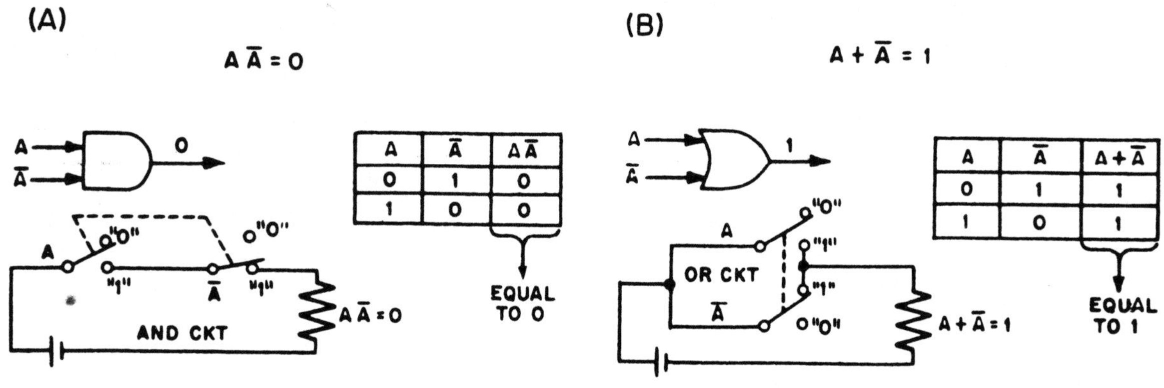

Figure 18.—Complementary Law.

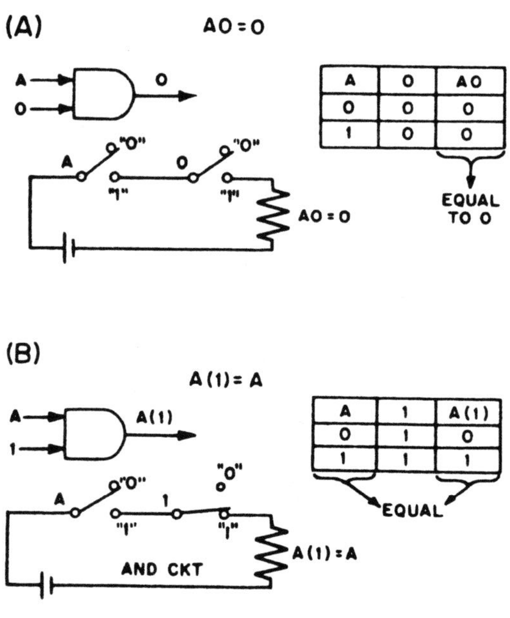

Figure 19.—Law of Intersection.

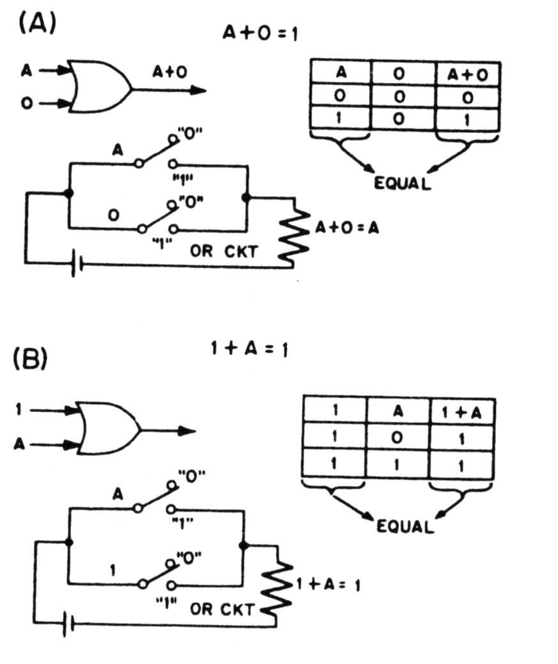

Figure 20.—Law of Union.

Whenever you split or join a vinculum, change the sign of operation. That is, AND to OR, or OR to AND.

In applying this theorem it should be remembered that when a vinculum covers part of an expression, the signs under the vinculum change and the signs outside the vinculum do not change; that is,

$$A\overline{BC} + \overline{D + E} = A(\overline{B} + \overline{C}) + \overline{D}\,\overline{E}$$

Notice that the grouping of letters must be maintained.

DISTRIBUTIVE LAW

There are two parts to the distributive law as shown in figure 22. The first identity is

$$A(B+C) = AB + AC$$

(A)

$\overline{(A+B)} = \overline{A}\,\overline{B}$

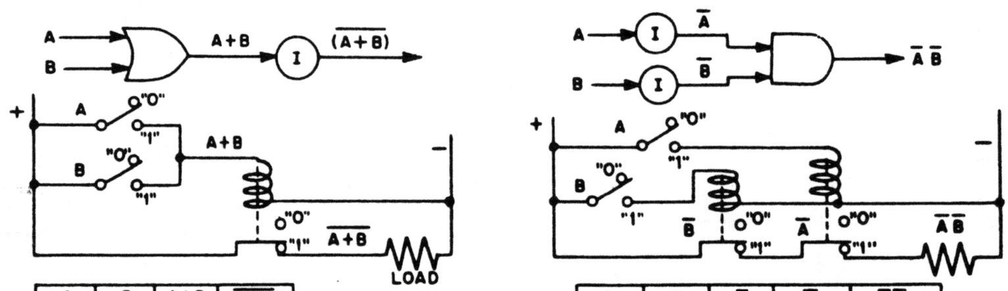

A	B	A+B	$\overline{(A+B)}$
0	0	0	1
0	1	1	0
1	0	1	0
1	1	1	0

A	B	$\overline{A}$	$\overline{B}$	$\overline{A}\,\overline{B}$
0	0	1	1	1
0	1	1	0	0
1	0	0	1	0
1	1	0	0	0

EQUAL

(B)

$\overline{AB} = \overline{A}+\overline{B}$

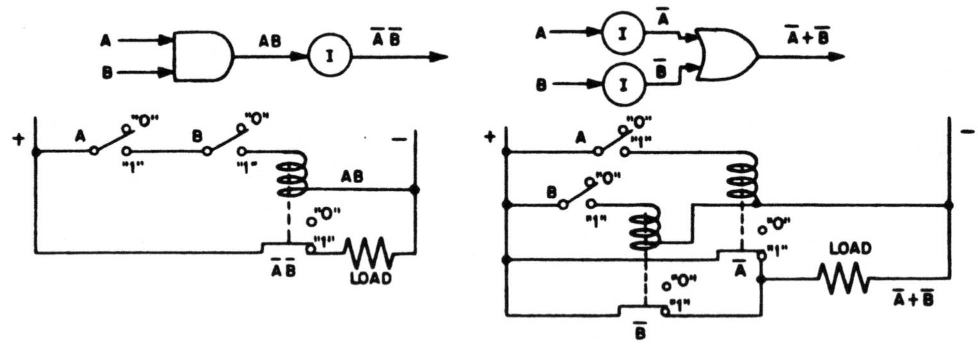

A	B	AB	$\overline{AB}$	$\overline{A}$	$\overline{B}$	$\overline{A}+\overline{B}$
0	0	0	1	1	1	1
0	1	0	1	1	0	1
1	0	0	1	0	1	1
1	1	1	0	0	0	0

EQUAL

Figure 21.—Law of Dualization.

and in order to obtain an output of 1 the A must be 1 and either B or C must be 1. This law is similar to the law of algebra which states that multiplication distributes over addition.

The second identity is

$$A + BC = (A+B)(A+C)$$

and in order to obtain an output of 1, at least one term in each of the parentheses must be 1. THIS LAW DOES NOT APPLY TO ORDINARY ALGEBRA. If this law did apply to ordinary algebra it would indicate that addition distributes over multiplication. In Boolean algebra this is true. Examples of the distributive law are as follows:

$$A(B+C+D) = AB + AC + AD$$

and

$$A + (B+C)(D+E) = (A+B+C)(A+D+E)$$

(A) A(B+C) = (AB) + (AC)

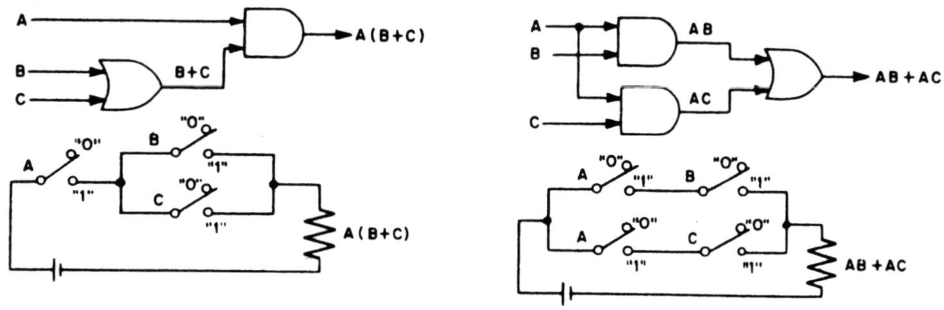

A	B	C	B+C	A(B+C)
0	0	0	0	0
0	0	1	1	0
0	1	0	1	0
0	1	1	1	0
1	0	0	0	0
1	0	1	1	1
1	1	0	1	1
1	1	1	1	1

A	B	C	AB	AC	AB+AC
0	0	0	0	0	0
0	0	1	0	0	0
0	1	0	0	0	0
0	1	1	0	0	0
1	0	0	0	0	0
1	0	1	0	1	1
1	1	0	1	0	1
1	1	1	1	1	1

EQUAL

(B) A+BC = (A+B)(A+C)

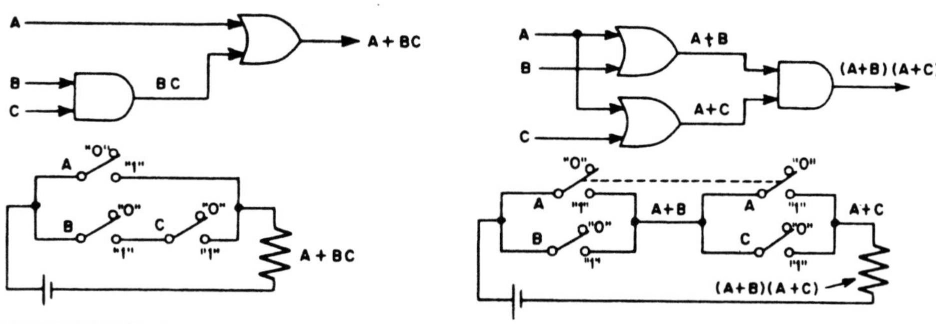

A	B	C	BC	A+BC
0	0	0	0	0
0	0	1	0	0
0	1	0	0	0
0	1	1	1	1
1	0	0	0	1
1	0	1	0	1
1	1	0	0	1
1	1	1	1	1

A	B	C	A+B	A+C	(A+B)(A+C)
0	0	0	0	0	0
0	0	1	0	1	0
0	1	0	1	0	0
0	1	1	1	1	1
1	0	0	1	1	1
1	0	1	1	1	1
1	1	0	1	1	1
1	1	1	1	1	1

EQUAL

Figure 22.—Distributive Law.

LAW OF ABSORPTION

The law of absorption is shown in figure 23. This law is written as

$$A(A + B) = A$$

and

$$A + AB = A$$

and indicates that the output is 1 whenever A is 1. Examples are

$$D(1 + E) = D \cdot 1 = D$$

and

$$A + AB + AC = A(1 + B + C)$$

$$= A \cdot 1$$

$$= A$$

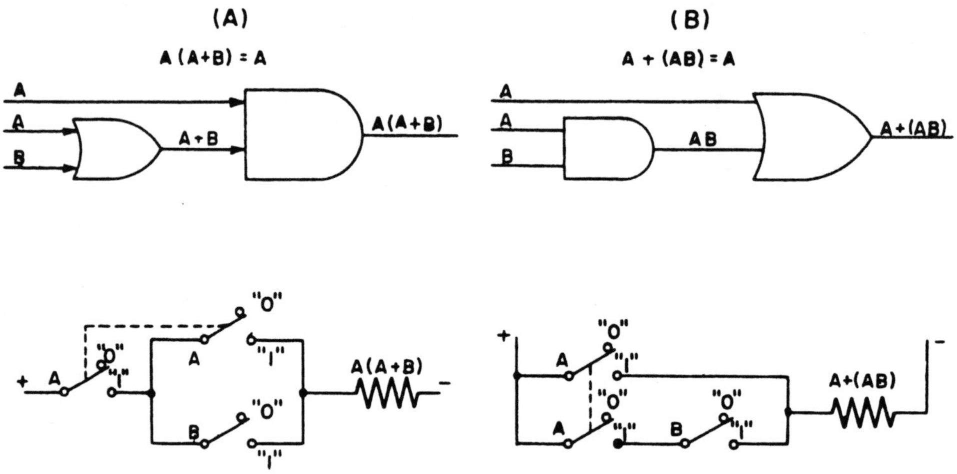

A	B	A + B	A(A + B)
0	0	0	0
0	1	1	0
1	0	1	1
1	1	1	1

EQUAL (A and A(A+B))

A	B	AB	A+(AB)
0	0	0	0
0	1	0	0
1	0	0	1
1	1	1	1

EQUAL (A and A+(AB))

Figure 23.—Absorption Law.

GLOSSARY OF ELECTRONIC TERMS

CONTENTS

ELECTRONIC SYMBOLS

CONTENTS

TRANSISTOR SYMBOLS

TUBE SYMBOLS

GLOSSARY OF ELECTRONIC TERMS

Acorn tube. An acorn-shaped vacuum tube designed for ultra-high-frequency circuits. The tube has short electron transit time and low interelectrode capacitance because of close spacing and small size electrodes.

Align. To adjust the tuned circuits of a receiver or transmitter for maximum signal response.

Alternation. One-half of a complete cycle.

Ammeter. An instrument for measuring the electron flow in amperes.

Ampere (amp). The basic unit of current or electron flow.

Amplification (A). The process of increasing the strength of a signal.

Amplification factor (μ). The ratio of a small change in plate voltage to a small change in grid voltage, with all other electrode voltages constant, required to produce the same small change in plate current.

Amplifier. A device used to increase the signal voltage, current, or power, generally composed of a vacuum tube and associated circuit called a stage. It may contain several stages in order to obtain a desired gain.

Amplitude. The maximum instantaneous value of an alternating voltage or current, measured in either the positive or negative direction.

Amplitude distortion. The changing of a waveshape so that it is no longer proportional to its original form. Also known as harmonic distortion.

Anode. A positive electrode; the plate of a vacuum tube.

Antenna. A device used to radiate or absorb r-f energy.

Aquadag. A graphite coating on the inside of certain cathode-ray tubes for collecting secondary electrons emitted by the screen.

Array (antenna). An arrangement of antenna elements, usually dipoles, which results in desirable directional characteristics.

Attenuation. The reduction in the strength of a signal.

Audio frequency (a-f). A frequency which can be detected as a sound by the human ear. The range of audio frequencies extends approximately from 20 to 20,000 cycles per second.

Autodyne circuit. A circuit in which the same elements and vacuum tube are used as an oscillator and as a detector. The output has a frequency equal to the difference between the frequencies of the received signal and the oscillator signal.

Automatic gain control (agc) A method of automatically regulating the gain of a receiver so that the output tends to remain constant though the incoming signal may vary in strength.

Automatic volume control (avc). *See* Automatic gain control.

Autotransformer. A transformer in which part of the primary winding is used as a secondary winding, or vice versa.

Azimuth. The angular measurement in a horizontal plane and in a clockwise direction, beginning at a point oriented to north.

Ballast resistance. A self-regulating resistance, usually connected in the primary circuit of a power transformer to compensate for variations in the line voltage.

Ballast tube. A tube which contains a ballast resistance.

Band of frequencies. The frequencies existing between two definite limits.

Band-pass filter. A circuit designed to pass with nearly equal response all currents having frequencies within a definite band, and to reduce substantially the amplitudes of currents of all frequencies outside that band.

Bazooka. *See* Line-balance converter.

Beam-power tube. A high vacuum tube in which the electron stream is directed in concentrated beams from the cathode to the plate. Variously termed beam-power tetrode and beam-power pentode.

Beat frequency. A frequency resulting from the combination of two different frequencies. It is numerically equal to the difference between or the sum of these two frequencies.

Beat note. *See* Beat frequency.

Bias. The average d-c voltage maintained between the cathode and control grid of a vacuum tube.

Biasing resistor. A resistor used to provide the voltage drop for a required bias.

Blanking. *See* Gating.

Bleeder. A resistance connected in parallel with a power-supply output to protect equipment from excessive voltages if the load is removed or substantially reduced; to improve the voltage regulation, and to drain the charge remaining in the filter capacitors when the unit is turned off.

Blocking capacitor. A capacitor used to block the flow of direct current while permitting the flow of alternating current.

Break-down voltage. The voltage at which an insulator or dielectric ruptures, or at which ionization and conduction take place in a gas or vapor.

Brilliance modulation. *See* Intensity modulation.

Buffer amplifier. An amplifier used to isolate the output of an oscillator from the effects produced by changes in voltage or loading in following circuits.

Buncher. The electrode of a velocity-modulated tube which alters the velocity of electrons in the constant current beam causing the electrons to become bunched in a drift space beyond the buncher electrode.

Bypass capacitor. A capacitor used to provide an alternating current path of comparatively low impedance around a circuit element.

Capacitance. The property of two or more bodies which enables them to store electrical energy in an electrostatic field between the bodies.

Capacitive coupling. A method of transferring energy from one circuit to another by means of a capacitor that is common to both circuits.

Capacitive reactance (X_c). The opposition offered to the flow of an alternating current by capacitance, expressed in ohms.

Capacitor. Two electrodes or sets of electrodes in the form of plates, separated from each other by an insulating material called the dielectric.

Carrier. The r-f component of a transmitted wave upon which an audio signal or other form of intelligence can be impressed.

Catcher. The electrode of a velocity-modulated tube which receives energy from the bunched electrons.

Cathode (K). The electrode in a vacuum tube which is the source of electron emission. Also a negative electrode.

Cathode bias. The method of biasing a tube by placing the biasing resistor in the common cathode return circuit, making the cathode more positive, rather than the grid more negative, with respect to ground.

Cathode follower. A vacuum-tube circuit in which the input signal is applied between the control grid and ground, and the output is taken from the cathode and ground. A cathode follower has a high input impedance and a low output impedance.

Characteristic impedance (Z_o). The ratio of the voltage to the current at every point along a transmission line on which there are no standing waves.

Choke. A coil which impedes the flow of alternating current of a specified frequency range because of its high inductive reactance at that range.

Chopping. *See* Limiting.

Clamping circuit. A circuit which maintains either amplitude extreme of a waveform at a certain level of potential.

Class A operation. Operation of a vacuum tube so that plate current flows throughout the entire operating cycle and distortion is kept to a minimum.

Class AB operation. Operation of a vacuum tube with grid bias so that the operating point is approximately halfway between Class A and Class B.

Class B operation. Operation of a vacuum tube with bias at or near cut-off so that plate current flows during approximately one-half cycle.

Class C operation. Operation of a vacuum tube with bias considerably beyond cut-off so that plate current flows for less than one-half cycle.

Clipping. *See* Limiting.

Coaxial cable. A transmission line consisting of two conductors concentric with and insulated from each other.

Coefficient of coupling (K). A numerical indication of the degree of coupling existing between two circuits, expressed in terms of either a decimal or a percentage.

Condenser. *See* Capacitor.

Conductance (G). The ability of a material to conduct or carry an electric current. It is the reciprocal of the resistance of the material, and is expressed in *ohms.*

Continuous waves. Radio waves which maintain a constant amplitude and a constant frequency.

Control grid (G). The electrode of a vacuum tube other than a diode upon which the signal voltage is impressed in order to control the plate current.

Control-grid-plate transconductance. *See* Transconductance.

Conversion transconductance (g_c). A characteristic associated with the mixer function of vacuum tubes, and used in the same manner as transconductance is used. It is the ratio of the i-f current in the primary of the first i-f transformer to the r-f signal voltage producing it.

Converter. *See* Mixer.

Converter tube. A multielement vacuum tube used both as a mixer and as an oscillator in a superheterodyne receiver. It creates a local frequency and combines it with an incoming signal to produce an intermediate frequency.

Counting circuit. A circuit which receives uniform pulses representing units to be counted and produces a voltage in proportion to their frequency.

Coupled impedance. The effect produced in the primary winding of a transformer by the influence of the current flowing in the secondary winding.

Coupling. The association of two circuits in such a way that energy may be transferred from one to the other.

Coupling element. The means by which energy is transferred from one circuit to another; the common impedance necessary for coupling.

Critical coupling. The degree of coupling which provides the maximum transfer of energy between two resonant circuits at the resonant frequency.

Crystal (*Xtal*). (1) A natural substance, such as quartz or tourmaline, which is capable of producing a voltage stress when under pressure, or producing pressure when under an applied voltage. Under stress it has the property of responding only to a given frequency when cut to a given thickness.

(2) A nonlinear element such as gelena or silicon, in which case the piezo-electric characteristic is not exhibited.

Crystal mixer. A device which employs the nonlinear characteristic of a crystal (nonpiezo-electric type) and a point contact to mix two frequencies.

Crystal oscillator. An oscillator circuit in which a piezoelectric crystal is used to control the frequency and to reduce frequency instability to a minimum.

Current (I). Flow of electrons; measured in amperes.

Cut-off (c.o.). The minimum value of negative grid bias which prevents the flow of plate current in a vacuum tube.

Cut-off limiting. Limiting the maximum output voltage of a vacuum-tube circuit by driving the grid beyond cut-off.

Cycle. One complete positive and one complete negative alternation of a current or voltage.

Damped waves. Waves which decrease exponentially in amplitude.

Decoupling network. A network of capacitors and chokes, or resistors, placed in leads which are common to two or more circuits to prevent unwanted interstage coupling.

Deflection sensitivity (*CRT*). The quotient of the displacement of the electron beam at the place of impact by the change in the deflecting field. It is usually expressed in millimeters per volt applied between the deflection electrodes, or in millimeters per gauss of the deflecting magnetic field.

Degeneration. The process whereby a part of the output signal of an amplifying device is returned to its input circuit in such a manner that it tends to cancel the input.

De-ionization potential. The potential at which ionization of the gas within a gas-filled tube ceases and conduction stops.

Demodulation. *See* Detection.

Detection. The process of separating the modulation component from the received signal.

Dielectric. An insulator; a term applied to the insulating material between the plates of a capacitor.

Dielectric constant. The ratio of the capacitance of a capacitor with a dielectric between the electrodes to the capacitance with air between the electrodes.

Differentiating circuit. A circuit which produces an output voltage substantially in proportion to the rate of change of the input voltage.

Diode. A two-electrode vacuum tube containing a cathode and a plate.

Diode detector. A detector circuit employing a diode tube.

Dipole antenna. Two metallic elements, each approximately one quarter wavelength long, which radiate r-f energy fed to them by the transmission line.

Directly heated cathode. A filament cathode which carries its own heating current for electron emission, as distinguished from an indirectly heated cathode.

Director (antenna). A parasitic antenna placed in front of a radiating element so that r-f radiation is aided in the forward direction.

Distortion. The production of an output waveform which is not a true reproduction of the input waveform. Distortion may consist of irregularities in amplitude, frequency, or phase.

Distributed capacitance. The capacitance that exists between the turns in a coil or choke, or between adjacent conductors or circuits, as distinguished from the capacitance which is concentrated in a capacitor.

Distributed inductance. The inductance that exists along the entire length of a conductor, as distinguished from the self-inductance which is concentrated in a coil.

Doorknob tube. A doorknob-shaped vacuum tube designed for ultra-high-frequency circuits. This tube has short electron transit time and low interelectrode capacitance, because of the close spacing and small size of electrodes.

Dropping resistor. A resistor used to decrease a given voltage to a lower value.

Dry electrolytic capacitor. An electrolytic capacitor using a paste instead of a liquid electrolyte. *See* Electrolytic capacitor.

Dynamic characteristics. The relation between the instantaneous plate voltage and plate current of a vacuum tube as the voltage applied to the grid is moved; thus, the characteristics of a vacuum tube during operation.

Dynatron. A negative resistance device; particularly, a tetrode operating on that portion of its i_p vs. e_p characteristic where secondary emission exists to such an extent that an increase in plate voltage actually causes a decrease in plate current, and, therefore, makes the circuit behave like a negative resistance.

Eccles-Jordan circuit (trigger circuit). A direct coupled multivibrator circuit possessing two conditions of stable equilibrium. Also known as a flip-flop circuit.

Effective value. The equivalent heating value of an alternating current or voltage, as compared to a direct current or voltage. It is 0.707 times the peak value of a sine wave. It is also called the rms value.

Efficiency. The ratio of output to input power, generally expressed as a percentage.

Electric field. A space in which an electric charge will experience a force exerted upon it.

Electrode. A terminal at which electricity passes from one medium into another.

Electrolyte. A water solution of a substance which is capable of conducting electricity. An electrolyte may be in the form of either a liquid or a paste.

Electrolytic capacitor. A capacitor employing a metallic plate and an electrolyte as the second plate separated by a dielectric which is produced by electrochemical action.

Electromagnetic field. A space field in which electric and magnetic vectors at right angles to each other travel in a direction at right angles to both.

Electron. The negatively charged particles of matter. The smallest particle of matter.

Electron emission. The liberation of electrons from a body into space under the influence of heat, light, impact, chemical disintegration, or potential difference.

Electronic switch. A circuit which causes a start-and-stop action or a switching action by electronic means.

Electronic voltmeter. *See* Vacuum tube voltmeter.

Electrostatic field. The field of influence between two charged bodies.

Equivalent circuit. A diagrammatic arrangement of coils, resistors, and capacitors, representing the effects of a more complicated circuit in order to permit easier analysis.

Farad (*f*). The unit of capacitance.

Feedback. A transfer of energy from the output circuit of a device back to its input.

Field. The space containing electric or magnetic lines of force.

Field intensity. Electrical strength of a field.

Filament. *See* Directly heated cathode.

Filter. A combination of circuit elements designed to pass a definite range of frequencies, attenuating all others.

Firing potential. The controlled potential at which conduction through a gas-filled tube begins.

First detector. *See* Mixer.

Fixed bias. A bias voltage of constant value, such as one obtained from a battery, power supply, or generator.

Fixed capacitor. A capacitor which has no provision for varying its capacitance.

Fixed resistor. A resistor which has no provision for varying its resistance.

Fluorescence. The property of emitting light as the immediate result of electronic bombardment.

Fly-back. The portion of the time base during which the spot is returning to the starting point. This is usually not seen on the screen of the cathode-ray tube, because of gating action or the rapidity with which it occurs.

Free electrons. Electrons which are loosely held and consequently tend to move at random among the atoms of the material.

Free oscillations. Oscillatory currents which continue to flow in a tuned circuit after the impressed voltage has been removed. Their frequency is the resonant frequency of the tuned circuit.

Frequency (f). The number of complete cycles per second existing in any form of wave motion; such as the number of cycles per second of an alternating current.

Frequency distortion. Distortion which occurs as a result of failure to amplify or attenuate equally all frequencies present in a complex wave.

Frequency modulation. See Modulation.

Frequency stability. The ability of an oscillator to maintain its operation at a constant frequency.

Full-wave rectifier circuit. A circuit which utilizes both the positive and the negative alternations of an alternating current to produce a direct current.

Gain (A). The ratio of the output power, voltage, or current to the input power, voltage, or current, respectively.

Gas tube. A tube filled with gas at low pressure in order to obtain certain desirable characteristics.

Gating (cathode-ray tube). Applying a rectangular voltage to the grid or cathode of a cathode-ray tube to sensitize it during the sweep time only.

Grid current. Current which flows between the cathode and the grid whenever the grid becomes positive with respect to the cathode.

Grid detection. Detection by rectification in the grid circuit of a detector.

Grid leak. A high resistance connected across the grid capacitor or between the grid and the cathode to provide a d-c path from grid to cathode and to limit the accumulation of charge on the grid.

Grid limiting. Limiting the positive grid voltage (minimum output voltage) of vacuum-tube circuit by means of a large series grid resistor.

Ground. A metallic connection with the earth to establish ground potential. Also, a common return to a point of zero r-f potential, such as the chassis of a receiver or a transmitter.

Half-wave rectification. The process of rectifying an alternating current wherein only one-half of the input cycle is passed and the other half is blocked by the action of the rectifier, thus producing pulsating direct current.

Hard tube. A high vacuum electronic tube.

Harmonic. An integral multiple of a fundamental frequency. (The second harmonic is twice the frequency of the fundamental or first harmonic.)

Harmonic distortion. Amplitude distortion.

Heater. The tube element used to indirectly heat a cathode.

Henry (h). The basic unit of inductance.

Helmholtz coil. A variometer having horizontal and vertical balanced coil windings, used to vary the angle of phase difference between any two similar waveforms of the same frequency.

Heterodyne. To beat or mix two signals of different frequencies.

High-frequency resistance. The resistance presented to the flow of high-frequency current. *See* Skin effect.

Horn radiator. Any open-ended metallic device for concentrating energy from a waveguide and directing this energy into space.

Hysteresis. A lagging of the magnetic flux in a magnetic material behind the magnetizing force which is producing it.

Image frequency. An undesired signal capable of beating with the local oscillator signal of a superheterodyne receiver which produces a difference frequency within the bandwidth of the i-f channel.

Impedance (Z). The total opposition offered to the flow of an alternating current. It may consist of any combination of resistance, inductive reactance, and capacitive reactance.

Impedance coil. See Choke.

Impedance coupling. The use of a tuned circuit or an impedance coil as the common coupling element between two circuits.

Impulse. Any force acting over a comparatively short period of time, such as a momentary rise in voltage.

Indirectly heated cathode. A cathode which is brought to the temperature necessary for electron emission by a separate heater element. Compare *Directly heated cathode.*

Inductance (L). The property of a circuit which tends to oppose a change in the existing current.

Induction. The act or process of producing voltage by the relative motion of a magnetic field across a conductor.

Inductive reactance (X_1). The opposition to the flow of alternating or pulsating current caused by the inductance of a circuit. It is measured in ohms.

Inductor. A circuit element designed so that its inductance is its most important electrical property; a coil.

Infinite. Extending indefinitely; having innumerable parts, capable of endless division within itself.

In phase. Applied to the condition that exists when two waves of the same frequency pass through their maximum and minimum values of like polarity at the same instant.

Instantaneous value. The magnitude at any particular instant when a value is continually varying with respect to time.

Integrating circuit. A circuit which produces an output voltage substantially in proportion to the frequency and amplitude of the input voltage.

Intensify. To increase the brilliance of an image on the screen of a cathode-ray tube.

Intensity modulation. The control of the brilliance of the trace on the screen of a cathode-ray tube in conformity with the signal.

Interelectrode capacitance. The capacitance existing between the electrodes in a vacuum tube.

Intermediate frequency (i-f). The fixed frequency to which r-f carrier waves are converted in a superheterodyne receiver.

Inverse peak voltage. The highest instantaneous negative potential which the plate can acquire with respect to the cathode without danger of injuring the tube.

Ion. An elementary particle of matter or a small group of such particles having a net positive or negative charge.

Ionization. Process by which ions are produced in solids, liquids, or gases.

Ionization potential. The lowest potential at which ionization takes place within a gas-filled tube.

Ionosphere. A region composed of highly ionized layers of atmosphere from 70 to 250 miles above the surface of the earth.

Kilo (k). A prefix meaning 1,000.
Kilocycle (kc). One thousand cycles; conversationally used to indicate 1,000 cycles per second.
Klystron. A tube in which oscillations are generated by the bunching of electrons (that is, velocity modulation). This tube utilizes the transit time between two given electrodes to deliver pulsating energy to a cavity resonator in order to sustain oscillations within the cavity.
Lag. The amount one wave is behind another in time; expressed in electrical degrees.
Lead The opposite of *lag.* Also, a wire or connection.
Leakage. The electrical loss due to poor insulation.
Lecher line. A section of open-wire transmission line used for measurements of standing waves.
Limiting. Removal by electronic means of one or both extremities of a waveform at a predetermined level.
Linear. Having an output which varies in direct proportion to the input.
Line-balance converter. A device used at the end of a coaxial line to isolate the outer conductor from ground.
Load. The impedance to which energy is being supplied.
Local oscillator. The oscillator used in a superheterodyne receiver the output of which is mixed with the desired r-f carrier to form the intermediate frequency.
Loose coupling. Less than critical coupling; coupling providing little transfer of energy.
Magnetic circuit. The complete path of magnetic lines of force.
Magnetic field (H). The space in which a magnetic force exists.
Magnetron. A vacuum-tube oscillator containing two electrodes, in which the flow of electrons from cathode to anode is controlled by an externally applied magnetic field.
Matched impedance. The condition which exists when two coupled circuits are so adjusted that their impedances are equal.
Meg (mega) (m). A prefix meaning one million.
Megacycle (M_c). One million cycles. Used conversationally to mean 1,000,000 cycles per second.
Metallic insulator. A shorted quarter-wave section of a transmission line which acts as an electrical insulator at a frequency corresponding to its quarter-wave length.
Mho. The unit of conductance.
Micro (μ). A prefix meaning one-millionth.
Microsecond (μs). One-millionth of a second.
Milli (m). A prefix meaning one-thousandth.
Milliampera (ma). One-thousandth of an ampere.
Mixer. A vacuum tube or crystal and suitable circuit used to combine the incoming and local-oscillator frequencies to produce an intermediate frequency. *See* Beat frequency.
Modulation. The process of varying the amplitude (amplitude modulation), the frequency (frequency modulation), or the phase (phase modulation) of a carrier wave in accordance with other signals in order to convey intelligence. The modulating signal may be an audio-frequency signal, video signal (as in television), or electrical pulses or tones to operate relays, etc.
Modulator. The circuit which provides the signal that varies the amplitude, frequency, or phase of the oscillations generated in the transmitter tube.
Multielectrode tube. A vacuum tube containing more than three electrodes associated with a single electron stream.
Multiunit tube. A vacuum tube containing within one envelope two or more groups of electrodes, each associated with separate electron streams.
Multivibrator. A type of relaxation oscillator for the generation of nonsinusoidal waves in which the output of each of its two tubes is coupled to the input of the other to sustain oscillations.
Mutual conductance (g_m). *See* Transconductance.
Mutual inductance. A circuit property existing when the relative position of two inductors causes the magnetic lines of force from one to link with the turns of the other.
Negative feedback. *See* Degeneration.
Neon bulb. A glass bulb containing two electrodes in neon gas at low pressure.
Network. Any electrical circuit containing two or more interconnected elements.
Neutralization. The process of nullifying the voltage fed back through the interelectrode capacitance of an amplifier tube, by providing an equal voltage of opposite phase; generally necessary only with triode tubes.
Node. A zero point; specifically, a current node is a point of zero current and a voltage node is a point of zero voltage.
Noninductive capacitor. A capacitor in which the inductive effects at high frequencies are reduced to the minimum.
Noninductive circuit. A circuit in which inductance is reduced to a minimum or negligible value.
Nonlinear. Having an output which does not vary in direct proportion to the input.
Ohm (ω). The unit of electrical resistance.
Open circuit. A circuit which does not provide a complete path for the flow of current.
Optimum coupling. *See* Critical coupling.
Oscillator. A circuit capable of converting direct current into alternating current of a frequency determined by the constants of the circuit. It generally uses a vacuum tube.
Oscillatory circuit. A circuit in which oscillations can be generated or sustained.
Oscillograph. *See* Oscilloscope.
Oscilloscope. An instrument for showing, visually, graphical representations of the waveforms encountered in electrical circuits.
Overdriven amplifier. An amplifier designed to distort the input signal waveform by a combination of cut-off limiting and saturation limiting.
Overload. A load greater than the rated load of an electrical device.
Parallel feed. Application of a d-c voltage to the plate or grid of a tube in parallel with an a-c circuit so that the d-c and a-c components flow in separate paths. Also called shunt feed.
Parallel-resonant circuit. A resonant circuit in which the applied voltage is connected across a parallel circuit formed by a capacitor and an inductor.
Paraphase amplifier. An amplifier which converts a single input into a push-pull output.

Parasitic suppressor. A resistor in a vacuum-tube circuit to prevent unwanted oscillations.

Peaking circuit. A type of circuit which converts an input to a peaked output waveform.

Peak plate current. The maximum instantaneous plate current passing through a tube.

Peak value. The maximum instantaneous value of a varying current, voltage, or power. It is equal to 1.414 times the effective value of a sine wave.

Pentode. A five-electrode vacuum tube containing a cathode, control, grid, screen grid, suppressor grid, and plate.

Phase difference. The time in electrical degrees by which one wave leads or lags another.

Phase inversion. A phase difference of 180° between two similar waveshapes of the same frequency.

Phase-splitting circuit. A circuit which produces from the same input waveform two output waveforms which differ in phase from each other.

Phosphorescence. The property of emitting light for some time after excitation by electronic bombardment.

Piezoelectric effect. The effect of producing a voltage by placing a stress, either by compression, by expansion, or by twisting, on a crystal, and, conversely, the effect of producing a stress in a crystal by applying a voltage to it.

Plate (P). The principal electrode in a tube to which the electron stream is attracted. *See* Anode.

Plate circuit. The complete electrical circuit connecting the cathode and plate of a vacuum tube.

Plate current (i_p). The current flowing in the plate circuit of a vacuum tube.

Plate detection. The operation of a vacuum-tube detector at or near cutoff so that the input signal is rectified in the plate circuit.

Plate dissipation. The power in watts consumed at the plate in the form of heat.

Plate efficiency. The ratio of the a-c power output from a tube to the average d-c power supplied to the plate circuit.

Plate impedance. *See* Plate resistance.

Plate-load impedance (R_L or Z_L). The impedance in the plate circuit across which the output signal voltage is developed by the alternating component of the plate current.

Plate modulation. Amplitude modulation of a class-C r-f amplifier by varying the plate voltage in accordance with the signal.

Plate resistance (r_p). The internal resistance to the flow of alternating current between the cathode and plate of tube. It is equal to a small change in plate voltage divided by the corresponding change in plate current, and is expressed in ohms. It is also called a-c resistance, internal impedance, plate impedance, and dynamic plate impedance. The static plate resistance, or resistance to the flow of *direct current* is a different value. It is denoted by R_p.

Positive feedback. *See* Regeneration.

Potentiometer. A variable voltage divider; a resistor which has a variable contact arm so that any portion of the potential applied between its ends may be selected.

Power. The rate of doing work or the rate of expending energy. The unit of electrical power is the watt.

Power amplification. The process of amlifying a signal to produce a gain in power, as distinguished from voltage amplification. The gain in the ratio of the alternating power output to the alternating power input of an amplifier.

Power factor. The ratio of the actual power of an alternating or pulsating current, as measured by a wattmeter, to the apparent power, as indicated by ammeter and voltmeter readings. The power factor if an inductor, capacitor, or insulator is an expression of the losses.

Power tube. A vacuum tube designed to handle a greater amount of power than the ordinary voltage-amplifying tube.

Primary circuit. The first, in electrical order, of two or more coupled circuits, in which a change in current induces a voltage in the other or secondary circuits; such as the primary winding of a transformer.

Propagation. *See* Wave propagation.

Pulsating current. A unidirectional current which increases and decreases in magnitude.

Push-pull circuit. A push-pull circuit usually refers to an amplifier circuit using two vacuum tubes in such a fashion that when one vacuum tube is operating on a positive alternation, the other vacuum tube operates on a negative alternation.

Q. The figure of merit of efficiency of a circuit or coil. Numerically it is equal to the inductive reactance divided by the resistance of the circuit or coil.

Radiate. To send out energy, such as r-f waves, into space.

Radiation resistance. A fictitious resistance which may be considered to dissipate the energy radiated from the antenna.

Radio frequency (r-f). Any frequency of electrical energy capable of propagation into space. Radio frequencies normally are much higher than sound-wave frequencies.

Radio-frequency amplification. The amplification of a radio wave by a receiver before detection, or by a transmitter before radiation.

Radio-frequency choke (RFC). An air-core or powdered iron core coil used to impede the flow of r-f currents.

Radio-frequency component. *See* Carrier.

Ratio. The value obtained by dividing one number by another, indicating their relative proportions.

Reactance (X). The opposition offered to the flow of an alternating current by the inductance, capacitance, or both, in any circuit.

Reciprocal. The value obtained by dividing the number 1 by any quantity.

Rectifier. A device used to change alternating current to unidirectional current.

Reflected impedance. *See* Coupled impedance.

Reflection. The turning back of a radio wave caused by reradiation from any conducting surface which is large in comparison to the wavelength of the radio wave.

Reflector. A metallic object placed behind a radiating antenna to prevent r-f radiation in an undesired direction and to reinforce radiation in a desired direction.

Regeneration. The process of returning a part of the output signal of an amplifier to its input circuit in such a manner that it reinforces the grid excitation and thereby increases the total amplification.

Regulation (voltage). The ratio of the change in voltage due to a load to the open-circuit voltage, expressed in per cent.

Relaxation oscillator. A circuit for the generation of nonsinusoidal waves by gradually storing and quickly releasing energy either in the electric field of a capacitor or in the magnetic field of an inductor.

Reluctance. The opposition to magnetic flux.

Resistance (*R*). The opposition to the flow of current caused by the nature and physical dimensions of a conductor.

Resistor. A circuit element whose chief characteristic is resistance; used to oppose the flow of current.

Resonance. The condition existing in a circuit in which the inductive and capacitive reactances cancel.

Resonance curve. A graphical representation of the manner in which a resonant circuit responds to various frequencies at and near the resonant frequency.

Rheostat. A variable resistor.

Ripple voltage. The fluctuations in the output voltage of a rectifier, filter, or generator.

rms. Abbreviation of root mean square. *See* Effective value.

Saturation. The condition existing in any circuit when an increase in the driving signal produces no further change in the resultant effect.

Saturation limiting. Limiting the minimum output voltage of a vacuum-tube circuit by operating the tube in the region of plate-current saturation (not to be confused with emission saturation).

Saturation point. The point beyond which an increase in either grid voltage, plate voltage, or both produces no increase in the existing plate current.

Screen dissipation. The power dissipated in the form of heat on the screen grid as the result of bombardment by the electron stream.

Screen grid (S_c). An electrode placed between the control grid and the plate of a vacuum tube to reduce interelectrode capacitance.

Secondary. The output coil of a transformer. *See* Primary circuit.

Secondary emission. The emission of electrons knocked loose from the plate, grid, or fluorescent screen of a vacuum tube by the impact or bombardment of electrons arriving from the cathode.

Selectivity. The degree to which a receiver is capable of discriminating between signals of different carrier frequencies.

Self-bias. The bias of a tube created by the voltage drop developed across a resistor through which either its cathode current or its grid current flows.

Self-excited oscillator. An oscillator depending on its resonant circuits for frequency determination. *See* Crystal oscillator.

Self-induction. The production of a counterelectromotive force in a conductor when its own magnetic field collapses or expands with a change in current in the conductor.

Sensitivity. The degree of response of a circuit to signals of the frequency to which it is tuned.

Series feed. Application of the d-c voltage to the plate or grid of a tube through the same impedance in which the alternating current flows. Compare *Parallel feed.*

Series resonance. The condition existing in a circuit when the source of voltage is in series with an inductor and capacitor whose reactances cancel each other at the applied frequency and thus reduce the impedance to a minimum.

Series-resonant circuit. A resonant circuit in which the capacitor and the inductor are in series with the applied voltage.

Shielding. A metallic covering used to prevent magnetic or electrostatic coupling between adjacent circuits.

Short-circuit. A low-impedance or zero-impedance path between two points.

Shunt. Parallel. A parallel resistor placed in an ammeter to increase its range.

Shunt feed. *See* Parallel feed.

Sine wave. The curve traced by the projection on a uniform time scale of the end of a rotating arm, or vector. Also known as a sinusoidal wave.

Skin effect. The tendency of alternating currents to flow near the surface of a conductor, thus being restricted to a small part of the total cross-sectional area. This effect increases the resistance and becomes more marked as the frequency rises.

Soft tube. A vacuum tube the characteristics of which are adversely affected by the presence of gas in the tube; not to be confused with tubes designed to operate with gas inside them.

Solenoid. A multiturn coil of wire wound in a uniform layer or layers on a hollow cylindrical form.

Space charge. The cloud of electrons existing in the space between the cathode and plate in a vacuum tube, formed by the electrons emitted from the cathode in excess of those immediately attracted to the plate.

Space current. The total current flowing between the cathode and all the other electrodes in a tube. This includes the plate current, grid current, screen-grid current, and any other electrode current which may be present.

Stability. Freedom from undesired variation.

Standing wave. A distribution of current and voltage on a transmission line formed by two sets of waves traveling in opposite directions, and characterized by the presence of a number of points of successive maxima and minima in the distribution curves.

Static. A fixed nonvarying condition; without motion.

Static characteristics. The characteristics of a tube with no output load and with d-c potentials applied to the grid and plate.

Superheterodyne. A receiver in which the incoming signal is mixed with a locally generated signal to produce a predetermined intermediate frequency.

Suppressor grid (Su). An electrode used in a vacuum tube to minimize the harmful effects of secondary emission from the plate.

Surge. Sudden changes of current or voltage in a circuit.

Surge impedance (ζ_o). *See* Characteristic impedance.

Sweep circuit. The part of a cathode-ray oscilloscope which provides a time-reference base.

Swing. The variation in frequency or amplitude of an electrical quantity.

Swinging choke. A choke with an effective inductance which varies with the amount of current passing through it. It is used in some power-supply filter circuits.

Synchronous. Happening at the same time; having the same period and phase.

Tank circuit. *See* Parallel-resonant circuit.

Tetrode. A four-electrode vacuum tube containing a cathode, control grid, screen grid, and plate.

Thermionic emission. Electron emission caused by heating an emitter.

Thermocouple ammeter. An ammeter which operates by means of a voltage produced by the heating effect of a current passed through the junction of two dissimilar metals. It is used for r-f measurements.

Thyratron. A hot-cathode, gas-discharge tube in which one or more electrodes are used to control electrostatically the starting of an unidirectional flow of current.

Tight coupling. Degree of coupling in which practically all of the magnetic lines of force produced by one coil link a second coil.

Trace. A visible line or lines appearing on the screen of a cathode-ray tube in operation.

Transconductance (G_m). The ratio of the change in plate current to the change in grid voltage producing this change in plate current, while all other electrode voltages remain constant.

Transformer. A device composed of two or more coils, linked by magnetic lines of force, used to transfer energy from one circuit to another.

Transient. The voltage or current which exists as the result of a change from one steady-state condition to another.

Transit time. The time which electrons take to travel between the cathode and the plate of a vacuum tube.

Transmission lines. Any conductor or system of conductors used to carry electrical energy from its source to a load.

Triggering. Starting an action in another circuit, which then functions for a time under its own control.

Triode. A three-electrode vacuum tube, containing a cathode, control grid, and plate.

Tuned circuit. A resonant circuit.

Tuning. The process of adjusting a radio circuit so that it resonates at the desired frequency.

Unbalanced line. A transmission line in which the voltages on the two conductors are not equal with respect to ground; for example, a coaxial line.

Unidirectional. In one direction only.

Vacuum-tube voltmeter (VTVM). A device which uses either the amplifier characteristic or the rectifier characteristic of a vacuum tube or both to measure either d-c or a-c voltages. Its input impedance is very high, and the current used to actuate the meter movement is not taken from the circuit being measured. It can be used to obtain accurate measurements in sensitive circuits.

Variable-u tube. A vacuum tube in which the control grid is irregularly spaced, so that the grid exercises a different amount of control on the electron stream at different points within its operating range.

Variocoupler. Two independent inductors, so arranged mechanically that their mutual inductance (coupling) can be varied.

Variometer. A variocoupler having its two coils connected in series, and so mounted that the movable coil may be rotated within the fixed coil, thus changing the total inductance of the unit.

Vector. A line used to represent both direction and magnitude.

Velocity modulation. A method of modulation in which the input signal voltage is used to change the velocity of electrons in a constant-current electron beam so that the electrons are grouped into bunches.

Video amplifier. A circuit capable of amplifying a very wide range of frequencies, including and exceeding the audio band of frequencies.

Volt (V). The unit of electrical potential.

Voltage amplification. The process of amplifying a signal to produce a gain in voltage. The voltage gain of an amplifier is the ratio of its alternating-voltage output to its alternating-voltage input.

Voltage divider. An impedance connected across a voltage source. The load is connected across a fraction of this impedance so that the load voltage is substantially in proportion to this fraction.

Voltage doubler. A method of increasing the voltage by rectifying both halves of a cycle and causing the outputs of both halves to be additive.

Voltage regulation. A measure of the degree to which a power source maintains its output-voltage stability under varying load conditions.

Watt (w). The unit of electrical power.

Wave. Loosely, an electromagnetic impulse, periodically changing in intensity and traveling through space. More specifically, the graphical representation of the intensity of that impulse over a period of time.

Waveform. The shape of the wave obtained when instantaneous values of an a-c quantity are plotted against time in rectangular coordinates.

Wavelength (λ). The distance, usually expressed in meters, traveled by a wave during the time interval of one complete cycle. It is equal to the velocity divided by the frequency.

Wave propagation. The transmission of r-f energy through space.

Wien-bridge circuit. A circuit in which the various values of capacitance and resistance are made to balance with each other at a certain frequency.

X. The symbol for reactance.

Z. The symbol for impedance.

ELECTRONICS SYMBOLS

AMPLIFIER (2)

general

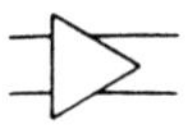

with two inputs

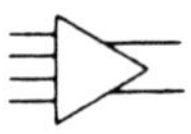

with two outputs

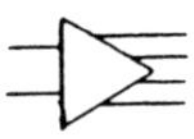

with adjustable gain

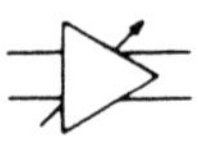

with associated power supply

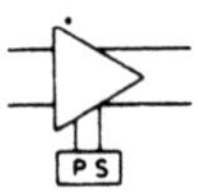

with associated attenuator

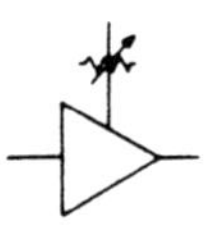

with external feedback path

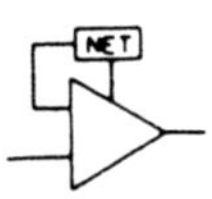

Amplifier Letter Combinations (amplifier-use identification in symbol if required)

BDG	Bridging
BST	Booster
CMP	Compression
DC	Direct Current
EXP	Expansion
LIM	Limiting
MON	Monitoring
PGM	Program
PRE	Preliminary
PWR	Power
TRQ	Torque

ANTENNA (3)

general

dipole

loop

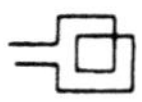

counterpoise

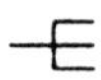

ARRESTER, LIGHTNING (4)

general

carbon block

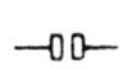

electrolytic or aluminum cell

horn gap

protective gap

sphere gap

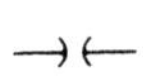

valve or film element

multigap

ATTENUATOR, FIXED

(see PAD) (57)
(same symbol as variable attenuator, without variability)

ATTENUATOR, VARIABLE (5)

balanced

unbalanced

AUDIBLE SIGNALING DEVICE (6)

bell, electrical; ringer, telephone

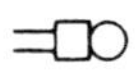

buzzer

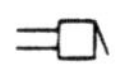

horn, electrical; loudspeaker; siren; underwater sound hydrophone, projector or transducer

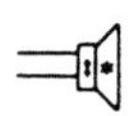

Horn, Letter Combinations (if required)

*HN	Horn, electrical
*HW	Howler
*LS	Loudspeaker
*SN	Siren
‡EM	Electromagnetic with moving coil
‡EMN	Electromagnetic with moving coil and neutralizing winding
‡MG	Magnetic armature
‡PM	Permanent magnet with moving coil

identification replaces (*) asterisk and (‡) dagger)

sounder, telegraph

BATTERY (7)

generalized direct current source; one cell

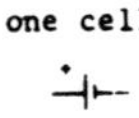

multicell

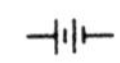

CAPACITOR (8)

general

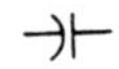

polarized

adjustable or variable

continuously adjustable or variable differential

phase-shifter

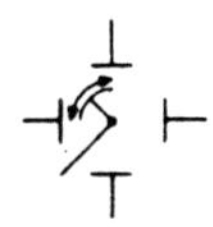

split-stator

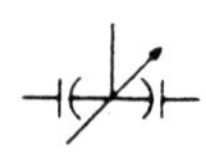

feed-through

CELL, PHOTOSENSITIVE (Semiconductor) (9)

asymmetrical photoconductive transducer

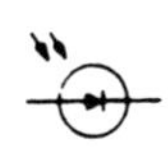

symmetrical photoconductive transducer

ELECTRONICS SYMBOLS

photovoltaic transducer; solar cell

CIRCUIT BREAKER (11)

general

with magnetic overload

drawout type

CIRCUIT ELEMENT (12)

general

Circuit Element Letter Combinations (replaces (*) asterisk)

EG	Equalizer
FAX	Facsimile set
FL	Filter
FL-BE	Filter, band elimination
FL-BP	Filter, band pass
FL-HP	Filter, high pass
FL-LP	Filter, low pass
PS	Power supply
RG	Recording unit
RU	Reproducing unit
DIAL	Telephone dial
TEL	Telephone station
TPR	Teleprinter
TTY	Teletypewriter

Additional Letter Combinations (symbols preferred)

AR	Amplifier
AT	Attenuator
C	Capacitor
CB	Circuit breaker
HS	Handset
I	Indicating or switch board lamp
L	Inductor
J	Jack
LS	Loudspeaker
MIC	Microphone
OSC	Oscillator
PAD	Pad
P	Plug
HT	Receiver, headset
K	Relay
R	Resistor
S	Switch or key switch
T	Transformer
WR	Wall receptacle

CLUTCH; BRAKE (14)

disengaged when operating means is de-energized

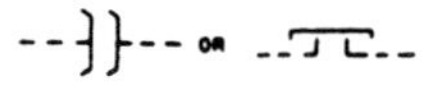

engaged when operating means is de-energized

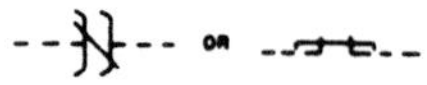

COIL, RELAY and OPERATING (16)

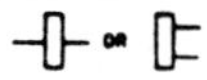

semicircular dot indicates inner end of wiring

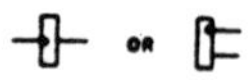

CONNECTOR (18)

assembly, movable or stationary portion; jack, plug, or receptacle

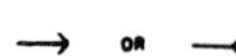

jack or receptacle

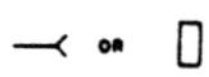

plug

separable connectors

two-conductor switchboard jack

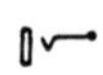

two-conductor switchboard plug

jacks normalled through one way

jacks normalled through both ways

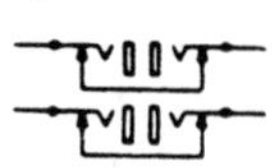

2-conductor nonpolarized, female contacts

2-conductor polarized, male contacts

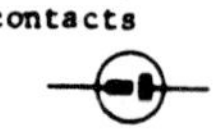

waveguide flange

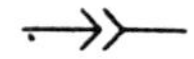

plain, rectangular

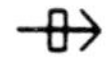

choke, rectangular

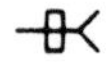

engaged 4-conductor; the plug has 1 male and 3 female contacts, individual contact designations shown

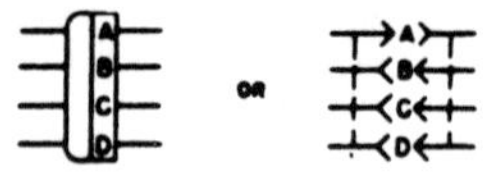

coaxial, outside conductor shown carried through

coaxial, center conductor shown carried through; outside conductor not carried through

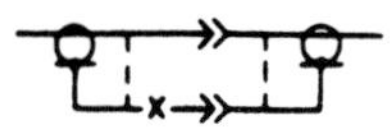

mated choke flanges in rectangular waveguide

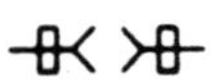

COUNTER, ELECTROMAGNETIC; MESSAGE REGISTER (26)

general

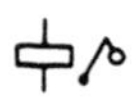

with a make contact

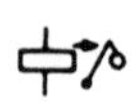

COUPLER, DIRECTIONAL (27) (common coaxial/waveguide usage)

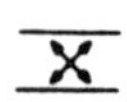

(common coaxial/waveguide usage)

E-plane aperture-coupling, 30-decibel transmission loss

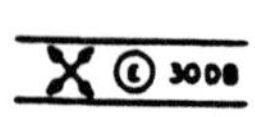

COUPLING (28)

by loop from coaxial to circular waveguide, direct-current grounds connected

CRYSTAL, PIEZO-ELECTRIC (62)

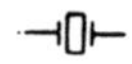

DELAY LINE (31)

general

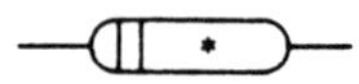

tapped delay

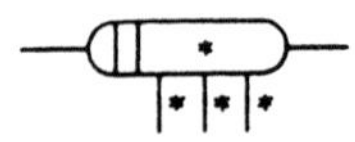

bifilar slow-wave structure (commonly used in traveling-wave tubes)

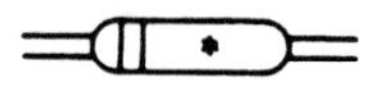

(length of delay indication replaces (*) asterisk)

DETECTOR, PRIMARY; MEASURING TRANSDUCER (30) (see HALL GENERATOR and THERMAL CONVERTER)

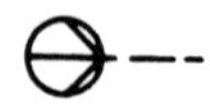

DISCONTINUITY (33) (common coaxial/waveguide usage)

equivalent series element, general

capacitive reactance

inductive reactance

inductance-capacitance circuit, infinite reactance at resonance

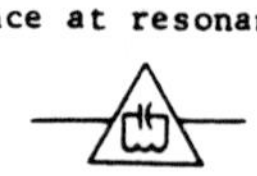

ELECTRONICS SYMBOLS

inductance-capacitance circuit, zero reactance at resonance

resistance

equivalent shunt element, general

capacitive susceptance

conductance

inductive susceptance

inductance-capacitance circuit, infinite susceptance at resonance

inductance-capacitance circuit, zero susceptance at resonance

ELECTRON TUBE (34)

triode

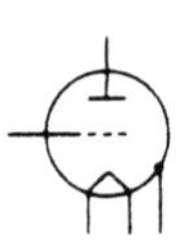

pentode, envelope connected to base terminal

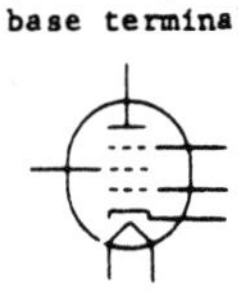

twin triode, equipotential cathode

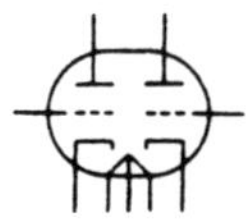

typical wiring figure to show tube symbols placed in any convenient positio[n]

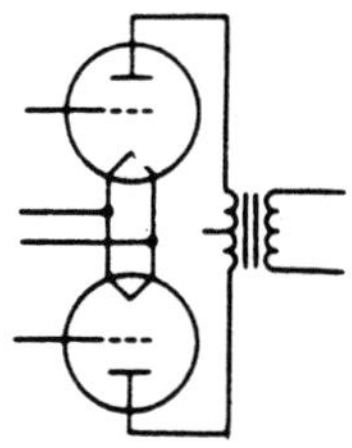

rectifier; voltage regulator
(see LAMP, GLOW)

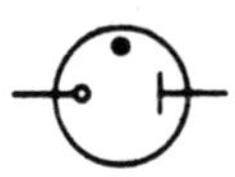

phototube, single and multiplier

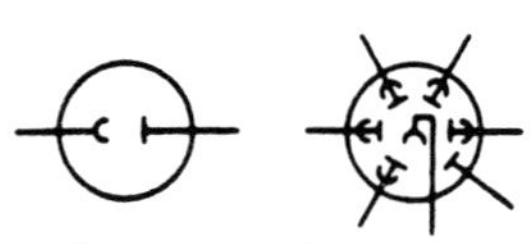

cathode-ray tube, electrostatic and magnetic deflection

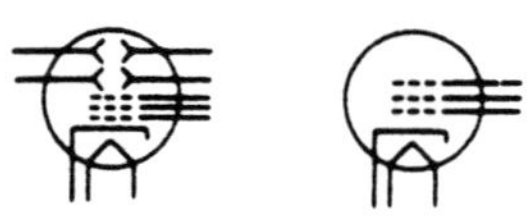

mercury-pool tube, ignitor and control grid (see RECTIFIER)

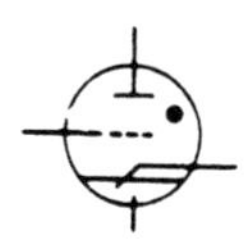

resonant magnetron, coaxial output and permanent magnet

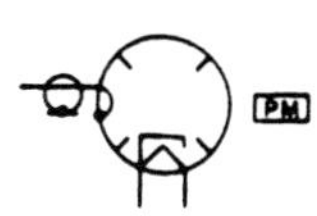

reflex klystron, integral cavity, aperture coupled

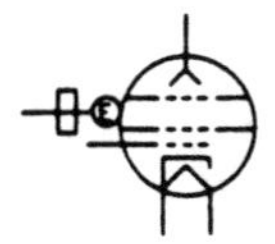

transmit-receive (TR) tube gas filled, tunable integral cavity, aperture coupled, with starter

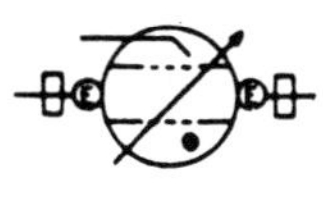

traveling-wave tube (typical)

forward-wave traveling-wave-tube amplifier shown with four grids, having slow-wave structure with attenuation, magnetic focusing by external permanent magnet, rf input and rf output coupling each E-plane aperture to external rectangular waveguide

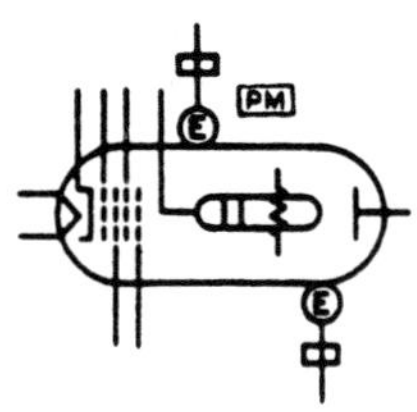

FERRITE DEVICES (100)

field polarization rotator

field polarization amplitude modulator

FUSE (36)

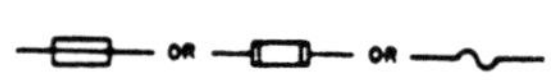

high-voltage primary cutout, dry

high-voltage primary cutout, oil

GOVERNOR (Contact-making) (37)

contacts shown here as closed

HALL GENERATOR (39)

HANDSET (40)

general

operator's set with push-to talk switch

HYBRID (41)

general

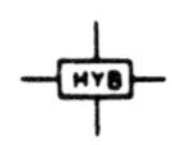

junction
(common coaxial/waveguide usage)

circular

(E, H or HE transverse field indicators replace (*) asterisk)

rectangular waveguide and coaxial coupling

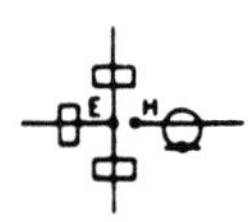

INDUCTOR (42)

general

ELECTRONICS SYMBOLS

magnetic core

tapped

adjustable, continuously adjustable

KEY, TELEGRAPH (43)

LAMP (44)

ballast lamp; ballast tube

lamp, fluorescent, 2 and 4 terminal

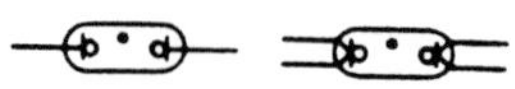

lamp, glow; neon lamp

a-c

d-c

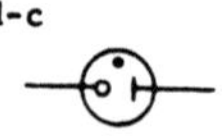

lamp, incandescent

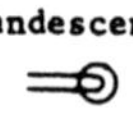

indicating lamp; switch-board lamp (see VISUAL SIGNALING DEVICE)

LOGIC (see 806B and Y32-14) (including some duplicate symbols; left and right-hand symbols are not mixed)

AND function

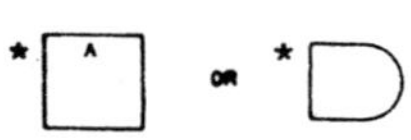

OR function

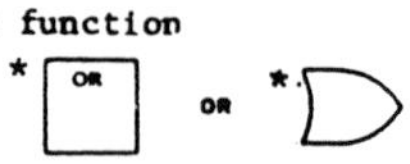

EXCLUSIVE-OR function

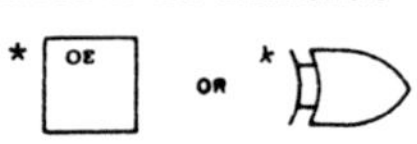

((*) input side of logic symbols in general)

condition indicators

state (logic negation)

a Logic Negation output becomes 1-state if and only if the input is not 1-state

an AND func. where output is low if and only if all inputs are high

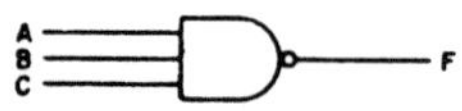

electric inverter

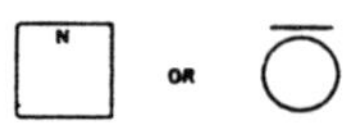

(elec. invtr. output becomes 1-state if and only if the input is 1-state) (elec. invtr. output is more pos. if and only if input is less pos.)

level (relative)

1-state is less +   1-state is more +

(symbol is a rt. triangle pointing in direction of flow)

an AND func. with input 1-states at more pos. level and output 1-state at less pos. level

single shot (one output)

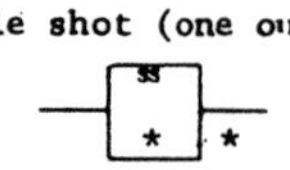

(waveform data replaces inside/outside (*))

schmitt trigger, waveform and two outputs

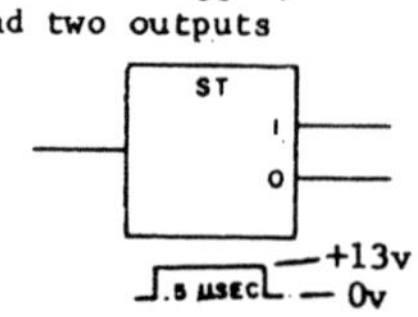

flip-flop, complementary

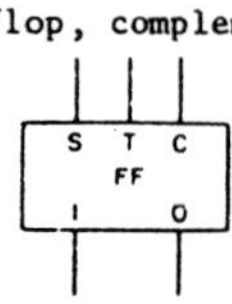

flip-flop, latch

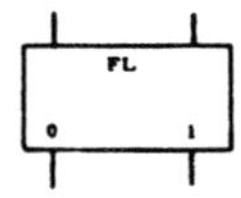

register

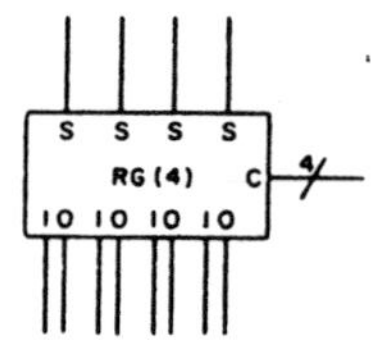

(binary register denoting four flip-flops and bits)

amplifier (see AMPLIFIER)

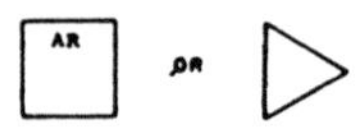

channel path(s) (see PATH, TRANSMISSION)

magnetic heads (see PICK-UP HEAD)

oscillator (see OSCILLATOR)

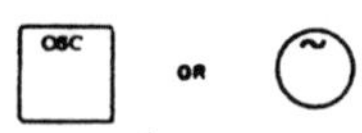

relay, contacts (see CONTACT, ELECTRICAL)

relay, electromagnetic (see RELAY COIL RECOGNITION)

signal flow (see DIRECTION OF FLOW)

time delay (see DELAY LINE)

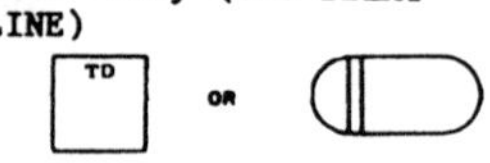

time delay with typical delay taps:

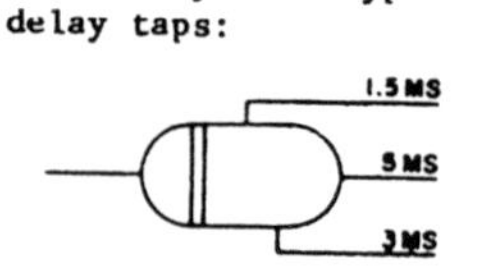

functions not otherwise symbolized

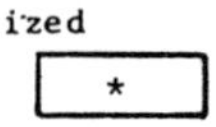

(identification replaces (*))

Logic Letter Combinations

S	set
C	clear (reset)
T	toggle (trigger)
(N)	number of bits
BO	blocking oscillator
CF	cathode follower
EF	emitter follower
FF	flip-flop
SS	single shot
ST	schmitt trigger
RG(N)	register (N stages)
SR	shift register

MACHINE, ROTATING (46)

generator

motor

METER, INSTRUMENT (48)

identification replaces (*) asterisk)

Meter Letter Combinations

A	Ammeter
AH	Ampere-hour
CMA	Contact-making (or breaking) ammeter
CMC	Contact-making (or breaking) clock
CMV	Contact-making (or breaking) voltmeter
CRO	Oscilloscope or cathode-ray oscillograph
DB	DB (decibel) meter
DBM	DBM (decibels referred to 1 milliwatt) meter
DM	Demand meter
DTR	Demand-totalizing relay
F	Frequency meter
G	Galvanometer
GD	Ground detector
I	Indicating
INT	Integrating
µA or UA	Microammeter
MA	Milliammeter
NM	Noise meter
OHM	Ohmmeter
OP	Oil pressure

ELECTRONICS SYMBOLS

MODE TRANSDUCER (53)

(common coaxial/waveguide usage)

transducer from rectangular waveguide to coaxial with mode suppression, direct-current grounds connected

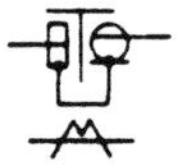

MOTION, MECHANICAL (54)

rotation applied to a resistor

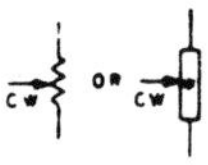

(identification replaces (*) asterisk)

NUCLEAR-RADIATION DETECTOR, gas filled; IONIZATION CHAMBER; PROPORTIONAL COUNTER TUBE; GEIGER-MULLER COUNTER TUBE (50) (see RADIATION-SENSITIVITY INDICATOR)

PATH, TRANSMISSION (58)

cable; 2-conductor, shield grounded and 5-conductor shielded

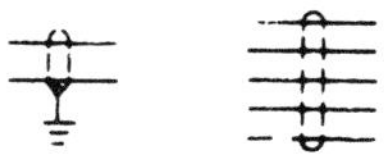

PICKUP HEAD (61)

general

writing; recording

reading; playback

erasing

writing, reading, and erasing

stereo

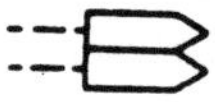

RECTIFIER (65)

semiconductor diode; metallic rectifier; electrolytic rectifier; asymmetrical varistor

mercury-pool tube power rectifier

fullwave bridge-type

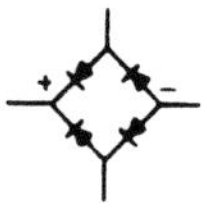

RESISTOR (68)

general

tapped

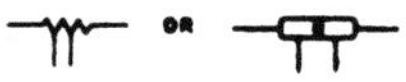

heating

symmetrical varistor resistor, voltage sensitive (silicon carbide, etc.)

(identification marks replace (*) asterisk)

with adjustable contact

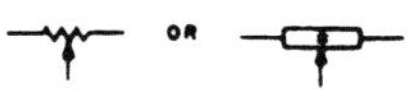

adjustable or continuously adjustable (variable)

(identification replaces (*) asterisk)

RESONATOR, TUNED CAVITY (71)

(common coaxial/waveguide usage)

resonator with mode suppression coupled by an E-plane aperture to a guided transmission path and by a loop to a coaxial path

tunable resonator with direct-current ground connected to an electron device and adjustably coupled by an E-plane aperture to a rectangular waveguide

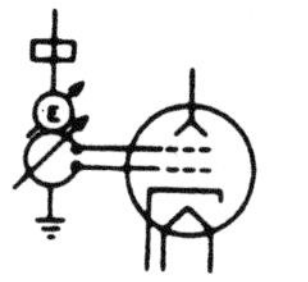

ROTARY JOINT, RF (COUPLER) (72)

general; with rectangular waveguide

(transmission path recognition symbol replaces (*) asterisk)

coaxial type in rectangular waveguide

circular waveguide type in rectangular waveguide

SEMICONDUCTOR DEVICE (73) (Two Terminal, diode)

semiconductor diode; rectifier

capacitive diode (also Varicap, Varactor, reactance diode, parametric diode)

breakdown diode, unidirectional (also backward diode, avalanche diode, voltage regulator diode, Zener diode, voltage reference diode)

breakdown diode, bidirectional and backward diode (also bipolar voltage limiter)

tunnel diode (also Esaki diode)

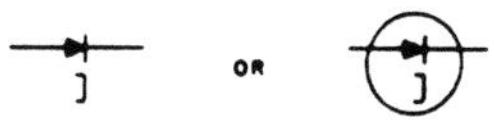

temperature-dependent diode

photodiode (also solar cell)

semiconductor diode, PNPN switch (also Shockley diode, four-layer diode and **SCR**).

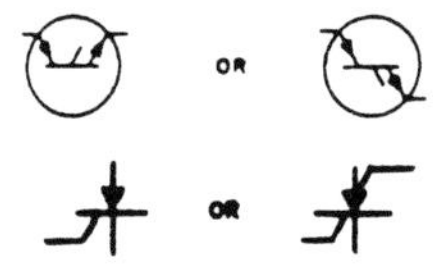

(Multi-Terminal, transistor, etc.)

PNP transistor

NPN transistor

unijunction transistor, N-type base

ELECTRONICS SYMBOLS

unijunction transistor, P-type base

field-effect transistor, N-type base

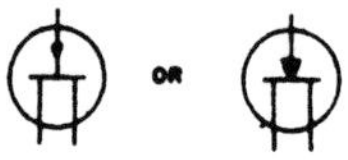

field-effect transistor, P-type base

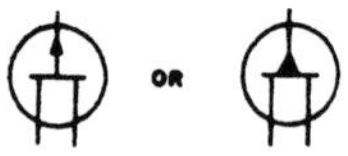

semiconductor triode, PNPN-type switch

semiconductor triode, NPNP-type switch

NPN transistor, transverse-biased base

PNIP transistor, ohmic connection to the intrinsic region

NPIN transistor, ohmic connection to the intrinsic region

PNIN transistor, ohmic connection to the intrinsic region

NPIP transistor, ohmic connection to the intrinsic region

SQUIB (75)

explosive

igniter

sensing link; fusible link operated

SWITCH (76)

push button, circuit closing (make)

push button, circuit opening (break)

nonlocking; momentary circuit closing (make)

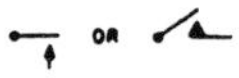

nonlocking; momentary circuit opening (break)

transfer

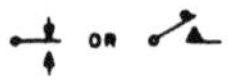

locking, circuit closing (make)

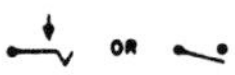

locking, circuit opening (break)

transfer, 3-position

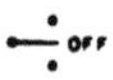

wafer

(example shown: 3-pole 3-circuit with 2 non-shorting and 1 shorting moving contacts)

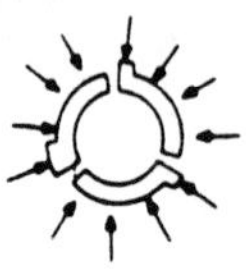

safety interlock, circuit opening and closing

2-pole field-discharge knife, with terminals and discharge resistor

(identification replaces (*) asterisk)

SYNCHRO (78)

Synchro Letter Combinations

CDX	Control-differential transmitter
CT	Control transformer
CX	Control transmitter
TDR	Torque-differential receiver
TDX	Torque-differential transmitter
TR	Torque receiver
TX	Torque transmitter
RS	Resolver
B	Outer winding rotatable in bearings

THERMAL ELEMENT (83)

actuating device

thermal cutout; flasher

thermal relay

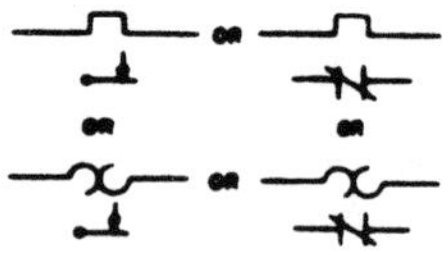

thermostat (operates on rising temperature), contact)

thermostat, make contact

thermostat, integral heater and transfer contacts

THERMISTOR; THERMAL RESISTOR (84)

with integral heater

THERMOCOUPLE (85)

temperature-measuring

current-measuring, integral heater connected

current-measuring, integral heater insulated

temperature-measuring, semiconductor

current-measuring, semiconductor

TRANSFORMER (86)

general

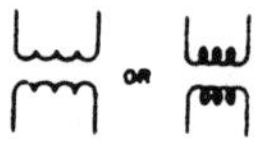

magnetic-core

one winding with adjustable inductance

separately adjustable inductance

adjustable mutual inductor, constant-current

ELECTRONICS SYMBOLS

autotransformer, 1-phase adjustable

current, with polarity marking

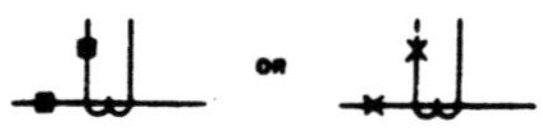

potential, with polarity mark

with direct-current connections and mode suppression between two rectangular waveguides

(common coaxial/waveguide usage)

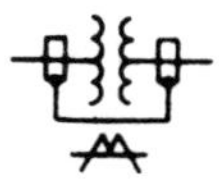

shielded, with magnetic core

with a shield between windings, connected to the frame

VIBRATOR; INTERRUPTER (87)

typical shunt drive (terminals shown)

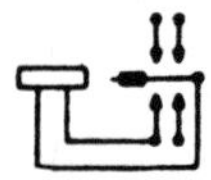

typical separate drive (terminals shown)

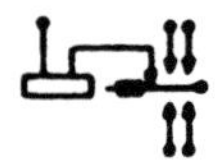

VISUAL SIGNALING DEVICE (88)

communication switchboard-type lamp

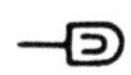

indicating, pilot, signaling, or switchboard light (see LAMP)

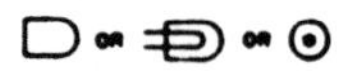

(identification replaces (*) asterisk)

indicating light letter combinations

A Amber
B Blue
C Clear
G Green
NE Neon
O Orange
OP Opalescent
P Purple
R Red
W White
Y Yellow

jeweled signal light

TRANSISTOR SYMBOLS

Semiconductor, General

BV . . . . . Breakdown voltage
T_A . . . . Ambient temperature
T_{ep} . . . . Operating temperature

Transistor

B, b . . . . Base electrode
C, c . . . . Collector electrode
C_{ib} . . . . Input capacitance (common base)
C_{ie} . . . . Input capacitance (common emitter)
C_{ob} . . . . Output capacitance (common base)
C_{oe} . . . . Output capacitance (common emitter)
E, e . . . . Emitter electrode
I_B . . . . . Base current (dc)
i_b . . . . . Base current (instantaneous)
I_C . . . . . Collector current (dc)
i_c . . . . . Collector current (instantaneous)
I_{CBO} . . . Collector cutoff current (dc) emitter open
I_{CEO} . . . Collector cutoff current (dc) base open
I_E . . . . . Emitter current
R_B . . . . . External base resistance
r_b . . . . . Base spreading resistance
r_i . . . . . . Input junction resistance
V_{BB} . . . . Base supply voltage
V_C . . . . . Collector voltage (with respect to ground or common point)
V_{BE} . . . Base to emitter voltage (dc)
V_{CB} . . . Collector to base voltage (dc)
V_{CE} . . . Collector to emitter voltage (dc)
V_{ce} . . . . Collector to emitter voltage (rms)
v_{ce} . . . . Collector to emitter voltage (instantaneous)
V_{CE} (sat) . . . . Collector to emitter saturation voltage
V_{EBO} . . Emitter to base voltage (static)
V_{CC} . . . Collector supply voltage
V_{EE} . . . Emitter supply voltage

TUBE SYMBOLS

A_{hf} . . . . High frequency gain
A_{lf} . . . . Low frequency gain
A_v . . . . Voltage gain
C_c . . . . Coupling capacitor
C_d . . . . Distributed capacitance
C_{gk} . . . . Grid-to-cathode capacitance
C_{gp} . . . . Grid-to-plate capacitance
C_i . . . . Input capacitance
C_k . . . . Cathode bypass capacitor
C_o . . . . Output capacitance
C_{pk} . . . . Plate-to-cathode capacitance
C_s . . . . Shunt capacitance ($C_d + C_i + C_o$)
E_b . . . . Plate volts (dc)
E_{bb} . . . . Supply volts (dc)
E_{bo} . . . . Quiescent plate voltage
E_{c1} . . . . Control grid voltage
E_{c2} . . . . Screen grid voltage
E_{cc} . . . . Control grid supply voltage
E_f . . . . Filament terminal voltage
e_b . . . . Instantaneous total plate volts (ac and dc)
e_{c1} . . . . Instantaneous total control grid volts (ac and dc)
e_{c2} . . . . Instantaneous total screen grid volts (ac and dc)
e_{g1} . . . . Instantaneous value of ac control grid volts
e_{g2} . . . . Instantaneous value of ac screen grid volts
e_{po} . . . . Instantaneous value of plate voltage above and below the quiescent value
E_g . . . . RMS value of grid volts
E_p . . . . RMS value of plate volts
gm . . . . Grid-plate transconductance (mutual conductance)
I_b . . . . DC value of plate volts
I_{bo} . . . . Quiescent value of plate current
I_{c1} . . . . DC value of control grid current
I_{c2} . . . . DC value of screen grid current
I_f . . . . Filament or heater current
I_{g1} . . . . RMS value of control grid current
I_{g2} . . . . RMS value of screen grid current
I_{gm1} . . . . . Crest values of ac current control grid
I_{gm2} . . . . . Crest values of ac current screen grid
I_p . . . . RMS values of plate current
I_{pm} . . . . . . Crest value of plate current
I_s . . . . Total electron emission
i_b . . . . Instantaneous total value of plate current
i_{c1} . . . . Instantaneous total value of control grid current
i_{c2} . . . . Instantaneous total value of screen grid current
i_{g1} . . . . Instantaneous ac value of control grid current
i_{g2} . . . . Instantaneous ac value of screen grid current
i_p . . . . Instantaneous ac value of plate current
i_{po} . . . . Instantaneous values of plate current above and below the quiescent value
R_b . . . . DC plate resistance
R_g . . . . DC grid resistance
R_k . . . . DC cathode resistance
RL . . . . Plate load resistance
r_p . . . . AC plate resistance
μ . . . . Amplification factor

ANSWER SHEET

TEST NO. ______________ PART ______ TITLE OF POSITION ______________________________
(AS GIVEN IN EXAMINATION ANNOUNCEMENT - INCLUDE OPTION, IF ANY)

PLACE OF EXAMINATION ______________________ (CITY OR TOWN) ______________ (STATE) DATE ______

RATING

USE THE SPECIAL PENCIL. MAKE GLOSSY BLACK MARKS.

1 A B C D E	26 A B C D E	51 A B C D E	76 A B C D E	101 A B C D E
2 A B C D E	27 A B C D E	52 A B C D E	77 A B C D E	102 A B C D E
3 A B C D E	28 A B C D E	53 A B C D E	78 A B C D E	103 A B C D E
4 A B C D E	29 A B C D E	54 A B C D E	79 A B C D E	104 A B C D E
5 A B C D E	30 A B C D E	55 A B C D E	80 A B C D E	105 A B C D E
6 A B C D E	31 A B C D E	56 A B C D E	81 A B C D E	106 A B C D E
7 A B C D E	32 A B C D E	57 A B C D E	82 A B C D E	107 A B C D E
8 A B C D E	33 A B C D E	58 A B C D E	83 A B C D E	108 A B C D E
9 A B C D E	34 A B C D E	59 A B C D E	84 A B C D E	109 A B C D E
10 A B C D E	35 A B C D E	60 A B C D E	85 A B C D E	110 A B C D E

Make only ONE mark for each answer. Additional and stray marks may be counted as mistakes. In making corrections, erase errors COMPLETELY.

11 A B C D E	36 A B C D E	61 A B C D E	86 A B C D E	111 A B C D E
12 A B C D E	37 A B C D E	62 A B C D E	87 A B C D E	112 A B C D E
13 A B C D E	38 A B C D E	63 A B C D E	88 A B C D E	113 A B C D E
14 A B C D E	39 A B C D E	64 A B C D E	89 A B C D E	114 A B C D E
15 A B C D E	40 A B C D E	65 A B C D E	90 A B C D E	115 A B C D E
16 A B C D E	41 A B C D E	66 A B C D E	91 A B C D E	116 A B C D E
17 A B C D E	42 A B C D E	67 A B C D E	92 A B C D E	117 A B C D E
18 A B C D E	43 A B C D E	68 A B C D E	93 A B C D E	118 A B C D E
19 A B C D E	44 A B C D E	69 A B C D E	94 A B C D E	119 A B C D E
20 A B C D E	45 A B C D E	70 A B C D E	95 A B C D E	120 A B C D E
21 A B C D E	46 A B C D E	71 A B C D E	96 A B C D E	121 A B C D E
22 A B C D E	47 A B C D E	72 A B C D E	97 A B C D E	122 A B C D E
23 A B C D E	48 A B C D E	73 A B C D E	98 A B C D E	123 A B C D E
24 A B C D E	49 A B C D E	74 A B C D E	99 A B C D E	124 A B C D E
25 A B C D E	50 A B C D E	75 A B C D E	100 A B C D E	125 A B C D E

ANSWER SHEET

TEST NO. ________ PART ______ TITLE OF POSITION ________________________________
(AS GIVEN IN EXAMINATION ANNOUNCEMENT - INCLUDE OPTION, IF ANY)

PLACE OF EXAMINATION ______________________________ (CITY OR TOWN) ______________ (STATE) DATE ________

RATIN

USE THE SPECIAL PENCIL. MAKE GLOSSY BLACK MARKS.

1 A B C D E	26 A B C D E	51 A B C D E	76 A B C D E	101 A B C D
2 A B C D E	27 A B C D E	52 A B C D E	77 A B C D E	102 A B C D
3 A B C D E	28 A B C D E	53 A B C D E	78 A B C D E	103 A B C D
4 A B C D E	29 A B C D E	54 A B C D E	79 A B C D E	104 A B C D
5 A B C D E	30 A B C D E	55 A B C D E	80 A B C D E	105 A B C D
6 A B C D E	31 A B C D E	56 A B C D E	81 A B C D E	106 A B C D
7 A B C D E	32 A B C D E	57 A B C D E	82 A B C D E	107 A B C D
8 A B C D E	33 A B C D E	58 A B C D E	83 A B C D E	108 A B C D
9 A B C D E	34 A B C D E	59 A B C D E	84 A B C D E	109 A B C D
10 A B C D E	35 A B C D E	60 A B C D E	85 A B C D E	110 A B C D

Make only ONE mark for each answer. Additional and stray marks may be counted as mistakes. In making corrections, erase errors COMPLETELY.

11 A B C D E	36 A B C D E	61 A B C D E	86 A B C D E	111 A B C D
12 A B C D E	37 A B C D E	62 A B C D E	87 A B C D E	112 A B C D
13 A B C D E	38 A B C D E	63 A B C D E	88 A B C D E	113 A B C D
14 A B C D E	39 A B C D E	64 A B C D E	89 A B C D E	114 A B C D
15 A B C D E	40 A B C D E	65 A B C D E	90 A B C D E	115 A B C D
16 A B C D E	41 A B C D E	66 A B C D E	91 A B C D E	116 A B C D
17 A B C D E	42 A B C D E	67 A B C D E	92 A B C D E	117 A B C D
18 A B C D E	43 A B C D E	68 A B C D E	93 A B C D E	118 A B C D
19 A B C D E	44 A B C D E	69 A B C D E	94 A B C D E	119 A B C D
20 A B C D E	45 A B C D E	70 A B C D E	95 A B C D E	120 A B C D
21 A B C D E	46 A B C D E	71 A B C D E	96 A B C D E	121 A B C D
22 A B C D E	47 A B C D E	72 A B C D E	97 A B C D E	122 A B C D
23 A B C D E	48 A B C D E	73 A B C D E	98 A B C D E	123 A B C D
24 A B C D E	49 A B C D E	74 A B C D E	99 A B C D E	124 A B C D
25 A B C D E	50 A B C D E	75 A B C D E	100 A B C D E	125 A B C D